GOVERNING MISINFORMATION IN EVERYDAY KNOWLEDGE COMMONS

Governing Misinformation in Everyday Knowledge Commons delves into the complex issue of misinformation in our daily lives. The book synthesizes three scholarly traditions – everyday life, misinformation, and governing knowledge commons – to present ten case studies of online and offline communities tackling diverse dilemmas regarding truth and information quality. The book highlights how communities manage issues of credibility, trust, and information quality continuously, to mitigate the impact of misinformation when possible. It also explores how social norms and intentional governance evolve to distinguish between problematic disinformation and little white lies. Through a coproduction of governance and (mis-)information, the book raises a set of ethical, economic, political, social, and technological questions that require systematic study and careful deliberation. This title is also available as Open Access on Cambridge Core.

Madelyn Rose Sanfilippo studies sociotechnical governance, broadly exploring privacy, inequality, and the political consequences of information technology. She has coedited *Governing Privacy in Knowledge Commons* (2021) and *Governing Smart Cities as Knowledge Commons* (2023), and coauthored three previous books: *Online Trolling and Its Perpetrators: Under the Cyberbridge* (2016), *Social Informatics Evolving* (2015), and *Multiculturalism and Information and Communication Technology* (2013).

Melissa G. Ocepek studies everyday information behavior, critical theory, copyright, and food studies. She has coauthored *Food in the Internet Age* (2013) and *Formal and Informal Approaches to Food Policy* (2014), and coedited *Deciding Where to Live: Information Studies on Where to Live in America* (2020).

CAMBRIDGE STUDIES ON GOVERNING KNOWLEDGE COMMONS

The mission of the series is to provide an authoritative space for high quality scholarship on the governance of knowledge commons. Following the path pioneered by Elinor Ostrom, recipient of the Nobel Prize in Economics for her work on institutional analysis of commons regimes in the natural environment, and the editors' work in *Governing Knowledge Commons*, this series uses a similar framework to systematically study knowledge commons in various sectors. Readers seeking more information on knowledge commons and this series can visit http://knowledge-commons.net, a repository for scholarship produced by an international, interdisciplinary group of knowledge commons researchers.

Series Editors
Katherine J. Strandburg
Brett M. Frischmann
Michael J. Madison

Other Books in the Series
Brett M. Frischmann, Michael J. Madison, and Madelyn Rose Sanfilippo, *Governing Smart Cities as Knowledge Commons*
Erwin Dekker and Pavel Kuchar, *Governing Markets as Knowledge Commons*
Madelyn Rose Sanfilippo, Brett M. Frischmann, and Katherine J. Strandburg, *Governing Privacy in Knowledge Commons*
Katherine J. Strandburg, Brett M. Frischmann, and Michael J. Madison, *Governing Medical Knowledge Commons*

Governing Misinformation in Everyday Knowledge Commons

Edited by

MADELYN ROSE SANFILIPPO
University of Illinois School of Information Sciences

MELISSA G. OCEPEK
University of Illinois School of Information Sciences

CAMBRIDGE
UNIVERSITY PRESS

CAMBRIDGE
UNIVERSITY PRESS

Shaftesbury Road, Cambridge CB2 8EA, United Kingdom

One Liberty Plaza, 20th Floor, New York, NY 10006, USA

477 Williamstown Road, Port Melbourne, VIC 3207, Australia

314–321, 3rd Floor, Plot 3, Splendor Forum, Jasola District Centre, New Delhi – 110025, India

103 Penang Road, #05-06/07, Visioncrest Commercial, Singapore 238467

Cambridge University Press is part of Cambridge University Press & Assessment,
a department of the University of Cambridge.

We share the University's mission to contribute to society through the pursuit of education, learning and research at the highest international levels of excellence.

www.cambridge.org
Information on this title: www.cambridge.org/9781009255158

DOI: 10.1017/9781009255165

When citing this work, please include a reference to the DOI 10.1017/9781009255165

First published 2025

A catalogue record for this publication is available from the British Library

A Cataloging-in-Publication data record for this book is available from the Library of Congress

ISBN 978-1-009-25515-8 Hardback

Contents

Figures

Tables

Contributors

Aparajita Bhandari, Assistant Professor, Department of English Language and Literature, University of Waterloo (Canada)

Jacquelyn Burkell, Professor, Faculty of Information and Media Studies, Western University

Gary Burnett, Professor, School of Information, Florida State University

Brett M. Frischmann, Charles Widger Endowed University Professor in Law, Business, and Economics at Villanova University Charles Widger School of Law, Pennsylvania

Alexandria Johnson, J.D. Candidate, School of Law, Duke University

Dominique Kelly, Doctoral Candidate, Faculty of Information and Media Studies, Western University

Kyung-Sun Kim, Professor, Information School, University of Wisconsin-Madison

William Marks, J.D. Candidate, Harvard Law School, Harvard University

Kate McDowell, Associate Professor, School of Information Sciences, University of Illinois at Urbana-Champaign

Smita Katherine Nair, Undergraduate Student, University of Illinois at Urbana-Champaign

David Nemer, Associate Professor, Department of Media Studies, University of Virginia

Rebecca Noone, Lecturer Information Studies, School of Humanities, University of Glasgow

Melissa G. Ocepek, Assistant Professor, School of Information Sciences, University of Illinois Urbana-Champaign

Britt Paris, Assistant Professor, Department of Library and Information Science, Rutgers University

Irene Pasquetto, Assistant Professor, School of Information, University of Michigan

Madelyn Rose Sanfilippo, Assistant Professor, School of Information Sciences, University of Illinois at Urbana-Champaign

Joanna Sin Sei-Ching, Associate Professor, Wee Kim Wee School of Communication and Information, Nanyang Technological University, Singapore

Elizabeth Wickes, Lecturer, School of Information Sciences, University of Illinois Urbana-Champaign

Shannon Williams, Doctoral Candidate, School of Information, Florida State University

Acknowledgments

We are grateful to the Governing Knowledge Commons (GKC) Workshop and the Governing Knowledge Commons Research Coordination Network (NSF 2017495) for their support. We also thank the publishing team at Cambridge University Press – including Matt Gallaway and Jadyn Fauconier-Herry – for making this book possible, along with the thoughtful contributors to this book: Aparajita Bhandari, Jacquelyn Burkell, Gary Burnett, Brett M. Frischmann, Alexandria Johnson, Dominique Kelly, Kyung-Sun Kim , William Marks, Kate McDowell, Smita Katherine Nair, David Nemer, Rebecca Noone, Britt Paris, Irene Pasquetto, Joanna Sin Sei-Ching, Elizabeth Wickes, and Shannon Williams.

Melissa would like to personally thank all the amazing students and colleagues who have discussed this work with her and encouraged her throughout this process, especially Julia Burns Petrella and Yibo Wang. This has been one of the most meaningful, challenging, and personal projects I have worked on, and unfortunately it collided with a difficult time in my life. You would not be reading this without the support I received from my colleagues Karen Wickett, Elizabeth Wickes, Matt Turk, and Sara Benson. I want to thank the friends and family who have always been by my side: Tina "Shorty" Henderson, Nathan Ingersoll, and Adam Vandenbrink; my coparent Ava Briley, and my muse Sydney Lynn Briley. To my coeditor, friend, and fellow foodie, the always thoughtful, brilliant, and oh-so-stylish Madelyn Rose Sanfilippo, thank you for embarking on this project with me and keeping me afloat.

Madelyn would like to personally thank the many thoughtful people who gave feedback on this project, including: Jack Brighton, Pnina Fichman, Tomer Kenneth, Christopher Leug, Yafit Lev-Aretz, Michael Madison, Jodi Schneider, and Katherine J. Strandburg. Their commentary and participation in workshopping chapters helped build our everyday misinformation knowledge commons. I also want to thank my family for their support and patience during this project, especially Joseph and Rose Sanfilippo. Finally, to my wonderful coeditor, Melissa G. Ocepek, thank you for being a considerate and creative collaborator as we brought this project to life. So many brilliant people helped to make this book possible.

1

Introduction

Governing Everyday Misinformation

Melissa G. Ocepek and Madelyn Rose Sanfilippo

Better to remain silent and be thought a fool than to speak and remove all doubt.

Abraham Lincoln
Mark Twain
Proverbs 17:28

Probably it was Maurice Switzer but there is not definitive proof.

O'Toole (2017)

IF BOOKS COULD KILL OR HOW LITTLE THINGS LEAD TO BIG PROBLEMS

The COVID-19 pandemic spread disruption and forced and/or allowed people to get out of their routines and see the world with new eyes, for good and for ill. Misinformation, while not new, became a term that was featured in headlines and became a part of family dinner conversations more than ever before throughout the pandemic in the US. In this edited volume you will read about dozens of examples of misinformation that you have been exposed to and that have affected your world. While a virus that kills and sickens hundreds of millions of people demonstrates the power and risk of misinformation, it also highlights the mundane ways our assumptions about the way the world works have been wrong and can have catastrophic consequences.

In the early aftermath of the global upheaval caused by the pandemic, one recent trend highlights the goals of this book exceedingly well, and that is calling bullshit on previously beloved and accepted cultural ideas. One example of this is the recent success of the podcast, *If Books Could Kill.* Recently ranked by Vulture as one of the top new podcasts, this series addresses "the airport bestsellers that captured our hearts and ruined our minds" (If Books Could Kill n.d.). The conceit that the most popular books that bored travelers grab on their way to their gate could be killing us may seem absurd, but of course the power of small, simple ideas lies in their subtlety and unobtrusiveness. No matter why you picked up and read *Freakonomics*, *The Secret*, or *Men Are from Mars, Women Are from Venus*, if you did, those ideas and the arguments behind them have penetrated your mind. Even if you never cracked the binding, many of the central ideas of these works have become embedded in our culture. As you will see throughout this book, those ideas, whether accurate or not, have an impact on the way information is considered and understood in our world.

In this Introduction, we will use similarly simple examples to highlight the main arguments and impetus behind this book to help us consider misinformation in a new light, away from COVID and election deniers. Instead we will highlight, in children's books and farmers markets, the three intellectual foundations of this work: the everyday, misinformation, and the governing knowledge commons (GKC) framework, beginning by defining and describing each. We believe that by bringing these three perspectives together we can see many of the big problems of our world in a new way that may help us to understand the nuanced realities of the spread of misinformation and help to prepare us and our institutions for the next misuse of information.

THE EVERYDAY

Much like the term information, the term "everyday" is used often but rarely defined. This is true throughout the media, scholarship, and everyday conversations (see what we did there). During the 2016 election Hillary Clinton and her campaign's use of the phrase "everyday Americans" evolved from a campaign mantra to a political story about what politicians call the masses, and how a candidate that seemed out of touch could try to appeal to her potential voters (Lerner 2015). We all know how that turned out. The everyday has been equated to average, or run of the mill, but of course every person's everyday experience is unique to their lives. Throughout this work we are interested in the small, typical, leisurely, and often overlooked parts of people's lives. The everyday is the essential part of all our lives that we rarely talk about. It isn't the Oscar-nominated film we can't stop discussing, it is the episode of *Friends* that we have seen twenty times and barely notice as we listen to it while falling asleep.

The everyday that we are using as the domain to explore misinformation is grounded in the critical theory of the everyday that sees immense power and import in the aspects of our lives that blend into the background, but make up the fundamental ways we see the world (Ocepek 2017). This provides a domain that is dynamic and emphasizes the goals of this work: to explore misinformation in places and ways that are often overlooked and understudied. Based on the work of Alfred Schütz and Henri Lefebvre, this domain was then expanded by Michel de Certeau, Donna Haraway, and Dorothy Smith (for more see Schutz 1973; de Certeau 1984;

Smith 1987; Haraway 2004; Lefebvre 2008). Their work denotes two important aspects of the everyday that we will explore throughout this volume: First, the everyday includes the totality of lived experience, and second, it is the quotidian aspects of life that are often the most powerful and ignored. These aspects initially seem at odds, as the totality of lived experience highlights that our everyday consists of all the big and little things that we experience, whether it is responding to your thirty-seventh email or going to Hawaii for the first time. This area of scholarship connects this part of life to highlight how the beauty of Hawaii can really only be understood by comparing it to the tedium of an Illinois winter. Our understanding of big abstract concepts like beauty, love, and success began to be formed in our lives before we had the power of speech, or writing, or poetry to share them, because the everyday world consists of hundreds and thousands of small moments that create our life-world and our worldview.

MISINFORMATION

Our society has many cultural markers of import, including the annual tradition of dictionary publishers proclaiming the word of the year. In 2018, the word selected by Dictionary.com was "misinformation" (Strauss 2018). The corresponding press release explains: "The rampant spread of misinformation poses new challenges for navigating life in 2018. As a dictionary, we believe understanding the concept is vital to identifying misinformation in the wild, and ultimately curbing its impact" (para. 4). The announcement goes on to share the Dictionary.com definition for misinformation as "false information that is spread, regardless of whether there is intent to mislead" (para. 5) followed by a description of how the online dictionary is dealing with what it sees as a worsening misinformation phenomenon. To fight misinformation, the dictionary notes that it is working to update and add related terms to its collection, including, "disinformation, echo chamber, confirmation bias, filter bubble, conspiracy theory, fake news, post-fact, post-truth, homophily, influencer, and gatekeeper" (para. 6; see Table 1.1). These are all important concepts that can help make visible many of the everyday and unnoticed structures of the information ecosystems that we all live in.

While there is no doubt that misinformation has become a much more popular topic of conversation in the last few years, the word, concept, and societal problem are not new. The term itself can trace its history back to the late 1500s and its impact has reverberated through history since long before then (Strauss 2018). Many types of misinformation are common and accepted facts of social life, whether it is office gossip or conspiracy theories or bad science that finds an audience. Human beings are incapable of living in a world where they do not ingest and spread misinformation; technology has changed the information ecosystems of all of us, but it hasn't changed our human nature to misinform.

TABLE 1.1 *Related misinformation terms*

Term	Definition	Examples
Disinformation	deliberately misleading or biased information; manipulated narrative or facts; propaganda	Special interest groups muddied the waters of the debate, spreading disinformation on social media.
Echo chamber	an environment in which the same opinions are repeatedly voiced and promoted, so that people are not exposed to opposing views	We need to move beyond the echo chamber of our network to understand diverse perspectives.
Confirmation bias	bias that results from the tendency to process and analyze information in such a way that it supports one's preexisting ideas and convictions	Confirmation bias is a major issue when we get all our news from social media sites.
Filter bubble	a phenomenon that limits an individual's exposure to a full spectrum of news and other information on the internet by algorithmically prioritizing content that matches a user's demographic profile and online history or excluding content that does not	My roommate streamed so many arthouse flicks on my account that she confused the filter bubble – the recommended movies page thinks I'm some kind of fancy-pants intellectual now.
Conspiracy theory	a theory that rejects the standard explanation for an event and instead credits a covert group or organization with carrying out a secret plot	One popular conspiracy theory accuses environmentalists of sabotage in last year's mine collapse.
Fake news	false news stories, often of a sensational nature, created to be widely shared or distributed for the purpose of generating revenue, or promoting or discrediting a public figure, political movement, company, etc.	It's impossible to avoid clickbait and fake news on social media.
Post-fact	see Post-truth.	We appear to be living in a post-fact society.
Post-truth	relating to or existing in an environment in which facts are viewed as irrelevant, or less important than personal beliefs and opinions, and emotional appeals are used to influence public opinion	Post truth politics.
Homophily	the tendency to form strong social connections with people who share one's defining characteristics, as age, gender, ethnicity, socioeconomic status, personal beliefs, etc.	Political homophily on social media.

Term	Definition	Examples
Influencer	a person who has the power to influence many people, as through social media or traditional media	Companies look for Facebook influencers who can promote their brand.
Gatekeeper	a person or thing that controls access, as to information, often acting as an arbiter of quality or legitimacy	An open internet allows innovators to bypass traditional gatekeepers and promote their work on its own merit.

All definitions and examples from Dictionary.com.

GOVERNING KNOWLEDGE COMMONS

One growing area of scholarship that helps us to see the invisible structures that move information around any social space is through the governing knowledge commons. The GKC framework comes to information science by way of legal scholars of intellectual property who were in turn inspired by the work of political economist Elinor Ostrom and her Institutional Analysis and Development (IAD) Framework (Frischmann, Madison, and Strandburg 2014). The IAD framework was used to gain useful insights into how commons arrangements and governance worked for natural resources. The GKC framework applies and extends IAD principles to information resources. It has been shown to be a useful framing for empirical studies on how innovation and creativity can be produced and encouraged. Previous work has used the GKC framework to gain useful insights into many of our world's biggest concerns and areas of exciting technological growth. These include previous tomes of case studies exploring open source software (Schweik and English 2012), genetic resources and research (Reichman, Uhlir, and Dedeurwaerdere 2016), medical knowledge (Strandburg, Frischmann, and Madison 2017), markets (Dekker and Kuchař 2022), privacy (Sanfilippo, Frischmann, and Strandburg 2021), and smart cities (Frischmann, Madison, and Sanfilippo 2023).

At its most basic, the GKC framework is a conceptual model to help us identify and explore how information is created, used, and moves through a context. Knowledge production and governance processes are iterative, as indicated in Figure 1.1, as communities work through action arenas which can be defined as contextual sets of challenges, and the outcomes in turn shape the context. The attributes of the framework help to identify the most relevant aspects of a knowledge commons to understand why some work well while others fail, and to help design better governance structures to increase the benefits of knowledge creation and sharing for a greater good (Frischmann, Madison, and Strandburg 2014).

From the GKC perspective, commons can be understood through the institutional arrangements around resources, community, and shared decision-making

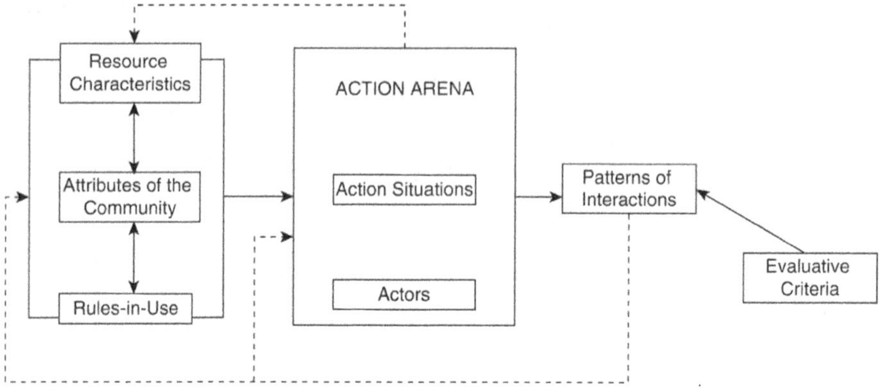

FIGURE 1.1 The knowledge commons framework.

over those resources and community. At the core, knowledge commons are the embodiment of coproduction of knowledge and community coupled with governance regarding knowledge throughout its lifecycle, wherein knowledge refers to a broad set of intellectual and cultural resources. The GKC framework offers a structured approach to understand how communities make decisions regarding the creation, sharing, use, and destruction of knowledge in different scenarios, as well as what their norms are regarding knowledge, as opposed to mapping exogenous expectations about what is right onto the community.

ILLUSTRATIVE EXAMPLES

Now that we have briefly defined the main concepts undergirding this work, we have selected a few everyday examples to begin our intellectual journey into everyday misinformation.

Telephone Game

There are few childhood pastimes as tried and true as the game of telephone, historically and xenophobically also named Chinese Whispers, mostly in the UK. The game is so commonplace to not even need description, but if we must, the game consists of a group of people, usually children, lined up whispering a message from person to person with the final participant sharing the message they heard and the first participant sharing how it started. As the message quickly and quietly gets passed along, elements are misheard, forgotten, and added until the final message typically bears minimal resemblance to its origin (Huang 2015). The game dates back to the era of the invention of the telephone and teaches children a valuable lesson, that hearsay is unreliable (Gamesver Team 2022).

While this example may seem too simple to warrant mention, the importance of understanding the telephone game in any work trying to understand misinformation is critical. It is not that the players are purposefully changing the message, though they certainly could be, it is that the message is only as reliable as the messenger we heard it from, and before that the person they heard it from, and that our minds and memory are only so good at receiving, encoding, and sharing information. The telephone game is a silly example that highlights how simple messages can easily be changed into misinformation, not out of ill intent, but because people are flawed and we can never assume as information passes from person to person it is not radically changed. For a strongly associated academic description of the communication theory behind the telephone game, see Shannon (1948) and Shannon and Weaver (1964).

Eat Ice Cream for Daily Happiness

Growing up in the former cheese capital of the world, Plymouth, WI, this chapter's first author, Melissa Ocepek, enjoyed dairy in all its forms and shares this personal example, which also strongly resonates with the chapter's second author, Madelyn Rose Sanfilippo, who grew up in a nearby Wisconsin community.

One of my favorite local landmarks is an old barn on the outskirts of town with a faded painted advertisement with a simple and powerful message that I internalized at an early age, "Eat Ice Cream for Daily Happiness" (Figure 1.2). As I became a food researcher, I was fascinated by the incredibly confusing and contradictory

FIGURE 1.2 Ice cream advertisement on a barn in Plymouth, WI.
©Terry Williamson.

nutrition advice that permeated my personal and professional life. Over the years I have interviewed many individuals who were so overwhelmed by the cacophony of advice they heard throughout their lives about what to eat and what to avoid, that they threw their hands in the air and gave up trying to feed their families the most "nutritious" food.

Recently, I was personally thrilled to see the nutrition community called out by recent reporting that their long-held fight against some of the most necessary elements of any human diet, fat and sugar, was in fact not empirically supported (Johns 2023). A public-health historian stumbled upon some exciting and bewildering research that demonstrates that ice cream has been found to have positive nutritional and health benefits. As someone who, since breastfeeding my child years ago, has been advocating that ice cream is one of our culture's most perfect foods, I felt vindicated and also angered. The writer who broke this story uncovered not only recent research that supports the health benefits of ice cream, but studies going back years that found similar results that were downplayed by their authors, likely due to fear that such results would not be well received by the nutrition community. Nutritional science, like all scientific communities, should be a space where research is shared and challenged in a way that allows results to speak for themselves, but like any other knowledge commons norms play a strong role in the information that is pursued, shared, and accepted. While counterintuitive information often makes for great headlines and article titles, there are some academic disciplinary norms so strong that to go against them can threaten a researcher's career. Anyone who has made the mistake of becoming a published author knows the fear that the words and arguments that we have written may be misinterpreted and lead to a negative outcome. You can imagine the nutrition researchers looking over their results, double-checking their math, and trying to explain a result that goes against not only their previous teachings and the field's epistemology, but also their deeply held beliefs about which foods are healthy. I am sure that some of the reasons someone chooses to become a nutritionist or nutrition scholar is to learn more about healthy food and convince people to make the healthiest choices for themselves and their families. Few nutrition scholars may look at the result that ice cream is a healthful food the way I did (as a very critical reader of nutrition research) and think "of course, so many of our understandings of health are based on anti-fat bias," but I digress. This example is meant to highlight how challenging it can be for true information to be shared when it flies in the face of the strongly held norms of an intellectual community.

Fraud at the Farmers Market

As a food researcher, this chapter's first author knows that there are few things in this world more misunderstood than the foods that we eat (another reason I am so skeptical of nutrition research). Trust is baked into much of our industrialized food processes. Most things we ingest daily were not grown, harvested, or even prepared by our own hands, and so we must trust that all the steps in the food production process are done with our safety in mind. Of course, anyone who pays attention to the news knows that our food systems allow harmful products to make it to store shelves occasionally. This has led many people to be wary of large-scale food production and industrialized farming and to prefer buying local and certified organic products. As a food researcher, I can also tell you that few consumers truly know and understand what terms like "local" and "organic" mean, let alone that only "organic" is actually defined in regulations, not just in marketing, but this issue is for another book (also, the only expiration dates that are regulated in the US are on infant formula; if it looks and smells fine, eat it).

Unfortunately, in the same ways that large-scale food production has been known to cut corners to increase profits and efficiency, so too are smaller producers misleading consumers. There are of course small levels of misinformation in local food production that we all assume and take for granted, marketing your dairy as coming from "happy" cows for instance, but in a multipart exposé, Laura Reiley, a food critic and writer from the *Tampa Bay Times*, uncovers multiple Tampa Bay food vendors and restaurant purveyors who are a part of several farm-to-fable schemes that sold misinformation along with produce and meat that rarely originated locally in Florida (Reiley 2016a).

In recent years farmers markets have exploded in popularity from limited agritourism to weekly occurrences in most cities and municipalities. And while it may seem like there are people growing and producing food all over the US, the truth is the amount of interest in local and especially organic produce wildly outpaces the availability (Reiley 2016b). Reiley's investigations found a variety of major and minor misinformation shared by multiple farmers, vendors, and even market managers. The following excerpt highlights how the ideas a shopper may have while perusing or even chatting with farmers market vendors may be far from accurate.

> When I (Reiley) notice asparagus and apples, which generally don't grow in Florida, I ask if it is resold produce from a broader radius. She (the vendor) says yes. And then I ask, specifically, which items are grown on Lee Farms. Her answer: "We are currently replanting." In 40 seconds, Lee Farms went from growing everything to nothing. I call the farmer, Christina Lee, whose Lee Farms is in Webster, and tell her about my experience. A lot of her crops were destroyed because it was a late winter, she says. She admits to visiting the wholesale market and selling Caribbean fruit and asparagus from Peru. Does she say all produce in their booth is their own? "In passing conversation, we say yes. If people stop and ask if it's ours, we say yes, it's ours, because most people don't have knowledge about what is grown in Florida." (Reiley 2016b, para 19)

Near this and several other Tampa Bay area farmers markets is the Publix wholesale market, where produce is sold to local markets and resellers, many of which pass off the products as their own. Some may have their own farm's products and sell some of the items that they grow, but without a thorough understanding of what local produce is typical or even possible and whether farms are even ready to be harvested, shoppers are left paying more for items they could easily buy at their supermarket.

The vendors have a clear vested interest in selling a good story of local and sustainable farm practices along with their produce, but they are not the only individuals allowing and even promoting misinformation at the market. While more and more cities and towns want to add farmers markets, they need more than just a space to set them up. Many market managers struggle to vet and verify sellers while they are also marketing and promoting a market with local "farmers." Much like the vendor who initially described her produce as all her own, the market manager of Tampa's downtown market quickly changed her tune from describing her market as "producer-only," and that she feels the "best part of the market" is "finding out where your food is coming from," to admitting that she had not vetted her sellers. "Verifying is a big job," she said. "I have to work other jobs as well. I just don't make enough money to vet vendors" (Reiley 2016b).

Farmers market fraud sadly does not end when the market closes. Reiley went on to investigate many local food claims made by restaurants, several of which advertise and highlight relationships with local farmers and sustainable producers. When claims from multiple restaurants were checked, most did not hold up. Reiley found everything from outdated signage about previous farm-to-restaurant relationships, inaccurate descriptions of frozen foods passed along as made from scratch, to veal schnitzel sold in a restaurant where no veal or veal invoices could be found. When pressed, many restaurant owners shared the hard truth about so much food advertising: "We try to do local and sustainable as much as possible, but it's not 100 percent," one restaurant owner said. "For the price point we're trying to sell items, it's just not possible." So, consumers are left misled about the quality and environmental impact of the foods they purchase and consume with little being done about it. The line between marketing and reality can be difficult to locate and it may move frequently.

In many ways food producers and sellers may see this as misinformation without a major impact. The claims of local and sustainable farming practices are in many ways a nice lie we all like to tell ourselves, even when we know that many of these claims are not or cannot be verified in the time we allot to acquiring food for our homes every week. There is a financial incentive in selling at farmers markets and marketing your food as local, as consumers typically expect to pay more for those products, but beyond the money it can also make everyone feel that they are participating in a marketplace that values social and environmental good, even if the claims vary from little white lies to full-on fabrications.

Fake Fine Dining: Ghost Kitchens of the Pandemic

While, sadly, farmers market fraud isn't that new, the pandemic highlighted a recent food trend that upset many consumers and restauranteurs alike, the recent creation of ghost kitchens throughout the US. When we began working on this book, we knew we wanted to discuss ghost kitchens. But like most Americans we only knew about the tip of the ghost kitchen iceberg. We learned about ghost kitchens during the COVID-19 pandemic when stay-at-home orders and concerns for public health drastically reshaped the restaurant industry and our very cultural understanding of "eating out." Like many foodies, we wanted to support local restaurants that were no longer able to welcome diners in by replacing normal eating out nights with ordering in. Our own experience aside, recent reporting and industry studies have shown that most restaurants changed their business practices including providing for the first time or increasing their off-premises sales channels, increasing outdoor dining, and simplifying their menus (Kelso 2021). These practices have shifted consumer behavior and provided new opportunities, leading all but 5 percent of restaurant operators to report that they plan to maintain at least some of these changes. Little did we or many consumers know that some of the restaurants that you would find on Grubhub, Doordash, or Uber Eats may not be the same brick-and-mortar establishments we knew, or even traditional restaurants at all.

Ghost kitchens did not start during the pandemic, but the dramatic increase in delivery and take-out orders supercharged this novel business model. "Ghost kitchen" is a term that can refer to any restaurant entity that sells food only for delivery. They are also known as dark kitchens, virtual brands, cloud kitchens, and delivery-only restaurants (Krishna 2021; Lucas 2021). Some of these kitchens were created when dine-in only restaurants shifted their business model during the pandemic. Other businesses were created by tech companies and hospitality groups to extend previous business ventures into the food space or to grow their delivery offerings; and a third group consists of passionate cooks looking for alternative ways to sell their food to consumers with lower overheads. With this variety of backgrounds, ghost kitchens have a range of ways to operate, some with full transparency, and others within a foggy haze of misinformation.

A delivery-only restaurant is not inherently cause for concern, but some hide their corporate ownership and others fully try to confuse and even defraud consumers into thinking they are purchasing food from an established brick-and-mortar restaurant. The following examples exemplify what we see as the ghost kitchen misinformation continuum.

The most honest ghost kitchens operate like Guy Fieri's Flavortown Kitchen, a "delivery-only restaurant featuring real-deal flavors from Chef Guy Fieri ... with over 170 locations in 34 states." Mr. Fieri and his business partners are making his food available without the cost and work of managing a restaurant with dine-in options (Guy Fieri's Flavortown Kitchen n.d.). This type of business is up front about its corporate ownership, its lack of a dine-in location, and the fact that it is operating largely online across the US.

The somewhat questionable type of ghost kitchen is represented by businesses such as "It's Just Wing," a ghost kitchen that Daniel Stamps found on his Uber Eats app that delivered sub-par wings to him in a box labeled Chili's Grill and Bar

(Dodds 2022). Whereas on the app "It's Just Wings" appeared to be an independent and new wing restaurant, it was actually a ghost kitchen owned by Chili's Grill and Bar and operating out of their dine-in restaurant kitchens. Causing Stamp to inadvertently order wings from a restaurant that he didn't want to order from based on his previous experience with their food, Stamp's story represents just one of the thousands of people represented by a recent National Restaurant Industry survey that found that 72 percent of consumers find it important that their delivery orders are made at a location that they can also visit (Kelso 2021). Uber Eats has pledged to remove companies that are trying to misinform customers by, for example, listing the same menu under fourteen different business names, while allowing other ghost kitchens that may be hiding some aspects of their business from consumers, such as who their corporate owners are, to still lightly misinform on their app (Perry 2023).

The most egregious type of ghost kitchen are those pretending to be known restaurants that people love. These kitchens hide the fact that they are restaurants that people may not feel so fond of. In 2015 the delivery app DoorDash was sued by the burger chain In-N-Out for delivering their food without the restaurant's permission (Barber 2020). While consumers may have still received food from an In-N-Out, the restaurant was concerned about the safety and quality of the food without their oversight over the timing of the deliveries. Two Japanese restaurants in San Francisco were being used by imposter ghost kitchens on multiple delivery apps (Campbell 2021). After it was reported that these ghost kitchens were using the name and likeness of well-known restaurants that they were not affiliated with, they were removed from the delivery platforms.

New innovations, whether they be in the food industry or tech industry, or as in the case of many ghost kitchens, both, take time to be well understood and accepted by consumers. Unfortunately, several bad actors and bad experiences have left an unpleasant taste in the mouth of many delivery app users when it comes to ordering from ghost kitchens. We all have an image in our mind of our favorite local restaurant, and of the cooks and staff that work there to make it a special place where we nourish more than just our bodies. Ghost kitchens are fundamentally changing what it means to be a restaurant; they often try to evoke the same positive feelings with well-known local establishments or hip new spots. Ghost kitchens lack the dine-in options that give consumers the personal interactions that build trust and loyalty. Consumer behavior has changed a lot when it comes to online shopping and food ordering in the last decade, but bad experiences could make ghost kitchens demonstrate that internet exceptionalism, the belief that the convenience of the internet can make any business thrive, only applies to positive consumer experiences.

Sam Bankman-Fried as the Slovenly Straw Man

Every new cultural era is defined by trends in popular culture, fashion, and social norms, and are often notable for great leaders (historically men) of the age. A recent great man we have seen quickly rise and fall, while changing what many think greatness looks like, is Sam Bankman-Fried. An out-of-nowhere billionaire who regularly showed up to important meetings and interviews in wrinkled t-shirts and shorts, with a rarely before seen amount of "I don't give a damn," Bankman-Fried was celebrated not only for his claimed causes of creating a "safer" crypto currency exchange, and donating heaps of money to causes for social good, but also for being a new kind of business leader. He wasn't polished and marketable; he seemed almost an anti-Elizabeth Holmes (the disgraced former founder of the blood testing company Theranos). Instead, he appeared to be all about the math, the tech, the money. For a brief window before his fraud made headlines, he was seen as the new type of ultra-logical "math bro," someone who could be trusted for all the reasons the leaders of the past couldn't. He seemingly didn't want power, or influence, or even money, he just wanted the math to work and to use his genius to see things no one else could. Of course, like so many of the narratives of the current moment, glitzed up with tech, anti-swagger, and other emblems of cultural clout, he was just like many other CEOs who are too focused on their success and not enough on their larger impact on the world. A frumpy CEO is marketing himself as much as one in a black turtleneck. Both are playing on our ideas of genius and success and representing themselves as a new kind of CEO who won't make the mistakes of the past.

During his fraud trial, "reams of evidence" were shared that the jury used to come to a guilty verdict, that Bankman-Fried mishandled, solicited, misrepresented, and spent around $10 billion of his clients and other people's money (Baker 2023). All this occurred in less than three and half years from cofounding FTX to its collapse. In that time, in addition to engaging in multiple forms of fraud, Bankman-Fried was also compared to past successful financial savants as varied and notable as J. P. Morgan and Bernie Maddoff, while building FTX into a company valued at $32 billion. Many people had speculated that Bankman-Fried could bring the world into a new financial era as the world's first trillionaire (Lewis 2023).

During Bankman-Fried's meteoric rise in the financial sector, many writers and journalists tried to understand this overnight billionaire who claimed to have little interest in the billionaire lifestyle, but instead wanted to use his financial gains to make the world a better place as one of the most famous members of the Effective Altruism movement (Lewis 2023). The media was soon fascinated by this new kind of billionaire, focusing on what made him different, rather than the things that in hindsight clearly demonstrate a long history of valuing winning and breaking the rules of the game rather than making thoughtful financial decisions. Great con artists, like great magicians, use storytelling and theatrics to make the audience look at what they want the audience to see and not at the truth of the situation. Sam Bankman-Fried was a master at distracting investors, consumers, and journalists by appearing unkempt and talking often about social good.

Looking beyond the specific crimes of Bankman-Fried is a story we have heard many times before about someone who appears different, but who builds a new platform to make money for himself and the people around him. After Bankman-Fried's conviction, Attorney General Merrick Garland put out a statement: "This case should send a clear message to anyone who tries to hide their crimes behind a shiny new thing they claim no one else is smart enough to understand" (Baker 2023). Bankman-Fried did not outsmart his investors, the media, or regulators with his technology, he did it by making everyone around him decide who he was and what he wanted. He played with our understandings of greed, power, and success while following much of the playbook of scammers and con artists who have been defrauding customers for centuries.

Theranos and Selling Misinformation

On May 30, 2023, Elizabeth Holmes, the disgraced former founder of the blood testing company Theranos, reported to federal prison to begin her eleven year and three-month sentence (Griffith 2023). She was convicted of four counts of wire fraud and conspiracy for deceiving investors. Beyond the legal crimes, Holmes has become a lightning rod for topics as varied as ethics in tech, domestic violence as rationale for criminal defense, and how the criminal justice system treats new mothers. In some ways, Holmes has become a turducken (deboned chicken stuffed into a deboned duck, further stuffed into a deboned turkey) of misinformation, where the mythology of Silicon Valley, the clichés of con artists, and the girl bossification of late-stage capitalism all have encased themselves around one woman.

As someone who came of age during the Enron debacle at the turn of the century, the chapter's first author has always been fascinated by the line between marketing and fraud, a line that plays with many of the key attributes that we will explore throughout this volume. Holmes, unfortunately, found herself on the wrong side of that line. I have been fascinated about her as a person and symbol, since I first learned of Elizabeth Holmes and Theranos long before the company's collapse and Holmes' conviction. I am not the only one fascinated, as can be seen by the popularity of the best-selling nonfiction book about the fraud, *Bad Blood: Secrets and Lies in a Silicon Valley Startup* by John Carreyrou; the hit documentary *The Inventor: Out for Blood in Silicon Valley*, directed by Alex Gibney; the hit podcast series, *The Dropout*, and the television mini-series of the same name, all of which I have ravenously consumed, discussed, and dissected. The story is so popular because it is a great example of reality being wilder than fiction, while also playing on so many of our favorite tropes about technology, female power, young genius, and scams.

In our view, long before she committed actual fraud, Holmes' biggest mistake was combining two industries that have so little in common, a Silicon Valley tech firm with a healthcare company. Almost nothing that Holmes did would have been a major problem in most tech startups. In fact, most of her "crimes" are commonplace. For example, she had her CFO, Henry Mosley, create a rosier projection of Theranos' projected earnings when she went out to venture capitalists to fund her company. Presenting numbers like this in Silicon Valley is so common it is known as a "hockey stick forecast" (Carreyrou 2020). Even her decision to fake her technology, rolling out a demo that looks good but does not actually work, was so routine that the term "vaporware" was coined in the 1980s to describe hardware or software that is rolled out with much acclaim only to take years to materialize or never to see the light of day. Holmes' company's demise and criminal prosecution largely occurred because her over-promised technology had real-world negative implications on the lives of hundreds of people with real medical needs. When software doesn't work as well as it is advertised, or even breaks, people get mad, but rarely do people get incorrect medical diagnoses. Our society and corporate regulators are generally fine with bombastic claims and elaborate marketing schemes, but there a few spaces where accuracy and truth really matter to people, and prosecutors will punish bad actors much more harshly for typical corporate behavior in another field. While misinformation exists in all areas of our world, there are some places where it is tolerated and even celebrated and some where it is tracked down and stopped. By trying to bring the Silicon Valley mindset coined at Facebook, of "move fast and break things" to healthcare, Holmes made a ruinous error that has impacted her clients, employees, investors, and her family forever.

CONTEXT AND RISK

What the spectrum of examples above is meant to highlight is the omnipresence of misinformation and the wildly different scales of impact they can have. No one stops children from playing the telephone game and scolds them for their inability to accurately share messages, but Elizabeth Holmes and Sam Bankman-Fried are both in prison at the time of publication. The difference, while obvious, is also really important when we think about and design strategies to deal with misinformation. The context and the risk of damage matters greatly. When beginning this project, we knew we were fascinated not by the novelty of misinformation, but by its ancient history and inevitability. As information science scholars and policy nerds, we are always questioning how new the latest societal problem or solution is, and how much the technologies in our pockets, backpacks, and covering our eyes have really changed the world. Being trained in this field has helped us see that few things are truly new. This volume will question many of the unspoken aspects of misinformation by exploring the most everyday uses, problems, and spaces to try and gain a nuanced understanding of something that may not be new, but definitely will shape our future.

REFERENCES

Baker, Katie. 2023. "'A Clear Message': Sam Bankman-Fried Is Found Guilty on All Seven Counts." The Ringer. November 3, 2023. www.theringer.com/tech/2023/11/3/23945301/sam-bankman-fried-sbf-verdict-guilty-all-counts-ftx-cryptocurrency-trial.

Barber, Gregory. 2020. "When 'Ghost Kitchens' Become Mystery Grubhub Listings." *Wired*, January 29, 2020. www.wired.com/story/ghost-kitchens-mystery-grubhub-listings.

Campbell, Ian Carlos. 2021. "Imposters Steal Restaurants' Names in Delivery App Deception." *The Verge*. April 9, 2021. www.theverge.com/2021/4/8/22374341/imposter-restaurants-doordash-grubhub-ghost-kitchen-wagyumafia-blowfish-sushi.

Carreyrou, John. 2020. *Bad Blood: Secrets and Lies in a Silicon Valley Startup*. 1st ed. New York: Vintage.

Certeau, Michel de. 1984. *The Practice of Everyday Life*. Berkeley: University of California Press.

Dekker, Erwin, and Pavel Kuchař. 2022. *Governing Markets as Knowledge Commons*. Cambridge Studies on Governing Knowledge Commons. Cambridge: Cambridge University Press.

Dodds, Io. 2022. "Behind the 'Fake' Uber Eats and DoorDash Restaurants That Are Infuriating Customers." *The Independent*, November 3, 2022, sec. World. www.independent.co.uk/news/world/americas/doordash-uber-eats-ghost-kitchens-virtual-restaurants-b2215680.html.

Frischmann, Brett M., Michael J. Madison, and Madelyn Rose Sanfilippo, eds. 2023. *Governing Smart Cities as Knowledge Commons*. Cambridge Studies on Governing Knowledge Commons. Cambridge: Cambridge University Press. https://doi.org/10.1017/9781108938532.

Frischmann, Brett M., Michael J. Madison, and Katherine J. Strandburg, ed. 2014. "Governing Knowledge Commons: Introduction & Chapter 1." In *Governing Knowledge Commons*. Oxford: Oxford University Press. https://papers.ssrn.com/abstract=2490622.

Gamesver Team. 2022. "Telephone (Chinese Whispers): History, Origins, Metaphor, Meaning …" *Gamesver* (blog). March 2, 2022. www.gamesver.com/telephone-chinese-whispers-history-origins-metaphor-meaning.

Griffith, Erin. 2023. "Elizabeth Holmes Reports to Prison to Begin More Than 11-Year Sentence." *The New York Times*, May 30, 2023, sec. Technology. www.nytimes.com/2023/05/30/technology/elizabeth-holmes-theranos-prison.html.

Guy Fieri's Flavortown Kitchen. n.d. Website. Accessed January 24, 2024. www.guyfieri.com/eat-at-guys/flavortown-kitchen.

Haraway, Donna Jeanne. 2004. *The Haraway Reader*. New York: Routledge.

Huang, Yunte. 2015. "Chinese Whispers." *Verge: Studies in Global Asias* 1 (1): 66–69. https://doi.org/10.5749/vergstudglobasia.1.1.0066.

If Books Could Kill, n.d. "Those Shoes Look Really Comfortable." Patreon audio podcast. Accessed February 7, 2023. www.patreon.com/IfBooksPod.

Johns, David Merritt. 2023. "Nutrition Science's Most Preposterous Result: Could Ice Cream Possibly Be Good for You?" *The Atlantic*, April 13, 2023. www.theatlantic.com/magazine/archive/2023/05/ice-cream-bad-for-you-health-study/673487.

Kelso, Alicia. 2021. "U.S. Restaurant Industry Finished the Year $240 Billion Below Pre-pandemic Sales Estimates." *Forbes*, January 26, 2021. www.forbes.com/sites/aliciakelso/2021/01/26/the-us-restaurant-industry-finished-the-year-240-billion-below-pre-pandemic-sales-estimates.

Krishna, Priya. 2021. "So, What's the Deal with Ghost Kitchens?" *Bon Appétit*, April 19, 2021. www.bonappetit.com/story/ghost-kitchens.

Lefebvre, Henri. 2008. *Critique of Everyday Life*. Special ed. London: Verso.

Lerner, Adam B. 2015. "The Calculus behind Hillary Clinton and 'Everyday Americans.'" *Politico*. April 18, 2015. www.politico.com/story/2015/04/the-calculus-behind-hillary-clinton-and-everyday-americans-117099.

Lewis, Michael. 2023. *Going Infinite: The Rise and Fall of a New Tycoon*. New York: W.W. Norton.

Lucas, Amelia. 2021. "Ghost Kitchens and Delivery-Only Restaurants May Be Too Popular for Their Own Good." *CNBC*. February 1, 2021. www.cnbc.com/2021/02/01/why-a-reckoning-may-be-ahead-for-ghost-kitchens-delivery-only-brands-.html.

Ocepek, Melissa G. 2017. "Bringing Out the Everyday in Everyday Information Behavior." *Journal of Documentation* 74 (2): 398–411. https://doi.org/10.1108/JD-10-2016-0119.

O'Toole, Garson. 2017. *Hemingway Didn't Say That: The Truth behind Familiar Quotations*. 1st ed. New York: Little A.

Perry, Alex. 2023. "Uber Eats Wants to Shut Down Those Fake 'Ghost Kitchens' in Your Neighborhood." Mashable. March 28, 2023. https://mashable.com/article/uber-eats-ghost-kitchens-shut-down.

Reichman, Jerome H., Paul F. Uhlir, and Tom Dedeurwaerdere. 2016. *Governing Digitally Integrated Genetic Resources, Data, and Literature: Global Intellectual Property Strategies for a Redesigned Microbial Research Commons*. Cambridge: Cambridge University Press. https://doi.org/10.1017/CBO9781139128957.

Reiley, Laura. 2016a. "At Tampa Bay Farm-to-Table Restaurants, You're Being Fed Fiction." *Tampa Bay Times*, April 13, 2016. www.tampabay.com/projects/2016/food/farm-to-fable/restaurants.

2016b. "Tampa Bay Farmers Markets Are Lacking in Just One Thing: Local Farmers." *Tampa Bay Times*, April 13, 2016. www.tampabay.com/projects/2016/food/farm-to-fable/farmers-markets.

Sanfilippo, Madelyn Rose, Brett M. Frischmann, and Katherine J. Strandburg, eds. 2021. *Governing Privacy in Knowledge Commons*. Cambridge Studies on Governing Knowledge Commons. Cambridge: Cambridge University Press. https://doi.org/10.1017/9781108749978.

Schutz, Alfred. 1973. *The Structures of the Life-World*. Northwestern University Studies in Phenomenology & Existential Philosophy. Evanston, IL: Northwestern University Press.

Schweik, Charles M., and Robert C. English. 2012. *Internet Success: A Study of Open-Source Software Commons*. Cambridge, MA: MIT. https://doi.org/10.7551/mitpress/9780262017251.001.0001.

Shannon, Claude Elwood. 1948. "A Mathematical Theory of Communication." *The Bell System Technical Journal* 27 (3): 379–423. https://doi.org/10.1002/j.1538-7305.1948.tb01338.x.

Shannon, Claude Elwood, and Warren Weaver. 1964. *The Mathematical Theory of Communication*. Urbana: University of Illinois Press.

Smith, Dorothy E. 1987. *The Everyday World as Problematic: A Feminist Sociology*. Northeastern Series in Feminist Theory. Boston: Northeastern University Press.

Strandburg, Katherine J., Brett M. Frischmann, and Michael J. Madison. 2017. "The Knowledge Commons Framework." In *Governing Medical Knowledge Commons*, ed. Brett M. Frischmann, Katherine J. Strandburg, and Michael J. Madison, 9–18. Cambridge Studies on Governing Knowledge Commons. Cambridge: Cambridge University Press. https://doi.org/10.1017/9781316544587.002.

Strauss, Valerie. 2018. "Analysis | Word of the Year: Misinformation. Here's Why." *Washington Post*, December 10, 2018.

2

Storytelling and/as Misinformation

Storytelling Dynamics and Narrative Structures for Three Cases of COVID-19 Viral Misinformation

Kate McDowell

INTRODUCTION

How do people encounter misinformation through storytelling in their everyday lives during a pandemic? This chapter focuses on the dynamics of storytelling in misinformation as a problematic aspect of the COVID-19 pandemic in three cases of misinformation storytelling. Storytelling offers a framework for researching collective experiences of information as a process that is inherently based in communities, with knowledge commons that are instantiated by the telling and retelling of stories, temporarily or permanently. To understand how difficult information is to govern in story form and through storytelling dynamics, this chapter uses new storytelling theory to explore three recent cases of COVID-19 misinformation related to drug misuse, exploiting vaccine hesitancy, and community wisdom in resisting medical racism. Understanding what goes wrong with these stories may be key to public health communications that engage effectively with communities' everyday misinformation challenges.

Kate McDowell regularly teaches both storytelling and data storytelling courses and is the 2022 recipient of the ASIS&T Outstanding Information Science Teacher Award. She researches and publishes in the areas of storytelling as information research, social justice storytelling, and what library storytelling can teach the information sciences about data storytelling. Her projects engage contexts such as libraries, non-profit fundraising, health misinformation, social justice in libraries, and others. Dr. McDowell has worked with regional, national, and international non-profits including the Pan-American Health Organization (PAHO, part of WHO) and the Public Library Association (PLA). Her nationally funded project Data Storytelling Toolkit for Librarians with co-PI Dr. Matthew Turk is under development (https://imls.gov/grants/awarded/re-250094-ols-21). Her storytelling research has involved training collaborations with advancement with both the University of Illinois at Urbana Champaign and the University of Illinois system (Chicago, Springfield), storytelling consulting work for multiple nonprofits including the 50th anniversary of the statewide Prairie Rivers Network that protects Illinois water, and storytelling workshops for the Consortium of Academic and Research Libraries in Illinois (CARLI). She formerly served as Interim Associate Dean for Academic Affairs and Assistant Dean for Student Affairs and has led multiple transformative projects for the School.

FIGURE 2.1 Visual themes from storytelling and/as misinformation: storytelling dynamics and narrative structures for three cases of COVID-19 viral misinformation.

INTRODUCTION AS STORY

Another way of introducing this chapter is to compare two historical stories of community experiences related to health and medicine. The first story comes from my family. My mother heard a story from her grandmother about being a young mother herself during the 1918 influenza epidemic, when the whole household came down with the flu. Her parents (my great-great-grandparents) left food on the porch and visited through an open window, using much the same "social distancing" practices in Wetumpka, Alabama in 1918 as in the COVID-19 pandemic today. The wisdom of these practices left an impression, and she recognized them immediately when they were called into use globally once again in spring of 2020. Continuity of stories in families can provide a kind of storytelling wisdom that accompanies the stories we inherit (McDowell 2021).

Also in Alabama, fourteen years later and 51 miles away in Tuskegee, the US Public Health Service launched the Tuskegee Syphillis study in 1932. The history of this study involves racism, torture, and genocidal practices against Black men who were unknowingly subjects of untreated syphilis infections. For human subjects researchers, this history is repeated at every basic Institutional Review Board (IRB) training, and this unethical study is one of the main reasons for the 1974 National Research Act, which requires "voluntary informed consent" from all human research subjects (Centers for Disease Control and Prevention 2021). Families who experienced the pain, suffering, and in many cases the death of their loved ones later learned that they had gone untreated because of deception, as a "scientific" experiment. Here as well, information was inherited in the form of storytelling wisdom and mistrust of federal medical programs.

Communities that have been subjected to this and other kinds of systemic persecution are reasonably skeptical of new health interventions. When medicine enacts genocide, it does terrible damage at the level of the commons, with impacts for generations to come, and this is only one story in a larger history of federally supported medical racism. The resistance of communities impacted should be seen not as an information problem, but as historical and storytelling-informed wisdom. After histories like these, people react "not with defensiveness, but with comprehension" (Howell 2005, 221). The routine misunderstandings of medical professionals deepen the harm. When current medical programs ignore or dismiss the wisdom of community survival, they commit an act of epistemcide, defined as the "killing, silencing, annihilation, or devaluing of a knowledge system" (Patin et al. 2021b).

In cases like these and many more, storytelling can yield theoretical insights into commons governance of shared knowledge and the tragedy of misinformation, information, and shared wisdom. Storytelling is an informal and yet ubiquitous means of information circulation. This chapter focuses on the case of storytelling and retelling of misinformation as a problematic aspect of the COVID-19 pandemic through an infodemics lens. Misinformation is false, inaccurate, misleading, or decontextualized so that it misinforms audiences, with strong similarities to disinformation which is designed to misinform or mislead, and malinformation which is intended to cause harm (Wardle and Derakhshan 2017). Infodemiology is the study of surges of information (including misinformation) that lead to confusion and risk during public health crises (Mavragani 2020). Put another way, "an infodemic is an overflow of information of varying quality that surges across digital and physical environments during an acute public health event. It leads to confusion, risk-taking, and behaviors that can harm health and lead to erosion of trust in health authorities and public health responses" (Calleja et al. 2021, 2). Infodemic issues related to COVID-19 are the focus here, but nonetheless they serve as just one kind of theorizing storytelling as an everyday means of information circulation. Understanding both the dynamics of storytelling and how they produce stories can help to identify what has gone wrong in infodemic contexts.

Efforts to address complex social issues require understandings of storytelling and retelling as a fundamental process of everyday information circulation. This chapter uses two storytelling theory frameworks to analyze three cases: medicine misuse, exploiting vaccine hesitancy, and aftermath of medical racism. Understanding what goes wrong with these stories may be key to public health communications that engage effectively with communities' everyday misinformation challenges.

LITERATURE REVIEW

Storytelling

Story is a fundamental information form that has been overlooked in the information sciences (IS), although it has a rich history of over 130 years of practice in library and information science (LIS) in youth services and librarianship (Agosto 2016; Bishop and Kimball 2006; Colón-Aguirre 2015; Del Negro 2017; Hearne 1999; Sturm and Nelson 2016). Storytelling is central to understanding social, collective, and community meaning-making because it intertwines informational and emotional elements. It is also a strong vantage point from which to engage everyday information epistemology and knowledge commons. IS as a field tends to swing like a pendulum between the implicit positivism of the computational and the social constructionism of humanistic approaches, but storytelling (as well as everyday information and knowledge commons) requires an epistemology that bridges these divides. Storytelling dynamics produce stories that inform while being deeply contextually relevant. "In story, wisdom often means discovering a way beyond the ways that seem obvious" (McDowell 2021). Of course, other human cultural and wisdom traditions are interwoven with storytelling practices that have lasted much longer, and both fields have "neglected to treat people as epistemic agents who are embedded in cultures, social relations, and identities" (Patin et al. 2021a)

Storytelling is here defined as telling a story within the dynamic triangle of the story, the teller, and the audience. Whether permanent familial relationships or temporary relationships among strangers at a shared event, the story emerges in the connection between the teller and the audience. Storytelling roles in a culture include who may tell as story as well as who is not permitted to tell or hear a story, as in cases of indigenous cultures encountering folklorists who are allowed to study a tradition but asked to destroy recorded stories (Toelken 1998).

Storytelling may be necessary to bridge divides between indigenous epistemologies and science, as Kimmerer demonstrates in her use of standard scientific paper headings to tell the story of a doctoral student who demonstrated that, contrary to most of her committee's beliefs, harvesting does not have to damage sweetgrass; indigenous knowledge of symbiotic harvesting stimulated more vigorous growth (Kimmerer 2013). The stories of indigenous lives in the Americas have been difficult to tell because of these and other limiting academic beliefs. In her article establishing the "importance of felt experiences as community knowledges," Million writes:

> The mainstream white society read Native stories through thick pathology narratives. Yet the same stories collectively witnessed the social violence that was and is colonialism's heart. Individually or collectively, these stories were hard to "tell." They were neither emotionally easy nor communally acceptable. Women (and men) who organized against family violence and politically sanctioned sex discrimination in their communities balanced the necessity to change things and constraints to "silence" their pain and experience. To "tell" called for a reevaluation of reservation and reserve beliefs about what was appropriate to say about your own family, your community. (Million 2009, 56)

All academic disciplines and endeavors have been formed around, absorbed, continued, and reified these storytelling inequities. As have science and political science, "LIS has undermined knowledge systems falling outside of Western traditions" (Patin et al. 2021a).

And yet the LIS tradition of storytelling as a dynamic relationship and process holds within it the possibility of a different kind of listening. In the storytelling triangle, the audience's relationship to the teller hinges in part on how they understand the teller's own relationship to the story as well as which story the teller chooses to tell that audience (Figure 2.2).

For storytelling to occur, some trust must exist between the teller and the audience. This trust is contextual and depends on demonstrating that the teller wants this audience to know this story. Whether a story is received as worthy of believing and retelling depends in part upon this trust. Second, the teller has a relationship to the story, whether as creator or reteller. The storyteller (and audience) cannot be neutral as they inevitably inhabit a worldview. Third, the audience has a relationship to the story, primarily through their interpretation. That relationship is informed by everything the teller says (gestures, performs, writes, records, etc.)

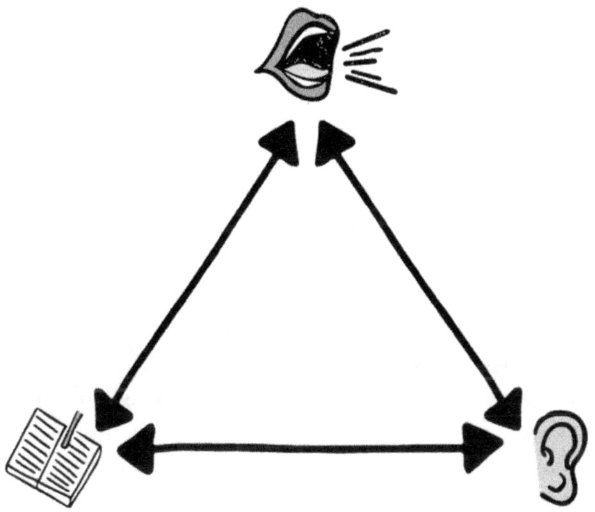

FIGURE 2.2 The storytelling triangle. Courtesy of Hilary Pope, artcoopshop.com.

in the course of telling the story and much more. The audience's interpretation can influence the meaning of the story for the teller as well, through listening to collective responses. Audiences interpret stories as groups, creating socially constructed meanings in the moment and sustaining cultural meanings over long spans of time. Retellings demonstrate how a story can be told so that others also recall and retell, and the story persists and moves on.

Storytelling has been studied extensively in other contexts, including organizations, with a strong emphasis on its role in creating and sustaining organizational cultures (Boje 2008; Denning 2011; Marek 2011). It has also been mobilized in counter-narrative contexts, as an important part of the process of hearing silenced voices and diversifying LIS (Cooke 2016). However, it is important to consider that storytelling is also a powerful noninstitutional information conduit. It is ubiquitous, flexible, and difficult to track. Folklore studies offer some insights into why this is. When considering the dynamics of folklore, retellings or reiterations of folk traditions are always simultaneously governed by cultural norms and inventive. "We may say that folklore itself is characterized by (1) certain cultural rules that determine strongly what gets articulated and how and when, and (2) by a looseness, an informality, an inclination toward rephrasing and change" (Toelken 1979, 31).

Storytelling and Commons

Storytelling relates to ideas of knowledge commons as resources. Stories are circulated through telling and retelling, carrying implicit cultural knowledge, and yet the specific form that a story takes depends on the teller, the audience, and the relationships between them. The cultural rules of a folk community swapping stories relate to the concept of knowledge commons, which "refers to an approach (commons) to governing the management or production of a particular type of resource (knowledge)." Commons is the institutional arrangement of resources in relation to a community, and knowledge means "cultural, intellectual, scientific, and social resources (and resource systems) that we inherit, use, experience, interact with, change, and pass on to future generations" (Frishmann et al. 2014). Storytelling offers a framework for researching collective experiences of information as a process that is inherently based in communities. However, stories circulate in ways that may make it challenging to clearly identify the commons, readily crossing boundaries implied by institutional arrangements. While stories do not require specific locations to circulate, their circulation relates to the concept of "inherently public property" (Rose 1986). Stories that have been retold again and again – from urban legends to cautionary tales to jokes – with no identifiable origin (or endpoint) may seem to belong to everyone who repeats them and no one at the same time. Storytelling, as defined in the library tradition, is cooperation that operates in a cultural context of public service.

Storytelling is also fundamental to the coproduction of community and of commons themselves. Communities build around stories of who we are, of what we share, and stories are ways of building, determining, and discerning who belongs within a particular commons. Storytelling is present in "remember when" recollections or "that's not how it happened" arguments among community members. Negotiating, repeating, celebrating, or bickering over stories and the realities they represent occur routinely, in both formal and informal ways, in a community that shares a commons. In some sense a community's stories are not just a part of its commons, they are simultaneously the signs of its borders and the production and property of its participants.

There is a productive contrast to be made between LIS storytelling and stand-up comedy. Some aspects are similar to storytelling, including the need for the audience: "a joke is only as good as its ability to make audiences laugh, which can only be gauged through public performance" (Bolles 2011, 241–242). However, LIS storytelling is predicated on retelling adaptations of folktales that are in the public domain (with multiple print versions) and are shared as original and spontaneous (transformative) adaptations in educational settings, and so rely on fair use. By contrast, as recent cases of "joke theft" have shown, stand-up comedy jokes and routines can be protected as intellectual property, such that "two entertainers can tell the same joke, but neither entertainer can use the other's combination of words" (239).

Folklorist and acclaimed librarian storyteller Betsy Hearne lauded the aesthetics of folkloric stories, with their "fast-moving, highly structured elemental plots" and "clearly delineated archetypal characters," for allowing each listener "to glean different emotional, socio-cultural, intellectual, spiritual, and physical connections with a tale" (Hearne 2011, 214). In the LIS storytelling tradition, staying true to the story means capturing its essence while telling it anew, emphasizing spontaneous expression in service of the audience. Early children's librarians underwent rigorous training for storytelling, practicing in front of mirrors (Shedlock 1915, 144), to learn their stories aurally, or through imagery, but not to memorize the words. Augusta Baker, the first to hold the position of Storytelling Specialist at the New York Public Library, directed new storytellers to emphasize "the story rather than upon the storyteller, who is, for the time being, simply a vehicle through which the beauty and wisdom and humor of the story comes to the listeners" (Baker and Greene, 1977, 58).

In LIS storytelling and other areas, knowledge commons can be enriched by use rather than depleted, as each retelling creates an original adaptation in the moment. Similar enrichment of commons is found in much longer cultural traditions. For example, in describing indigenous knowledge, Joranson proposes looking "inside the commons" to see that there is an understanding of knowledge "that does not become depleted through use." "As we continue to develop language to describe the knowledge commons, it is important to explore language that does not keep these resources in opposition or as separate, but makes their mutuality visible" (Joranson 2008, 70–71). In the same way, stories are not depleted by LIS storytelling, but instead may grow richer with retelling over time. When tales are consensually swapped (and not stolen from one culture by another), there may in fact be continual renewal and replenishment of the commons.

There are more reasons to be interested in story and storytelling as simultaneously informational and emotional processes based on recent research from other fields. For example, recent neurological research finds that neural story processing involves "mirroring process of embodied subjectivity" or experiences of "narrative emotions" predicated upon story's "ability to intertwine our experience of time" (Armstrong 2020). Specialized "mirror neurons" in the brain contribute to experiencing empathy through story (Rizzolatti 2008), and contextual empathy cues increase the potential for empathetic experience through story (Roshanaei et al. 2019).

Storytelling as Methodology and Metaphor

Epistemologically and methodologically, storytelling is a fruitful metaphor for the commons in our time. In our physical, virtual, and hybrid worlds, an exchange of story can take place anywhere, and the retelling of a story may change (enhance or degrade) its content; there is no shortage of story created by the use or reuse of story. Baker emphasized that "storytelling at its best is mutual creation." (Baker and Greene 1977, xii). This focus on relationships is key to understanding storytelling dynamics. Because stories are constituted through narrative experience, and audiences are partly constitutive of the stories told to and with them, storytelling offers a framework for researching collective experiences of information as a process.

To understand infodemics, for example, we need nuanced competencies to unpack and understand misinformation. The following section define storytelling dynamics and narrative structures that are key to understanding misinformation storytelling.

Engaging storytelling can help to reveal "how individuals really interact with information on a fundamental level," by prodding us to "look to the relationship individuals have with information." Part of that relationship entails when an individual is part of an audience, when they are listening to a storyteller, and how each story emerges as an aspect of that relationship. In considering everyday life information behavior Ocepek (2018) urges scholars to look to "narrative, lived experience, and other non-traditional forms of information as valuable resources and means for understanding everyday life" (404–405). Storytelling dynamics are critical to making sense of how people use and experience information, but they are not the only important aspect. Equally important is a fundamental understanding of story as information in narrative structures.

Storytelling Narrative Structures

Narrative structures are story patterns, and their recurrence across eons of time and multiple human cultures signal their robustness for simultaneously conveying information and emotion in story form. Their persistence demonstrates what has worked with audiences previously. Based on a broad overview of narratology and related narrative theories, three simple and relatively intuitive structures were developed and refined over eight years of workshops and teaching, with input from audiences through discussion and interviews (Table 2.1).

The *continuity* narrative structure is about cycles, journeys from stability to disruption and on to a new stability. The emotional impact of the equilibrium structure is resilience or stability despite challenge. This structure is often useful for stories about rebuilding and reorganizing in order to sustain continuity. For example, when antivax rhetoric surges in an infodemic, stories about the history,

TABLE 2.1 *Narrative structures*

Narrative structure	Emotional impact	Originating theory
Continuity	Stability and resilience despite challenges, reassurance of continuity	Tsvetan Todorov's *The Fantastic*, and the story "The Stonecutter" by Andrew Lang, attributed as Japanese but actually Dutch in origin
Transformation	Awe at transformation, joy of watching a hero triumph	Joseph Campbell's *The Hero With a Thousand Faces*
Discovery	Mystery, suspense, intrigue, and satisfaction of coming to understanding	Roland Barthes' hermeneutic or enigma code, one of five semiotic codes in *S/Z*

invention, and successes of vaccines and vaccination campaigns communicate that, despite challenges and change, the practice of vaccination evolves and endures (Lang 1903, 192–197; Sturm 2008; Todorov 2004).

In the *transformation* structure, based on the hero's journey, a protagonist encounters obstacles and is transformed b the process of overcoming them (Campbell 1949). This structure conveys a feeling of awe at the hero's triumph over obstacles, but it is not a universal structure and should be used with caution, because self-aggrandizing stories can backfire and risk losing the audience's interest or trust. In an infodemic context, this structure characterizes the global transformation from the pandemic of COVID-19 to the availability of effective vaccines.

The *discovery* narrative structure comes from the hermeneutic or enigma code, which is one of five codes defined by Roland Barthes that define ways that meaning drives a narrative (Barthes 1970). The discovery narrative structure is a recurring cycle of suspense and discovery, intrigue and information, curiosity and satisfaction. Emotionally, it is like a mystery story, where the audience follows the investigation of a detective-as-narrator in coming to understand what happened, why, and how things could be different. The story of Alexander Fleming's discovery of a mold that killed bacteria to the development of penicillin is a classic discovery narrative.

Although story plots often communicate causality, they also elide strict definitions of causality, and in fact the plot of a story may range from true causation to loose correlations in story sequence. In an aspect of confabulation, events that do not constitute fully causal explanations can seem to be associated in a story plot. Because narrative structures are flexible, any of these structures can be used to make compelling stories, whether true or confabulated or someplace in between.

WHERE STORYTELLING GOES WRONG: COVID-19 MISINFORMATION

Some of the most complex issues related to social justice, civil rights, racism, and more have been understood as relating to storytelling and narrative (Bell 2010; Polletta 2006). Victims of misinformation may lack resources to fact-check the stories they hear. Even if they can fact-check, they may not be able to describe the emotional effects of an infodemic, which may be, as Million says, "hard to 'tell'." Many people – from the most vulnerable populations to the most accomplished social scientists – have needed time to understand the social media disinformation strategies at play in confusing and disorienting audiences to create even more vulnerability.

We know very well in these times that stories that are retold and believed by millions of people are not always true. People want to understand the world around them, and that desire is so strong that, when things don't fit, human brains confabulate, finding and making stories out of whatever is available. Some psychological experiments have revealed that confabulation can be reinforced, so that participants develop false memories (Zaragoza et al. 2021). Storytelling dynamics may, in some cases, increase the propensity to share misinformation. The COVID-19 infodemic has brought new ways of understanding misinformation and its harms (Hansson et al. 2021). Further research is needed.

The present analysis focuses on understanding how storytelling dynamics (relationships between teller, audience, story) and narrative structures (continuity, transformation, discovery) play out in three example cases of misinformation. While each of these three examples – *medicine misuse, exploiting vaccine hesitancy, and aftermath of medical racism* – could be studied (and in some cases, have been) with a focus on social media, taking a storytelling approach offers some fresh insights about what goes wrong. The predominance of methods using metric analysis of shares (likes, retweets, etc.) has a tendency to obscure the ways that individuals frequently share misinformation as storytellers, adding their own comments, captions, messages, and more. In this sense, as storytellers, they repeat misinformation from their own point of view and tailor misinformation to their audiences. Sometimes the false or wrong information is amplified in content or in emotional urgency through these processes. While understanding the metric magnitude of viral misinformation is important, so too is understanding the qualitative aspects that speak to how and why specific tellers and audiences engage in listening to and retelling misinforming stories. Some misinformation may not be in strict narrative form, but it is part of a narrative experience – whether accurately explanatory or confabulatory – which is how storytelling operates in everyday life, and it evokes and invites story making. The same dynamics of surprise or outrage that lead to shares or retweets operate around the (now perhaps fabled) workplace watercooler, with the "you'll never believe this" stories. And audiences may lack the resources – time, access to peer-reviewed literature, knowledge, and more – to fully investigate whether something is true themselves, which may compound the spread and harms of misinformation.

In considering storytelling dynamics and narrative structures, it is important to establish a way of accounting for storytelling and/as information. This refinement of the previously published S-DIKW framework provides a tool for this analysis. Storytelling is here conceptualized as a set of abilities that relate to various levels of the DIKW hierarchy. Rather than each level being a descriptor, in the S-DIKW framework each level relates to human abilities to derive stories from data, to interpret data with context as information stories, to take action based on those information stories, and to enact wisdom in the selection of stories. (McDowell 2021)

- **S-Data** Basis of information in story
- **S-Information** Data interpretation with context as story
- **S-Knowledge** Actionable information in story
- **S-Wisdom** Which story to tell when, how, to whom, and more

It is important to distinguish between the broader definition of knowledge as a defining element of knowledge commons and s-knowledge as used here, which is derived from the narrower definition in the DIKW framework of actionable information (Rowley 2007). In relation to S-Wisdom, this definition is likely only partial, but provides a launching point for information fields to more actively and deliberately grapple with wisdom as a concept.

Storytelling analysis, then, can shed light on misinformation examples by (1) analyzing the relationships between teller, audience, and story, and (2) exploring what narrative structures are used to shape the meaning of the story. In the three examples that follow, a combination of these approaches helps to unpack how storytelling went wrong in three viral stories that exemplify various aspects of the COVID-19 infodemic.

Medicine Misuse: Desperation and Ivermectin

In fall of 2021, more than a year into the COVID-19 pandemic, stories about Ivermectin as a possible treatment circulated rapidly. This occurred at a point when it was still too early to tell whether it would be an effective COVID treatment, a speculation which clinical trials later disproved. Ivermectin was readily available at stores that sold veterinary supplies, since it is commonly used to treat parasites such as heartworms in dogs and worms and other parasites in livestock. Unfortunately, this misuse of medicine, particularly in horse-sized doses, led to both human illnesses and a temporary shortage of this previously common veterinary medication.

Unlike other drugs that are difficult to obtain without prescription, Ivermectin was commonly available and well known to veterinarians and animal caretakers who work with horses. Storytellers who retold, shared, tweeted, and otherwise spread this misinformation were, at best, helpless and willing to try anything available due to widespread panic. At worst, they were exploiting this panic to sell a "treatment" that was worthless or harmful or to boost their metrics on social media. In many cases, this story was retold as part of conspiracy theories. Tellers falsely claimed that they were engaged in righteous revelation of "the truth" from those "in power" who were hiding it. This "what they're not telling you" rhetoric is a recognizable aspect of populist rhetoric, which uses digital media to "stage intense dramas of dark forces and brave rebels fighting against them, led by charismatic rhetoricians revealing secrets and telling truths" (Finlayson 2022, 82). For audiences, this rhetoric further heightens reactivity, enlists them as "truth-tellers" who are actually spreading misinformation, and contributes to the rapid circulation of fake stories online.

From a narrative structure perspective, this kind of story engages both transformation and discovery. As a *transformation* story, the teller is promoting themselves as a truth-telling "hero," overcoming purported obstacles of "them" who don't want "you/us" to know – albeit one who gives no evidence of who the mysterious "them" might be that are trying to hide the truth. Audiences are readily swept up in the sense of drama, which might connect to their own senses of isolation or victimization, hoping that a populist hero will save them. Information literacy approaches might consider adding to the usual questions about authority and authoritativeness something more specific about teller–audience dynamics, such that audiences are encouraged to look for these rhetorical heroics as a cue to be skeptical of the associated information.

This rhetoric also amplifies fear that audiences are excluded from knowing that a disease treatment has been discovered. Purporting to be a *discovery* narrative, this trope of revealing something everyday with unexpected and hidden uses is familiar online, where household tips abound for the use of everyday substances for other purposes. Unfortunately, some of these "tips" are toxic. For example, in May 2022, viral TikTok videos promoted a "tip" storing avocados in water to make them last longer, which prompted the United States Food and Drug Administration to issue a warning against the practice (Fowler 2022). More broadly, the discovery narrative is familiar from so many stories of scientific advancement, from the discovery of penicillin to trips to the moon. From an information storytelling perspective, we might ask why are these stories of a cure – right in front of our noses but simultaneously hidden – so compelling? This may be related to brands which market themselves as "exclusive" or with "secret ingredients," and why behind-the-scenes stories are so compelling in marketing and fundraising (McDowell and Miller 2021). When considering knowledge commons, it is important to consider what might defuse the intensity associated with this aura of mystery.

Exploiting Vaccine Hesitancy: Faked Magnetism

In summer of 2021, a number of stories went viral about another purported mystery, a claim of causality between magnetism and vaccination. Specifically, rumors abounded that vaccines contained metals, microchips, or other substances that caused the human body to become magnetic at the site of the vaccination injection. Some people even produced video "evidence" that vaccinations caused bodily magnetism. These videos were bogus and scientifically laughable, and yet they tapped directly into audiences' fears of negative vaccine side-effects.

Deceptive YouTubers bumped up their metrics with likes, shares, and follows with video theatrics showing metal "sticking" to their skin where they were vaccinated. Audiences already primed by fears of a global pandemic and the complex conglomeration of safety concerns that it brought were especially vulnerable to these manipulations. At best, such videos were confabulation or delusion from antivaccination (antivax) perspectives; extensions of conspiracy theories about vaccination. At worst, they were performances that deliberately shocked audiences with false information and magic tricks of simple visual deception. Audiences in this context operate as social media currency, meant to spread this story because of its shock value. At the same time, the potential negative impact of this deception took on greater weight with the rapid spread of COVID and clear evidence that vaccines save lives.

Conspiracy theories present an intriguing set of dynamics from a storytelling perspective. First, tellers and audiences quickly become locked into an ever-increasing set of interrelated stories with an ever-narrowing window for contrary evidence. If audiences trust the source and are shocked or otherwise emotionally activated by the story, they are more likely to retell it. Over time, the tight connection between teller and audience inside this trust widens into a gap between insider audiences who share beliefs and anyone else as outsiders. Conspiracy theories become a system of stories through which tellers and audiences focus on a limited set of information, confirming each other's perspectives to the complete exclusion of any contrary evidence or interpretation.

From a narrative structure standpoint, this is most like a *continuity* story, in that the underlying message is to focus on maintaining the status quo despite disruption, in the COVID-19 case from a global pandemic. The implied argument is to rely on "natural immunity" rather than vaccination, to resist public health and governmental guidelines and treat them as part of the problem rather than part of the solution. At the same time, oversimplified public health rhetoric has tended to elide the complexities of vaccination. Yes, there were (and are) side effects of vaccination, and with a new vaccine not all side effects are immediately known (or knowable). When so much is unknown, storytellers readily confabulate, and those confabulations become stories that feed into conspiracy theories and faked performances.

Aftermath of Medical Racism: Vaccine Questioning

Many studies have now shown that COVID-19 vaccines were met by significant hesitancy even by those who are not generally opposed to vaccination, and that disproportionately, communities of people of color were more likely to pause before seeking vaccination. For example, as documented in a fall 2021 report, "in most states where data is available, Black people are receiving a smaller percentage of vaccines relative to their overall population, despite them accounting for a much larger share of COVID-19 deaths" (Dodson et al. 2021). This pervasive mistrust has been termed "vaccine hesitancy," but the term hesitancy is incomplete at best and misleading at worst. Like so many terms, it is decontextualized from the broader history of the United States.

It is vital to consider, from a storytelling perspective, what inherited information resulting from traumatic medical racism might mean for everyday information behavior. If two storytellers present contradictory stories, the factor of trust between teller and audience can tip the balance one way or the other. For the survivors of the victims of the Tuskegee atrocities, for example, the federal government and public health authorities that are the direct successors to the authorities that killed and tortured their relatives might not be trustworthy, to put it mildly. For their families, relatives who survived these atrocities would likely be more trustworthy than public health officials. BIPOC[1] family stories of Japanese "internment" camps to

[1] Black, Indigenous, and people of color.

"Indian schools" reveal precisely how and when systemic racism was enacted, while white families are routinely taught to elide, overlook, or forget these atrocities (Ali-Khan 2022). These are just three examples among hundreds in US history. Why would a person whose family was decimated by a government medical program within living memory (the Tuskegee Study ended only in 1972) be enthusiastic to comply with new government medical directives? Why would scientists and public health officials expect otherwise? Only a profound disconnect between (predominately white) public health storytellers and BIPOC audiences could lead to disdain for "vaccine hesitancy" instead of respect for vaccine questioning.

Narrative structures operating in this context may also be at odds. On the one hand, the *transformation* structure turns those who survived despite terrible obstacles into heroes, with lessons learned that include caution about new government medical programs. On the other hand, public health officials hope to tell a *discovery* narrative about the invention and value of vaccines against COVID-19, in the urgent context of pandemic. They expect to be trusted as experts and believe that more information is the answer. This attitude fails because they are outsiders to communities who have previously suffered because of such experts. If transformative survival has required caution while a discovery story demands immediate action, some audiences would understandably require time to consider those narrative contradictions.

This relates to the complexities of disconnects between knowledge commons, particularly when information travels through family stories. Families are frequent sources of information. This information is both explicit and implicit, instructional and cultural, and informs all aspects of life including which knowledge commons are available to, unavailable to, or even systematically obscured from particular individuals.

Unfortunately, there is evidence that white audiences cannot always hear the stories that BIPOC people tell. In a series of storytelling for social justice workshops, facilitators developed guidelines and a "pausing" practice to combat unconscious racism in the moment after a BIPOC individual tells their story of "discovering the degree to which racial advantage and disadvantage are constructed":

> Before she can finish, a white participant interrupts and asserts that she too has experienced disadvantage and that the problem is really class not race. One can feel the entire room hold its collective breath. Clearly this is a crucial moment. Will we move on with the discussion, as often happens, smoothing over tension and disagreement? Instead, one of the facilitators quietly asks the white participant to repeat what she heard her African American colleague say. When she cannot do so, it becomes clear that she had tuned out the first speaker. (Bell 2010, 103)

Any knowledge commons is constructed without attention to the complex ways that white audiences routinely fail to listen to BIPOC storytellers may perpetuate systemic racism. In the case of COVID-19 vaccination, the statistics have shifted over time as more people receive at least some vaccination, but disparities remain in

TABLE 2.2 *Storytelling insights*

Cases	Insights from triangle	Insights from narrative structure
Medicine misuse: desperation and Ivermectin	Teller allies with audience against "them" and "what they're not telling you" "Secret" revealed to audience, aura of mystery	*Transformation* story, teller as "hero" revealing "truth" *discovery* narrative, revealing something everyday with hidden uses
Exploiting vaccine hesitancy: faked magnetism	Teller uses shock based on trickery to activate audience emotionally to resist If audience trusts teller, then likelihood of retelling increases	*Continuity* story, in that the underlying message is to focus on maintaining the status quo despite disruption
Aftermath of medical racism: vaccine questioning	Inherited information: teller and audience are insiders together as families who survived governmental medical officials expect trust because of expertise and believe that more information is the answer	*Transformation* structure makes those survived despite terrible obstacles heroes Public health officials hope to tell a *discovery* narrative about the invention and value of vaccines

vaccination availability, job-related viral exposure, and the health effects of systemic racism that increase risk of poor health outcomes from COVID-19 infection, even as much remains unknown (Table 2.2).

STORYTELLING ABILITIES AND MISINFORMATION

Understanding the storytelling abilities required to spread misinformation may help us to specifically examine where in a storytelling process does information become inaccurate. In the context of the COVID-19 infodemic, the S-DIKW framework can be adapted to provide a structure for understanding when storytelling goes wrong, through examining how each of these levels of human storytelling abilities relates to creating or spreading misinformation as story. This can in turn help to explore and explain where and how misinformation amplifies social justice and civil rights or perpetuates racism.

With many compelling forms of misinformation circulating, coupled with the disappointing research discovery that fake news circulates faster than real news (Vosoughi et al. 2018), how can storytelling help to combat these issues? For example, how can public health leaders serve as storytellers to propagate accurate information stories while also acknowledging the historical injustices committed by public health work? For examples like misused Ivermectin or faked vaccine magnetism, it would be good to have a more rigorous understanding of how audiences come to adapt and retell these stories as their own, further misinforming communities.

One approach is to consider the S-DIKW framework from the perspective of misinformation, by asking: How has the story gone wrong? It might be false from the start (bad s-data); misunderstood (bad s-information); or wrongly put into action (bad s-knowledge); or wisdom, containing stories of important community knowledge but limited, misapplied, or risky (bad s-wisdom). The four-part framework below revises the S-DIKW framework for understanding the storytelling abilities necessary for misinformation storytelling.

Misinformation S-DIKW

- **Bad S-Data:** Evoking cultural cues that imply factuality for data that is false
- **Bad S-Information:** False data with context that misinforms in story
- **Bad S-Knowledge:** Stories based on false information that lead to ineffective or harmful actions
- **Bad S-Wisdom:** Reactivity that leads to retelling misinformation as story without checking sources in ways that amplify harm

Like the DIKW hierarchy, these levels typically build on one another. Bad S-Data, for example, frequently appears on platforms like Twitter in the use of manipulated or patently false (and unsourced) graphs or charts that look like data visualizations. A common example is faked data showing grossly exaggerated negative effects of vaccination. Bad S-Information would connect such a faked visualization with a caption addressing audiences and often encouraging further sharing. Bad S-Knowledge would be stories from people who use bad S-Data and bad S-Information as justification for harmful or negligent actions, such as refusing to be vaccinated despite public health and doctors' advice. Bad S-Wisdom is at least (and there might be more to consider here as well) telling a bad story at a time that will cause or reinforce harm to human health outcomes.

The three cases above each provide an interesting lens for considering the Misinformation S-DIKW framework. For the first case, Ivermectin was briefly considered a potentially useful antiviral drug, so at a particular moment prior to clinical trials, there was real (albeit partial and inconclusive) s-data about its potential. However, once that was disproven, then the lingering interest in Ivermectin was based on bad s-data that attempted to revive a theory that had been shown to be false. The way that time – for example, the process and outcome of clinical trials – factors into medical discoveries and the scientific process in general is not easy to tell as a story. The urgency of pandemic threats make this more difficult and complicated.

In the second case, faked magnetism was based on bad s-data that was produced as falsehood from the outset. Bad s-data became bad s-information by spreading false fears about vaccination, bad s-knowledge by discouraging vaccination, and bad s-wisdom through retelling. In essence, this is thoroughly bad S-DIKW.

The third case, however, is again more complex. The histories of medical racism injustices are real. In other words, they are based on good s-data and s-information, because the harm caused was real. In considering what to do, the s-knowledge to act cautiously and skeptically based on prior abuses is reasonable. The problem occurs at the level of s-wisdom. It is bad s-wisdom to avoid vaccination in a situation where vaccination can save lives and reduce harm. But it is also bad s-wisdom for medical and public health professionals to expect there to be no lasting consequences of governmentally sponsored medical abuse.

Ultimately, by looking at what goes wrong, a storytelling analysis may reveal more about the design strategies used by disinformation sources as they craft stories intended to go viral.

WHERE IS THE WISDOM?

Knowing how to tell the truth is easy with accurate data. Knowing how to tell what information is accurate is also relatively easy. However, telling a story that persuades audiences to act by conveying knowledge is more difficult, and conveying wisdom is more difficult still. As T. S. Elliot wrote: "Where is the wisdom we have lost in knowledge? Where is the knowledge we have lost in information?" (Eliot 1934). For example, debunking false information can be tricky, because reiterating the fake message in order to correct it may serve to reinforce it through repetition.

What part might wisdom have in discerning whether to retell such a story? Recent storytelling theory defines wisdom in two ways: (1) knowing which story to tell when and to whom, and (2) seeing a way beyond the ways that are obvious (McDowell 2021). Knowing which story to tell involves understanding what audiences will find enticing, convincing, and repeatable. From a public health knowledge commons perspective, it will mean making true stories as interesting as they are accurate.

One kind of trouble comes in the combination of false transformation narratives and conspiracy dynamics between tellers and audiences. Let's imagine a storyteller who, skeptical of COVID-19 news, speculates about the urgency of vaccination. That storyteller connects with audience members who share their stories expressing fears in falsehoods that the government is adding chips or tracking devices to people through vaccines. Together, reacting with fear and righteous indignation rooted in emotional responses linked to victimization, these stories escalate, moving quickly across the internet and amplifying from skeptical speculations to claims of conspiracy. Once a conspiracy theory is activated, the door may have already closed, dividing believers who form an increasingly tight-knit circle from anyone else who might hold contrary evidence or opinion (Hansson et al. 2021).

In thinking about a way beyond the ways that seem obvious, it is currently easier to imagine banding together to create misinformation than to stop it. Recent internet memes about "camping" as a ruse for traveling to access abortion are an example, in that groups rapidly made up and spread a cover story for people whose reproductive rights had been stripped by the Supreme Court. In a much older example, from a Haitian folktale, the girl Tipingee's stepmother tries to give her away as a servant to a stranger. The stepmother dresses her in red to mark her for the taking, but Tipingee overhears and shares the plan with all of the village girls. The next day, when the stranger comes to steal Tipingee away, they are all wearing red, hiding her in plain sight, and repeating the (infectious) chorus of "I'm Tipingee, she's Tipingee, we're Tipingee too." In the ending, justice prevails, and the stranger takes the stepmother instead (Ragan 1998).

Folktales may have an important place in considering the commons. Stories handed down through generations in a culture contain the wisdom of that way of living. And many Indigenous perspectives focus on knowledge as a shared resource. From the perspectives of "overstudied others," "Western knowledge and knowledge production is perceived as supporting and reproducing settler colonialism and the reification of knowledge as a commodity. This is directly opposed to Indigenous perceptions of knowledge as a shared community resource" (Tuck and Yang 2014, cited in Oliphant 2021, 960).

Of course, as the Toelken example mentioned earlier shows, the wisdom of a commons might involve hiding information as well as informing. Since it is patently unethical to manipulate, injure, and withhold treatment from humans in scientific experiments, stories like Tipingee beg the question: When is it ethical and perhaps even necessary to deceive in order to do the right thing? Communities which have been subject to settler colonialists have sometimes wisely hidden their knowledge in order to survive state-sponsored genocide.

Everyday misinformation and knowledge commons alike need to engage actively and rigorously with the concept of wisdom, particularly in considering how to forge a way beyond the ways that seem obvious. One aspect of wisdom should be a broad consideration of how challenging everyday circumstances can be for information sharing and knowledge commons. Sharing accurate information is difficult when the information is changing quickly, as with the emergence of a novel coronavirus that sparked the COVID-19 pandemic. Scientific advances continue on a monthly if not weekly basis, as do announcements of new variants discovered, and so it is difficult even for adept scientifically literate audiences to keep up. Even though the teams that invented vaccines moved incredibly quickly in a story that will become one of the scientific triumphs of our era, the rapid movement of information and misinformation remains complex and challenging.

CONCLUSION

Storytelling wisdom may be a key tool for engaging deep or complex social problems. Some of the most complex issues related to social justice, civil rights, racism, and more have been understood as relating to storytelling and narrative (Bell 2010; Polletta 2006). Survival in a pandemic can be supported by inherited family stories like the one my great-grandmother told to my mother about safely visiting through an open window. Survival against medical racism can be supported by family stories that tell of the horrors of Tuskegee and the questioning of governmental medical programs. Recently, scholars have begun to name some of the information harms of racism and its connection to epistemicide, or the "killing, silencing, annihilation, or devaluing of a knowledge system" (Patin et al. 2021b). As much damage as misinformation storytelling can cause, equally damaging are the forces that silence voices and still stories that could support survival if only they could be known.

Cultivating storytelling wisdom in the face of an infodemic requires great care in thinking about the ways that a particular version of a story emerges between a teller and an audience. Understanding how storytelling dynamics can spread misinformation stories may be key to public health communications that engage effectively with communities' everyday misinformation challenges. Storytelling offers a framework for researching collective experiences of information as a process that is inherently based in communities, with knowledge commons that are instantiated by the telling and retelling of stories, temporarily or permanently.

In other areas of misinformation research, similar approaches may be helpful in revealing what is happening and how to intervene to support accurate information sharing. In any field involving a commons orientation – public health, law, sociology, and many more – these frameworks could be productive for analyzing the interplay and layers of narrative dynamics and the narrative structures in order to more systematically understand how storytelling is contributing to misinformation. This may be especially important when competing contradictory narratives are at play between those governing and those governed, even within the same community. At the same time, there is no one form of analysis of information that will solve all of the social information problems today. Storytelling is one among many ways of analyzing information, but it has been an overlooked one until recently, and it deserves to be included in discussions about everyday misinformation and knowledge commons.

With its person-to-person form of circulation and retelling as recreation, it may be that storytelling is one of the most ungovernable aspects of knowledge commons, but it should be considered. It is worth analyzing complexities of exchanges among tellers and audiences and the stories that emerge and evolve between them. Misinformation moves at great speed, and understanding storytelling may help us to understand part of what creates that speed.

ACKNOWLEDGMENTS

Many thanks to colleagues Ian Brooks, Rebecca Kyser, and Jodi Schneider for fruitful exchanges of ideas related to storytelling and misinformation.

REFERENCES

Agosto, Denise E. 2013. "If I Had Three Wishes: The Educational and Social/Emotional Benefits of Oral Storytelling." *Storytelling, Self, Society* 9 (1): 53–76. https://doi.org/10.13110/storselfsoci.9.1.0053.

2016. "Why Storytelling Matters: Unveiling the Literacy Benefits of Storytelling." *Children and Libraries* 14 (2): 21–26. https://doi.org/10.5860/cal.14n2.21.

Ali-Khan, Sophia. 2022. *A Good Country*. Penguin Random House. www.penguinrandomhouse.com/books/645807/a-good-country-by-sofia-ali-khan/.

Armstrong, Paul B. 2020. *Stories and the Brain: The Neuroscience of Narrative*. Baltimore: Johns Hopkins University Press.

Baker, Augusta, and Ellin Greene. 1977. *Storytelling Art and Technique*. New York: Bowker.

Barthes, Roland. 1970. "Sémiologie et urbanisme," 53: (1–13): 141.

Bell, Lee Anne. 2010. *Storytelling for Social Justice: Connecting Narrative and the Arts in Antiracist Teaching*. 2nd ed. New York: Routledge. https://doi.org/10.4324/9780203852231.

Bishop, Kay, and Melanie A. Kimball. 2006. "Engaging Students in Storytelling." *Teacher Librarian* 33 (4): 28–31.

Boje, David M. 2008. *Storytelling Organizations*. London: Sage.

Bolles, Elizabeth Moranian. 2011. "Stand-Up Comedy, Joke Theft, and Copyright Law." *Tulane Journal of Technology & Intellectual Property* 14: 237–260.

Calleja, Neville, AbdelHalim AbdAllah, and Julienne N. Anoko et al. 2021. "A Public Health Research Agenda for Managing Infodemics: Methods and Results of the First WHO Infodemiology Conference." *JMIR Infodemiology* 1 (1): e30979. https://doi.org/10.2196/30979.

Campbell, Joseph. 1949. *The Hero with a Thousand Faces*. Princeton: Princeton University Press.

Centers for Disease Control and Prevention. 2021. "Tuskegee Study – Research Implications – CDC – NCHHSTP." April 26, 2021. www.cdc.gov/tuskegee/after.htm.

Colón-Aguirre, Mónica. 2015. "Organizational Storytelling among Academic Reference Librarians." *Portal: Libraries and the Academy* 15 (2): 233–250.

Cooke, Nicole A. 2016. "Counter-Storytelling in the LIS Curriculum." In *Perspectives on Libraries as Institutions of Human Rights and Social Justice*, 41: 331–348. Advances in Librarianship. Emerald. https://doi.org/10.1108/S0065–283020160000041014.

Del Negro, Janice. 2017. *Engaging Teens with Story: How to Inspire and Educate Youth with Storytelling*. Santa Barbara, CA: Libraries Unlimited, an imprint of ABC-CLIO.

Denning, Stephen. 2011. *The Leader's Guide to Storytelling: Mastering the Art and Discipline of Business Narrative, Revised and Updated*. San Francisco: Jossey-Bass.

Eliot, Thomas Stearn. 1934. "Choruses from the Rock." *The Complete Poems and Plays of T. S. Eliot*.

Eliot, T. S. 1962. *The Waste Land, and Other Poems*. A Harvest/HBJ Book. New York: Harcourt, Brace, Jovanovich.

Finlayson, Alan. 2022. "Brexit, YouTube and the Populist Rhetorical Ethos." In *Populist Rhetorics: Case Studies and a Minimalist Definition*, ed. Christian Kock and Lisa Villadsen. Rhetoric, Politics and Society. Cham: Springer, 81–106. https://doi.org/10.1007/978-3-030-87351-6_4.

Fowler, Kate. 2022. "FDA Warns Popular Avocado Hack Could Be Breeding Salmonella." *Newsweek*. May 20, 2022. www.newsweek.com/fda-warning-avocado-hack-salmonella-bacteria-1708562.

Frischmann, Brett M., Michael J. Madison, and Katherine J. Strandburg. 2014. "Governing Knowledge Commons." In *Governing Knowledge Commons*. Oxford University Press, 1–43.

Hansson, Sten, Kati Orru, and Ala Pigrée et al. 2021. "COVID-19 Information Disorder: Six Types of Harmful Information during the Pandemic in Europe." *Journal of Risk Research* 24 (3–4): 380–393. https://doi.org/10.1080/13669877.2020.1871058.

Hearne, Betsy. 1999. "Swapping Tales and Stealing Stories: The Ethics and Aesthetics of Folklore in Children's Literature." *Library Trends* 47 (3): 509–528.

2011. "Folklore in Children's Literature: Contents and Discontents." In *Handbook of Research on Children's and Young Adult Literature*, 209–223. Routledge.

Howell, Joel D. 2005. "Trust and the Tuskegee Experiments." In *Clio in the Clinic: History in Medical Practice*, ed. Jacalyn Duffin. 213–223. Oxford University Press.

Joranson, Kate. 2008. "Indigenous Knowledge and the Knowledge Commons." *International Information & Library Review* 40 (1): 64–72. https://doi.org/10.1080/10572317.2008.10762763.

Kimmerer, Robin Wall. 2013. *Braiding Sweetgrass: Indigenous Wisdom, Scientific Knowledge, and the Teachings of Plants*. 1st ed. Minneapolis, MN: Milkweed.

Lang, Andrew. 1903. *Social Origins*. Longmans, Green.

Marek, Kate. 2011. *Organizational Storytelling for Librarians: Using Stories for Effective Leadership*. Chicago: American Library Association.

Mavragani, Amaryllis. 2020. "Infodemiology and Infoveillance: Scoping Review." *Journal of Medical Internet Research* 22 (4): e16206. https://doi.org/10.2196/16206.

McDowell, Kate. 2020. "Storytelling, Young Adults, and Three Paradoxes." In *Transforming Young Adult Services*, 2nd ed., ed. Anthony Bernier, 93–109. Chicago, IL: American Library Association-Neal Schuman.

2021. "Storytelling Wisdom: Story, Information, and DIKW." *Journal of the Association for Information Science and Technology* 72 (10 [Special Issue: Paradigm Shift in the Field of Information]): 1223–1233. https://doi.org/10.1002/asi.24466.

McDowell, Kate, and Deborah Miller. 2021. "Great Storytelling: Best Practices for Advancement." *Journal of Education Advancement & Marketing* 6 (3): 257–267.

Million, Dian. 2009. "Felt Theory: An Indigenous Feminist Approach to Affect and History." *Wicazo Sa Review* 24 (2): 53–76.

Ocepek, Melissa G. 2018. "Bringing out the Everyday in Everyday Information Behavior." *Journal of Documentation* 74 (2): 398–411. https://doi.org/10.1108/JD-10-2016-0119.

Oliphant, Tami. 2021. "Emerging (Information) Realities and Epistemic Injustice." *Journal of the Association for Information Science & Technology* 72 (8): 951–962. https://doi.org/10.1002/asi.24461.

Patin, Beth, Tami Oliphant, and Kayla Lar-Son et al. 2021a. "At the Margins of Epistemology: Amplifying Alternative Ways of Knowing in Library and Information Science." *Proceedings of the Association for Information Science & Technology* 58 (1): 630–633. https://doi.org/10.1002/pra2.515.

Patin, Beth, Melinda Sebastian, Jieun Yeon, Danielle Bertolini, and Alexandra Grimm. 2021b. "Interrupting Epistemicide: A Practical Framework for Naming, Identifying, and Ending Epistemic Injustice in the Information Professions." *Journal of the Association for Information Science and Technology* 72 (10): 1306–1318. https://doi.org/10.1002/asi.24479.

Polletta, Francesca. 2006. *It Was Like a Fever: Storytelling in Protest and Politics*. Chicago: University of Chicago Press. https://doi.org/10.1177/009430610803700327.

Ragan, Kathleen. 1998. *Fearless Girls, Wise Women, and Beloved Sisters: Heroines in Folktales from around the World*. 1st ed. New York: W.W. Norton.

Rizzolatti, Giacomo. 2008. *Mirrors in the Brain: How Our Minds Share Actions and Emotions*, ed. Corrado Sinigaglia. Oxford: Oxford University Press.

Rose, Carol. 1986. "The Comedy of the Commons: Custom, Commerce, and Inherently Public Property." *University of Chicago Law Review* 53 (3): 711–781. https://doi.org/10.2307/1599583.

Roshanaei, Mahnaz, Christopher Tran, Sylvia Morelli, Cornelia Caragea, and Elena Zheleva. 2019. "Paths to Empathy: Heterogeneous Effects of Reading Personal Stories Online." *Proceedings – 2019 IEEE International Conference on Data Science and Advanced Analytics, DSAA 2019*, no. 1: 570–79. https://doi.org/10.1109/DSAA.2019.00072.

Rowley, Jennifer. 2007. "The Wisdom Hierarchy: Representations of the DIKW Hierarchy." *Journal of Information Science* 33 (2): 163–180. https://doi.org/10.1177/0165551506070706.

Shedlock, Marie L. 1915. *The Art of the Story-Teller*. 1st ed. D. Appleton.

Sturm, Brian W. 1999. "The Enchanted Imagination: Storytelling's Power to Entrance Listeners." *School Library Media Research* 2 (6): 1–21.

2008. "The Process of Sharing Stories with Young People." *Knowledge Quest* 36 (5): 12.

Sturm, Brian W., and Sarah Beth Nelson. 2016. "With Our Own Words: Librarians' Perceptions of the Values of Storytelling in Libraries." *Storytelling, Self, Society* 12 (1): 4–4. https://doi.org/10.13110/storselfsoci.12.1.0004.

Todorov, Tzvetan. 2004. *Literatura não é Teoria, é Paixão*. Difel.

2004. *Literatura não é Teoria, é Paixão*. Difel.

Toelken, Barre. 1979. *The Dynamics of Folklore*. Boston: Houghton Mifflin.

1998. "The Yellowman Tapes, 1966–1997." *Journal of American Folklore* 111 (442): 381–391. https://doi.org/10.2307/541046.

Tuck, Eve, and Wayne Yang. 2014. "R-Words: Refusing Research." In *Humanizing Research: Decolonizing Qualitative Inquiry with Youth and Communities*, 223–247. Thousand Oaks, CA: Sage. https://dx.doi.org/10.4135/9781544329611.

Vosoughi, Soroush, Deb Roy, and Sinan Aral. 2018. "The Spread of True and False News Online." *Science* 359 (6380): 1146–1151. https://doi.org/10.1126/science.aap9559.

Wardle, Claire, and Hossein Derakhshan. 2017. "Information Disorder: Toward an Interdisciplinary Framework for Research and Policy Making." DGI(2017)09, 162317GBR. F-67075 Strasbourg Cedex: Council of Europe. www.coe.int.

3

It's Not (All) about the Information

The Role of Cognition in Creating and Sustaining False Beliefs

Dominique Kelly and Jacquelyn Burkell

INTRODUCTION

There is currently a great deal of talk about "misinformation," "disinformation," and "fake news." There are widespread calls to limit the promotion of false information, particularly by the press, and to correct false information when and where it appears. These discussions, and these concerns, focus on the *content* of communications and on the need to ensure that communications are factually *correct* and *accurate*. Both are, indisputably, critical goals in the battle against "fake news" and other types of misleading information.

In the final analysis, however, it is not the false information per se that is of concern – rather, it is the *false beliefs* that can arise from that information that are the problem. Christopher Fox (1983) makes this issue eminently clear in his careful analysis of information and misinformation. According to Fox, to "misinform" someone is intentionally to cause them to believe something that is untrue, typically by promulgating false or inaccurate information. Critical to this definition is the end state – the state of false belief – and consistent with this focus, the term "misinformed" is widely used as an adjective to describe someone who holds incorrect or false beliefs. Thus, teens are said to be misinformed about the proper use of condoms (Rosenberg 2001), patients are identified as misinformed about the causes and treatment of the common cold (Braun et al. 2000), and prior to the 2016 United States election, a surprisingly high proportion of eligible voters were said to be misinformed about Barack Obama's birthplace (Holman and Lay 2019).

In this chapter, we focus on the *state* of holding false beliefs, and examine one particular *cause* of this state – specifically the biased cognitive processing of objectively true information. Our selective focus adds to a rich body of literature discussing and distinguishing information, misinformation, and disinformation (e.g., Fox 1983). We recognize that false beliefs can arise from objectively inaccurate information, especially where there is an intention to mislead. We also recognize that *trust* figures largely in the development of beliefs, and misplaced trust could lead

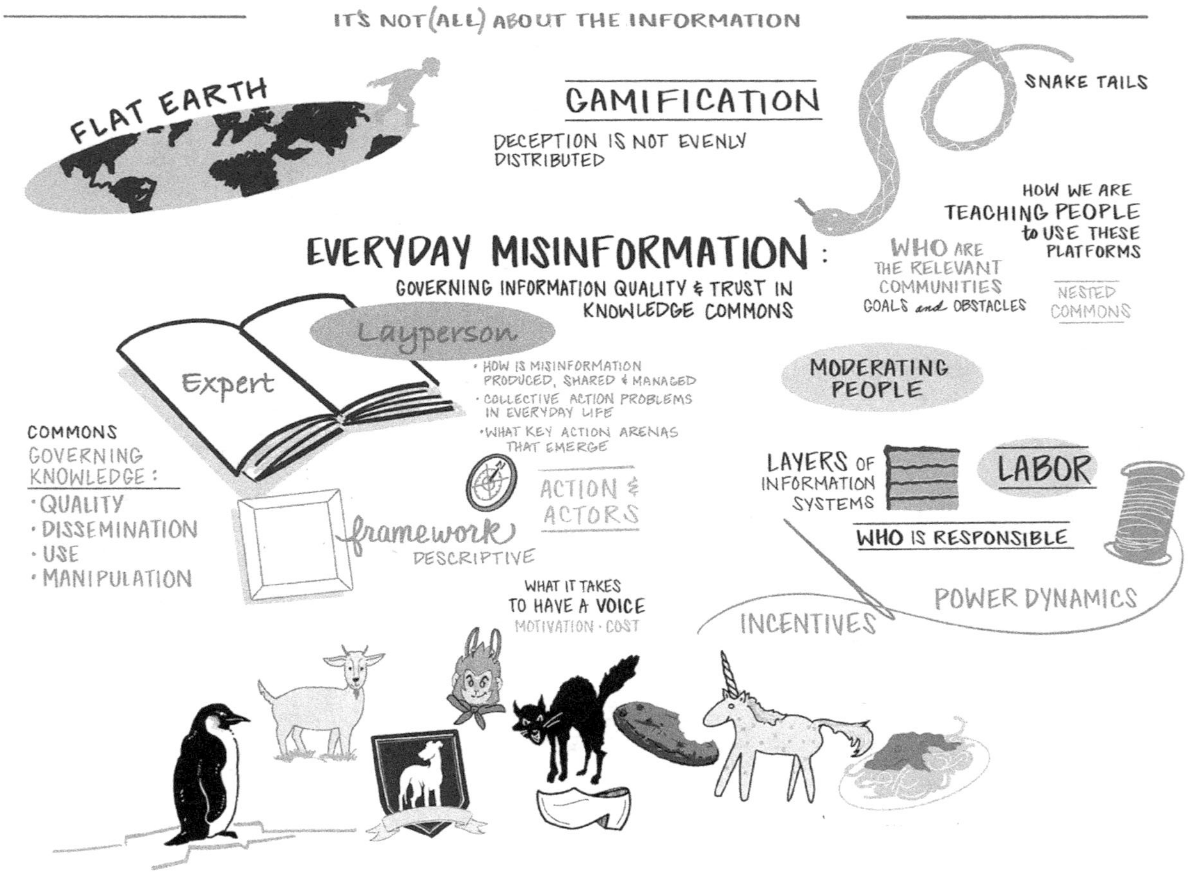

FIGURE 3.1 Visual themes related to everyday misinformation.

one to believe false information if the source of that information has in the past been reliable. More insidious is the impact of the increasing sophistication of false information (e.g., false deepfake videos depicting, with great apparent realism, events that did not happen) on our ability and willingness to trust, and thus believe, information that is in fact accurate (see Fallis 2020 for a discussion of this point). The development of true, or accurate, beliefs relies at least in part on the ability of the public to discern accurate information and reject false information, and to distinguish relevant from irrelevant information, and digital literacy training to improve these skills is an important aspect of the battle against misinformation (Scherer and Pennycook 2020). Our analysis adds a new perspective on the issue of false beliefs, suggesting an additional avenue through which such beliefs can arise, and additional strategies that can be used to minimize the development of false beliefs.

Setting the Context

To begin, a clarification: we are in no way denying the value of accurate information or the cost of false information. Both can and do tend to have the effects one would anticipate, in that accurate information tends to foster accurate beliefs, and false information tends to undermine them. Efforts to reduce the spread of misinformation (Bak-Coleman et al. 2022; Kim et al. 2018), to identify and correct false information (Rubin et al. 2019; Vraga and Bode 2017), and to assist information consumers to do the same (Sharon and Baram-Tsabari 2020; Vraga, Tully, and Bode 2020) are now and will remain tremendously important.

Nevertheless, we contend that even if these goals are perfectly achieved, misapprehensions will persist – because people are not simply passive receivers of information. The act of [mis]informing does not end with [mis]information – a great deal happens in the mind after information is encountered. When someone receives information, it is filtered through cognitive processes, including strategies and heuristics that can introduce bias, and incorporated into existing belief structures. In other words, the information is *used.* Our focus is on those moments, and processes, of information use – what Savolainen (2009) terms the typically "unspecified 'appendix' of information seeking" (116) – and the impact that these have on beliefs.

Thus, in this chapter, we examine the impact of information use in developing and sustaining misapprehensions – *even when the information being processed is not itself misleading or inaccurate.* Examples of this type of false belief abound. For example, many people incorrectly estimate the frequency of murder to be much higher than that of suicide (Fischoff, Slovic, and Lichtenstein 1977), and travelers often express greater concern about flying as opposed to driving. There are, however, no false news reports or other statements that assign a higher probability to murder as compared to suicide, or a greater risk to flying as compared to driving. The misapprehensions arise instead from the ways in which decision makers cope with the limitations of what Herbert Simon termed "bounded rationality" (Wheeler 2018). These limitations lead us to rely on mental shortcuts, or *heuristics,* that support the best possible decisions with the limited resources – attention, memory, and information – that we have at our disposal. Although these heuristics generally work well, they lead to severe and systematic errors in thinking, or *cognitive biases* (Tversky and Kahneman 1974, 1124). These biases, in turn, can lead us to develop or persist in incorrect conclusions and false beliefs.

We explore these issues in the specific context of the COVID-19 pandemic, but the arguments, and conclusions, have wide application. We structure the chapter around specific misconceptions, exploring how cognitive biases and heuristics can influence these misperceptions, and how careful information design and education can ameliorate those effects.

MISCONCEPTIONS IN THE CONTEXT OF THE COVID-19 PANDEMIC

Since the onset of the coronavirus pandemic in 2020, much research has documented the existence and impact of COVID-related misinformation (e.g., Roozenbeek et al. 2020; Tasnim, Hossain, and Mazumder 2020). In this section, we examine how people's biased processing of *accurate* COVID-related information contributed to five misconceptions observed over the course of the pandemic. We explain how the misconceptions stem from different kinds of cognitive biases, and outline information design and educational strategies that could potentially reduce the biases' effects. We also show how the biases influenced people's willingness to take up preventive actions, such as wearing masks, complying with social distancing guidelines, and getting vaccinated. Within the governing knowledge commons framework (Frischmann, Madison, and Strandburg 2014), these preventive actions constitute distinct "action arenas" (Ostrom 2005) where participants make decisions with varying outcomes.

Misconception 1: "COVID-19 Isn't Really a Problem – There Are Only a Few Cases"

Even in the earliest stages of the COVID-19 crisis, there was widespread understanding that case counts were growing – but there was also widespread and severe underestimation of the rate of growth, and thus severe underestimation of the number of future COVID-19 cases (Banerjee and Majumdar 2020; Villanova 2022). Part of the explanation lies in *exponential growth bias*: a general tendency to underestimate the impact of exponential growth. This bias is well-demonstrated in a children's book by Demi entitled *One Grain of Rice*. According to the story, an Indian peasant girl was offered a reward of her choice for service to the king. Her request, which appeared modest to the sovereign, was for a single grain of rice on the first day, and for each of the following thirty days an additional amount of double the number of grains received the day before. The greedy king readily agreed, thinking he had struck a very good bargain. On the last day of the month, however, the girl received four *storehouses* full of rice in addition to all that had been provided up to that time. In the story, the king has fallen prey to *exponential growth bias*, and the young girl has used his misconception to her advantage to feed her entire community.

Unfortunately, the impact of exponential growth bias is not as positive when it comes to responding to the COVID-19 crisis. Especially in the initial stages, the spread of infectious diseases follows an exponential function, with a few positive cases exploding into a widespread pandemic if the disease is, as with COVID-19, sufficiently transmittable (Banerjee, Bhattacharya, and Majumdar 2021). Initially, the number of reported COVID-19 infections was low, and a focus on this low number, coupled with a predictable misapprehension of the impact of exponential growth, led many to underestimate the severity of the situation (Lammers, Crusius, and Gast 2020). During the outbreak of March 2020, for example, the number of coronavirus patients in the United States doubled about every three days; however, a study conducted during this period showed that American participants mistakenly perceived the growth of cases of the virus as linear (Lammers, Crusius, and Gast 2020). This misperception could, in turn, hinder the adoption of measures to fight and contain the pandemic. Research has shown that, in the comprehension of COVID-19 disease data, the exponential growth bias can predict noncompliance with safety measures such as handwashing, mask-wearing, and the use of sanitizers (Banerjee, Bhattacharya, and Majumdar 2021). Moreover, interventions that correct the misperception of exponential coronavirus growth have been shown to significantly increase support for social distancing (Lammers, Crusius, and Gast 2020).

Some instructional or educational strategies appear to reduce the effect of exponential growth bias. Lammers, Crusius, and Gast (2020) found that exponential growth bias could be reduced by asking audiences to read a few sentences that explained the bias (e.g., "keep in mind that many people forget that the speed by which the corona virus spreads, increases each day . . ."). Even greater reduction in the bias was observed in an experimental context where participants were encouraged to "step through" case number increases when asked to estimate a total number at a specific future date given an initial case count and doubling time (e.g., estimate cases at day fifteen given five initial cases and a doubling time of three days).

Exponential growth bias can also be reduced by careful information design. There is some evidence that presenting growth information in terms of doubling times rather than growth rates (i.e., "the number of cases is expected to double every fifteen days" as compared to "cases are expected to grow at a rate of 5 percent per day") can reduce exponential growth bias (Schonger and Sele 2021). Graphical representations of exponential growth are especially difficult to grasp (Berenbaum 2021), and the dominant mode of presenting COVID-19 case numbers in the print and online media has been graphical (Banerjee, Bhattacharya, and Majumdar 2021). Banerjee, Bhattacharya, and Majumdar (2021) experimentally demonstrated that the representation of past COVID-19 case numbers in numerical form (see Table 3.1) significantly decreased the exponential growth bias relative to graphical representation (see

TABLE 3.1 *Hypothetical growth of COVID-19 cases over fifteen days with a doubling time of three days, presented numerically*

Day	Number of COVID-19 cases
1	5
3	10
6	20
9	40
12	80
15	160

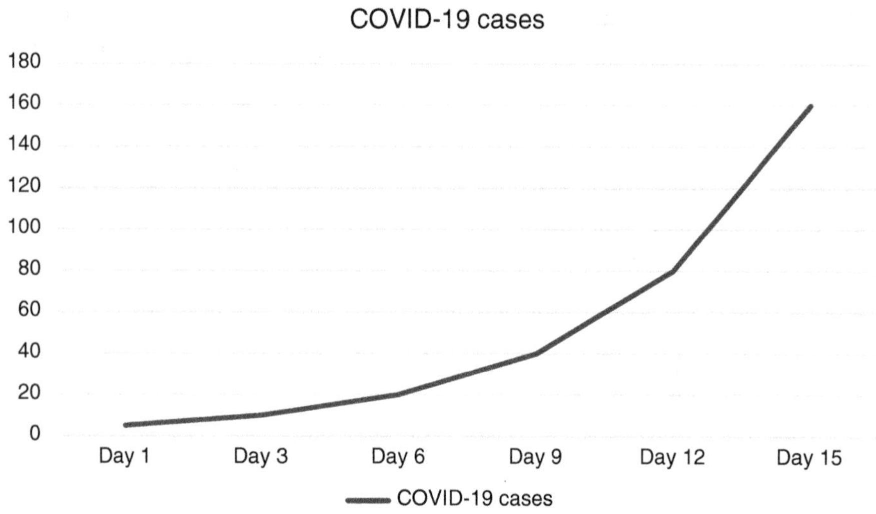

FIGURE 3.2 Hypothetical growth of COVID-19 cases over fifteen days with a doubling time of three days, presented graphically.

Figure 3.2). From this, they suggest that data should be shown via raw numbers (see Table 3.1) and presented "alongside familiar 'flatten-the-curve' style graphics" (8).

Misconception 2: "COVID-19 Won't Happen To Me"

Epidemiologists raised the threat of a global pandemic as early as January 2020 and "announced that more than 40–70 % of the world population could be infected within the end of the year" (Bottemanne et al. 2020, 2); in the same month, the Chinese city of Wuhan, with a population of 11 million people, went into full quarantine (Cai 2020). Despite this and other strong and mounting objective evidence of widespread infection, people underestimated their personal risk of contracting and/or transmitting COVID-19. Survey data collected in several

European countries in February 2020, for example, revealed that a large majority of participants believed that their risk of contracting the coronavirus was around 1 percent (Raude et al. 2020). (By comparison, as of February 2022, approximately 58 percent of the United States population – and 75 percent of US children – had contracted COVID-19, according to clinical testing of blood samples for SARS-CoV-2 antibodies (Ducharme 2022).) Many also viewed themselves as less likely than average, and less likely than others, to contract the disease. During the early phases of the outbreak in March 2020, German adults who had not yet tested positive for COVID-19 perceived family members and friends to be at a higher risk of infection than themselves (Gerhold 2020), and Polish university students estimated their likelihood of contracting the coronavirus as lower than that of their peers (Dolinski et al. 2020). Meanwhile, one respondent in an American poll claimed that they were "cautiously optimistic" the United States would "nip [the virus] in the bud" before it could spread as it had in Europe (Allyn and Sprunt 2020).

This unwarranted and elevated perception of personal safety is consistent with the *optimism bias* (Weinstein 1983). Optimism bias leads us to think ourselves *more likely than average* to experience positive outcomes, and *less likely than average* to experience negative outcomes. This bias appears to arise from an inappropriate comparison group – when thinking about our risk of serious consequences of disease, for example, we implicitly compare ourselves to others who we believe to be less healthy and more vulnerable than ourselves, rather than to a more representative "average" other – and by comparison, we estimate ourselves as less likely to experience the negative effects (Weinstein 1983).

The misapprehension of reduced personal risk could lead people to ignore public health recommendations and refrain from personal hygiene practices and precautions (Pascual-Leone et al. 2021; see Wise et al. 2020 for a discussion of the relationship between risk perception and protective behaviors). Indeed, research shows that individuals with high optimism bias engaged less in protective behavioral changes during the COVID-19 pandemic in 2020 (Fragkaki et al. 2021).

The delivery of objectively accurate risk information, even by trusted messengers, will not correct this bias (Felgendreff et al. 2021); however, early research on the optimism bias indicates that personalized risk comparators on a group (Weinstein 1983) or individual (Alicke et al. 1995) basis can help to correct risk perceptions. This targeted communication approach, with messages designed for specific individuals or subpopulations, will be effective in reducing optimism bias when audiences can be segregated. For general (untargeted) communications, there is specific research in the COVID-19 context, consistent with prior research in other domains, showing that unrealistic optimism is reduced by communications (video or text) that emphasize the risk-reducing activities of others (e.g., compliance with medical recommendations for social distancing; Dolinski et al. 2022). In general, interventions that explicitly or implicitly provide an appropriate comparison group for personal risk estimation will help to mitigate optimism bias.

Misconception 3: "Vaccines Don't Work"

Research shows that, in real-world settings, COVID-19 vaccines offer a high degree of protection against SARS-CoV-2-related diseases (Zheng et al. 2022), and the efficacy of these vaccines has been reported extensively and positively in the mainstream media (e.g., Hayes 2021; Thomas and Hanna 2021). Nonetheless, many remain "vaccine hesitant" (Kirzinger et al. 2021), choosing not to take the vaccination. One common argument among the vaccine hesitant is that "vaccines don't work" – and the cited evidence is that the vaccinated continue to make up "most people admitted to hospital with Covid-19" (Benedictus 2021),[1] and a large proportion of those dying from the disease (latest data from Canada indicates that from December 2020 to June 2022 there have been 10,385 COVID-related deaths among the unvaccinated, and 7,423 COVID-related deaths among the vaccinated: Statista 2022). If vaccines work, the reasoning goes, those who are vaccinated should not be contracting COVID-19, should not be hospitalized with COVID-19, and should not die from the disease. This is, of course, a misunderstanding of the math and science. No vaccine is 100 percent effective, but an effective vaccine reduces the likelihood of contracting a disease, and can also reduce the severity of the disease if it is contracted. According to the best available data, COVID-19 vaccines provide both types of protection (Centers for Disease Control and Prevention 2022). So, why the persistent belief that the vaccines don't work?

This misperception arises at least in part from a failure to take into account the vaccination base rate, or the proportion of people in the entire population who are vaccinated compared to those who are unvaccinated. Imagine a virus that will, without any protection, affect 30 percent of the population. Imagine also that 95 percent of the population have received a vaccine for the virus. If the vaccine provides *no* protection, then 30 percent of the unvaccinated and 30 percent of the vaccinated will be infected, and if we consider the positive cases only, 95 percent of those will be individuals who have received the vaccine (because the vaccine offers no protection, and 95 percent of the population has received the vaccine: Egger and Egger 2022). If the vaccine provides perfect protection (which never happens), then *all* cases will be among the unvaccinated. In general, however, reality falls somewhere in between. Say, for example, that the vaccine is 80 percent protective against disease; in other words, only 20 percent of those who are vaccinated and who would otherwise have been infected will actually get the virus. In this case, 30 percent of the unvaccinated population will still get the disease, but the rate among the vaccinated will fall to 6 percent (the 20 percent for whom the vaccine is ineffective, out of the 30 percent who would otherwise be expected to become infected).

So far, so good – but here comes the surprising part: under these circumstances, the large majority of infections will still occur in people who are vaccinated. In fact, *almost 80 percent of the people who get the condition will have received the*

[1] https://fullfact.org/health/economist-vaccination-status/.

vaccination. Put another way, of 1,000 randomly selected people in this population, 15 unvaccinated are expected to contract COVID-19 (30 percent of the 5 percent of the population who are unvaccinated), and 57 among the vaccinated are expected to get the disease (20 percent of the 30 percent who would otherwise get COVID-19, among the 95 percent of the population who are vaccinated). In general, if the proportion of those who have a condition is anything less than the proportion of the population vaccinated against the disease, then the vaccine is providing protection – but that doesn't *seem* right when the large majority of those we see with the condition have been vaccinated against it.

Base rates – in this case, the rate of vaccination in the population – have a surprisingly large and counterintuitive impact on outcomes, leading to inaccurate intuitions in a wide variety of circumstances. Thus, for example, we tend to over-interpret the results of accurate screening tests for rare conditions (Burkell 2004), incorrectly assuming that a positive screening test is a strong indicator that the condition is present. Screening tests, by definition, test for relatively rare conditions with a fast and low-cost initial test designed to capture the large majority of true positive cases. These tests are by definition imperfect, and are focused on avoiding false negative results (saying that someone does *not* have the condition when in fact they do) at the expense of an increased risk of false positive results (saying that someone *has* the condition when in fact they do not) that can only be resolved through further testing. As a consequence, and as illustrated in Table 3.2, a large proportion of positive screening test results are actually *false* (in the case presented, 95 of 150 positive results, or 66 percent) – that is, the individual does not in fact have the identified condition. This situation arises precisely because there are so many people without the condition – or, to put it another way, because the base rate of the condition in the population is low, with the result that even a small tendency to give a false positive result creates a large absolute number of such results. Base rate neglect may also play a role in the alarm recently expressed by bicycling advocacy groups over an increase in bicycling deaths (Advocacy

TABLE 3.2 *Hypothetical screening test results for a population of 1,000; condition present in 5% of the population, test results in 0% false negative results (sensitivity 100%), 10% false positive results (specificity 90%)*

	Screen negative	Screen positive	
Actual negative	True negative results: 855	False positive results: 95	Actual negative cases: 950
Actual positive	False negative results: 0	True positive results: 50	Actual positive cases: 50
	Screen negative results: 855	Screen positive results: 145	

Advance 2021).[2] While there are undoubtedly many explanations for this shift, one that is *not* regularly acknowledged is the increase in the number of cyclists on the road (Mazerolle 2021).[3] This represents an increase in the *base rate* of cycling as a mode of transportation, which would increase the number of observed cycling accidents even if no other risks to cyclists were to change.

Base rate neglect may explain, at least in part, why a sizable segment of unvaccinated people perceive COVID-19 vaccines to be ineffective (Kirzinger et al. 2021) despite credible proof to the contrary.[4] Evidence of this misperception can be easily observed, even among healthcare professionals (Kampf 2021). Governmental data showing a higher proportion of vaccinated than unvaccinated people in COVID-19 deaths has been used to argue for the inefficacy of vaccines online (e.g., The Exposé 2022).[5] Base rate neglect was also at the root of the February 2022 pronouncement by a high-ranking Canadian politician that the COVID-19 vaccine was no longer effective (Quon 2022).[6] In that case, data showing that the number of new COVID-19 cases in the province of Saskatchewan was "about the same in [*sic*] vaccinated and unvaccinated people" was presented as evidence of vaccine inefficacy. This conclusion ignored the critical fact that nearly 80 percent of the province's eligible population had received at least two vaccine doses, and even a small proportion of breakthrough cases would result in a relatively large number of vaccinated individuals testing positive for COVID-19.

At the same time, many recognize that these perceptions arise from base rate neglect, and efforts to correct this misperception have appeared in a variety of sources including news articles (e.g., Devis 2021; Ferreira 2022)[7][8] and social media posts (e.g., Rumilly 2021).[9] Approaches to addressing base rate neglect include graphics that "zoom out" to show the base rate of vaccinated and unvaccinated groups in a hypothetical population (Rumilly 2021), allowing viewers to visualize how the unvaccinated population is disproportionately hospitalized with COVID-19, even if more hospitalized patients are vaccinated than unvaccinated. Animations demonstrating changes in the percent of vaccinated (as opposed to unvaccinated) COVID-19 hospitalizations with changes in the base rate of vaccination in the population have also been used to achieve this outcome (Panthagani 2022).[10]

[2] www.advocacyadvance.org/2021/05/bicycle-injury-and-fatality-statistics-during-the-pandemic/.

[3] www.cbc.ca/news/business/bicycle-boom-industry-turmoil-covid-19-1.5956400.

[4] www.kff.org/coronavirus-covid-19/poll-finding/kff-covid-19-vaccine-monitor-july-2021/.

[5] https://expose-news.com/2022/04/09/italy-7-in-10-covid-deaths-vaccinated/.

[6] www.cbc.ca/news/canada/saskatchewan/covid-19-transmission-scott-moe-1.6336479.

[7] https://cosmosmagazine.com/health/covid/why-are-there-so-many-vaccinated-people-in-hospital/.

[8] www.ctvnews.ca/health/coronavirus/most-covid-19-hospitalizations-in-provinces-are-among-the-vaccinated-here-s-why-1.5770226.

[9] https://twitter.com/MarcRummy/status/1418672725686640643?s=20.

[10] https://dearpandemic.org/base-rate-fallacy/.

Many of these alternative visualizations are presented in the context of "corrective" or educational information (e.g., in a blog intended to "help you stay on the frontline of health information": Panthagani 2022). These are effective tools for motivated audiences focused on the accuracy of their own perceptions, but it is important also to ensure that information about case counts is initially presented in ways that minimize base rate neglect and thereby minimize the inaccurate interpretation that vaccines do not provide protection. Case counts presented with vaccination status but without drawing attention to the base rate of vaccination in the population invite misinterpretation arising from base rate neglect. At the very least, therefore, media and other public reports should include the relevant base rate information along with counts. A better approach is to take advantage of presentation formats that have been demonstrated to lead to better understanding. Specifically, frequency histograms (Burkell 2004) that show infected and uninfected members of vaccinated and unvaccinated populations have been demonstrated to reduce base rate neglect.

Misconception 4: "Vaccines Aren't Safe"

In general, vaccines are safe and effective at reducing disease, having undergone rigorous testing for both efficacy and negative side effects before approval. Although the COVID-19 vaccines were developed in a relatively short time in response to an acute health crisis, they are no exception to these general rules, and have been widely demonstrated to reduce incidence and severity of COVID-19 infection with few if any negative health consequences (Centers for Disease Control and Prevention 2022; Henry and Glasziou 2021). Nonetheless, and consistent with public response to other vaccines, there remains a portion of the population who refuse the vaccine, citing safety concerns (King et al. 2021; Monte 2021).[11] The issue of vaccine safety is complex, and indeed there are questions about the long-term impact, including unintended consequences, of the mRNA vaccines (Seneff and Nigh 2021). In many cases, however, hesitancy is linked to an overestimation of the incidence of vaccine-related adverse effects, including among unvaccinated health professionals (Ehrenstein et al. 2010).

This inaccurate risk perception can be attributed at least in part to the *availability bias*, which is a tendency to base frequency or probability estimates on the ease with which specific instances can be recalled or imagined (Tversky and Kahneman 1973). When less likely outcomes are more salient – for example, by virtue of the frequency with which they are reported – the probability of these events can be overestimated. Thus, for example, we will tend to overestimate the likelihood of an automobile accident immediately after witnessing one, and we will overestimate the likelihood of an airplane crash if one has recently been covered in the news. The availability

[11] www.census.gov/library/stories/2021/12/who-are-the-adults-not-vaccinated-against-covid.html.

bias is stronger for negative than for positive events, perhaps because those events are more salient (Stapel, Reicher, and Spears 1995). Media coverage is a significant factor in the availability bias, and has been shown to influence health concerns among the general public (Brezis, Halpern-Reichert, and Schwaber 2004) and even physician diagnostic decisions (Schmidt et al. 2014).

A rare serious adverse effect of a vaccine covered in the media provides "a vivid and emotionally compelling anti-vaccination message, likely to be recalled during decision making" that could cause one to overestimate the probability of an adverse effect following immunization (Azarpanah et al. 2021, 8). In March 2021, for example, reports emerged in the mainstream media of a possible link between AstraZeneca's COVID-19 vaccine and rare but potentially fatal blood clots (e.g., Gronholt-Pedersen 2021; Reuters staff 2021).[12,13] Some of these reports included descriptions of individual cases, such as that of a sixty-year-old Danish woman who developed a blood clot and died after vaccination (Gronholt-Pedersen 2021). In the wake of this media coverage, one March 2021 poll showed that Canadians were much warier of the AstraZeneca vaccine than other vaccines approved for use in Canada (Bryden 2021).[14] Although blood clots among those with low blood platelet counts were eventually confirmed to be a side effect of the AstraZeneca vaccine, the European Medicines Agency (2021) stressed that the complication was "very rare" and that the vaccine's benefits in preventing COVID-19 outweighed the risks of side effects.

In a context where low-probability events, such as vaccine side effects, are more likely to be reported than higher-probability outcomes, such as vaccination without incident, the availability bias will lead audiences to overestimate the risk of the low-probability event. Counteracting this bias requires that the alternative – safe and incident-free vaccinations – are made more salient and more easily recalled. Strategies to achieve these outcomes include multiple repetitions of the items judged less frequent, and/or increasing the memorability of the reports of incident-free vaccinations (Lewandowsky and Smith 1983). The latter could be supported, for example, by the intuitively appealing strategy of reporting stories of high-profile individuals who have received vaccination and not experienced side effects. Some strategies to reduce availability bias focus on instruction and training (e.g., Mamede et al. 2020). These approaches are most useful when specific individuals are engaged in repeated risk estimation tasks (e.g., physicians making a diagnosis), and often involve guided reflection on decision-making processes along with specific stepwise strategies including the consideration of alternative conclusions (Prakash, Sladek, and Schuwirth 2019).

12 https://globalnews.ca/news/7696802/astrazeneca-denmark-blood-clot-suspension/.

13 www.reuters.com/article/health-coronavirus-austria-nurse-idUSL8N2L506P.

14 www.cp24.com/news/canadians-far-more-wary-of-astrazeneca-than-other-covid-19-vaccines-poll-1.5367801?cache=yes%3FautoPlay%3Dtrue%3FclipId%3D89950.

Misconception 5: "We've Had the Pandemic – So We're Safe for a Long Time"

No one knows when the next pandemic will occur, and our expectations on this point *matter*. Belief that another pandemic is likely in the near future will spur prevention and resilience activities, while the assumption that it will be a long time before we have another pandemic will foster complacency and a disinterest in taking up risk mitigation activities. Like deciding whether to take actions to curb the spread of COVID-19, choosing to engage in or avoid measures that could prevent the *next* pandemic can be conceptualized as a distinct action arena. Although there is no "correct" prediction regarding when we will next see a pandemic, it is important that perceptions are unbiased and appropriately calibrated to objective and current information about risk. These perceptions, however, are influenced by biases and heuristics, and it is crucial that we understand what these are, how they operate, and how they will influence the perceived risk of a new pandemic.

There are numerous cognitive biases that could and likely do influence our expectations about when another pandemic will occur. One of these is the *gambler's fallacy*, well demonstrated in the history of the "Lion's Share" slot machine at the MGM Grand hotel-casino. In 2014, this slot machine paid out a progressive jackpot of $2.4 million after collecting money for twenty years from unsuccessful gamblers (CNN Wire 2014). Some, like the author of one blog post, believed that the long "dry" period meant the slot machine was "due for a win" (Best US Casinos, n.d.)[15] Assuming a fair machine with outcomes determined randomly, this intuition is unequivocally wrong. Players at that particular machine were no more likely than those at any other machine with the same odds to have a "winning" spin – because like all independent random events, slot machine spins have no "memory." For each new instance, the probability of winning is exactly the same. The contrary belief – essentially the belief that luck (good or bad) must change – is called the gambler's fallacy (Tversky and Kahneman 1974).

The gambler's fallacy isn't limited to the casino. The fallacy leads people to believe that improbable events operate on an implicit "schedule," and thus become increasingly likely as time passes without an event (in Eastern Canada, after a few years without a big snowstorm you'll hear the refrain "we're due for a big one") and, once they occur, are unlikely to be repeated, at least in the near future (think about the familiar adage "lightning never strikes twice"). Soccer goalkeepers fall prey to the gambler's fallacy when faced with penalty kicks (Wogan 2014), and asylum judges, loan officers, and baseball umpires show signs of the gambler's fallacy when making multiple decisions (Chen, Moskowitz, and Shue 2016).

The gambler's fallacy appears to be operating in at least some expectations regarding pandemics. COVID-19 has been described as a "once-in-a-lifetime" (e.g.,

[15] www.bestuscasinos.org/vegas/lions-share-jackpot-won/.

Guterres, n.d.)[16] or "once-in-a-century" (e.g., Gates 2020) event.[17] Even World Health Organization (WHO) Director General Tedros Adhanom Ghebreyesus characterized the pandemic as a "once-in-a-century health crisis" during a 2020 meeting of the WHO's emergency committee (Reuters staff 2020).[18] This belief might partially arise from the fact that the Spanish flu – which has attracted "unprecedented interest" due to COVID-19 and has clear parallels to the COVID-19 pandemic itself (Simonetti, Martini, and Armocida 2021, E613) – broke out approximately 100 years prior, in 1918. According to some:

> the gambler's fallacy is almost irresistible when we think about pandemics. We think the fact that a new disease has emerged from the natural world so recently, and caused such a terrible catastrophe, means that we're due some luck. Surely we must be due a long reprieve before the next one. Surely we will have time to prepare. (Birch 2021)

Even if pandemics were completely random events – and they are not – this reasoning would be flawed: the fact that we had a pandemic 100 years ago and another one just recently does not suggest that the next is due in another 100 years. Indeed, research that models pandemic risk "shows that the frequency and severity of spillover infectious disease – directly from wildlife host to humans – is steadily increasing" (Smitham and Glassman 2021). Contrary to the assumption that it will be a long time before another severe pandemic, one research team estimates "the annual probability of a pandemic on the scale of COVID-19 in any given year to be between 2.5 and 3.3 percent, which means a 47–57 percent chance of another global pandemic as deadly as COVID in the next 25 years" (Smitham and Glassman 2021). Moreover, and contrary to the gambler's fallacy, barring factors that *increase* the annual risk (e.g., increased zoonotic spillover; see Birch 2021), the chances of a new pandemic in each of the next 100 years is *identical*, and a new pandemic just as likely after one virus-free year as it is after 99 virus-free years.

The gambler's fallacy (also known as belief in the law of small numbers; Tversky and Kahneman 1971) is a remarkably persistent cognitive bias (see, e.g., Bishop, Thompson, and Parker 2022). Basic training regarding random events does not appear to reduce the bias (Beach and Swensson 1967). There is some evidence, however, that the perceptual grouping of events can influence the gambler's fallacy. In particular, if an event (say, for example, the upcoming year) is seen as grouped with or a continuation of a prior series of events (e.g., the past 100 years), then the gambler's fallacy is evident. If, however, the event is viewed as the beginning of a *new* series, the gambler's fallacy is reduced or even eliminated (Roney and Trick 2003). In the COVID-19 context, this would suggest that the underestimation of the risk of a new pandemic could be reduced if risk discussions

[16] www.un.org/en/un-coronavirus-communications-team/all-hands-deck-fight-once-lifetime-pandemic.

[17] www.nejm.org/doi/full/10.1056/nejmp2003762.

[18] www.reuters.com/article/us-health-coronavirus-who-idUSKCN24W27L.

were forward looking (e.g., "over the next 100 years"), rather than backward looking ("in the past 100 years").

INFORMATION IS NOT A PANACEA

By now, you should be convinced that inaccurate beliefs can and do arise in the face of accurate information – and that careful information design and/or education can reduce this possibility. At this point, we want to address a different question: specifically, is correct(ive) information a panacea for the misinformed?

The seemingly obvious approach to correcting misperceptions is to provide people with accurate information. This approach is consistent with the *knowledge deficit model* (or *information deficit model*) which suggests that public skepticism toward modern science and technology is mainly caused by insufficient information – or a "knowledge deficit" – which can be remedied by more information about these topics (Dickson 2005; Simis et al. 2016). In the case of COVID-19 vaccine hesitancy, for example, some have argued that clear and transparent communication about vaccine risks and benefits could increase vaccine uptake among the general public (Kerr et al. 2021).

The knowledge deficit model generally attributes misapprehension to a lack of appropriate information and assumes that "a thorough and accessible explanation of facts" (Ecker et al. 2022, 13) should overcome incorrect beliefs, including those arising from objectively false information and those that are the product of the types of cognitive biases outlined earlier in this chapter. Research indicates, however, that merely conveying accurate information is often insufficient to correct misperceptions, for two interrelated reasons. First, we have a tendency to selectively attend to and seek information that is consistent with preexisting beliefs, in a bias termed the *confirmation bias* (Jones and Sugden 2001; Wason 1960). Second, objectively false information that is encountered and accepted continues to influence our thinking, even in the face of new and accurate information, including when that corrective information has been accepted as true – a phenomenon known as the *continued influence effect* (CIE) (Ecker et al. 2022; Lewandowsky et al. 2012; Wittenberg and Berinsky 2020).

Confirmation Bias

Confirmation bias describes the tendency to selectively seek out and attend to information that is consistent with previously held beliefs or attitudes (Jones and Sugden 2001; Wason 1960); *evaluation bias*, which is associated, is the tendency to more positively evaluate information that is consistent with prior beliefs and attitudes. The effect of these two biases is to minimize *cognitive dissonance* (Festinger 1957), or the uncomfortable state of internal cognitive inconsistency that can arise when we are confronted with something inconsistent with our beliefs and/or attitudes. The effects of confirmation bias are exacerbated in the online social environment, where past information consumption choices influence future offerings, not only to the individual but also to others who are identified as "similar" based on complex algorithmic processing (Ling 2020). Confirmation bias contributes to opinion polarization, through a feedback loop wherein even a slight bias or tendency in thinking (e.g., belief that vaccines are unsafe) is exaggerated by selective attention to supporting information, which in turn creates an even stronger tendency to ignore inconsistent messages (Modgil et al. 2021; Xu et al. 2022). Surprisingly, there is some evidence that the tendency to select confirmatory messages and to perceive them as more convincing is higher among individuals with higher literacy levels (in the case of the research, *health* literacy levels: Meppelink et al. 2019). Evaluation biases similarly lead to discounting of disconfirming evidence. Information that is strongly inconsistent with previously held beliefs is evaluated as less credible, and is therefore less likely to be accepted, than information consistent with those beliefs (Christensen 2021).

Confirmation bias can be reduced (somewhat paradoxically) by making the corrective information *more* difficult to process – creating what some researchers have termed "disfluency" in information processing (Hernandez and Preston 2013; see also Rajsic, Wilson, and Pratt 2018). Disfluency is affected by features such as the visual clarity of text and is believed to promote a critical and analytical mindset (Hernandez and Preston 2013). Research has shown, for instance, that high school students score higher on exams when learning materials are presented in hard-to-read fonts (Diemand-Yauman, Oppenheimer, and Vaughan 2011). In two studies, Hernandez and Preston (2013) found that participants' preexisting attitudes toward certain issues became less extreme when they read arguments on the issues in a disfluent format (e.g., light gray, bolded and italicized Haettenschweiler font vs. standard 12-point Times New Roman). The second study further determined that participants were only able to disconfirm their prior biases when they were not under "cognitive load" (i.e., distraction or time pressure) (Hernandez and Preston 2013).

In some cases, simply increasing the salience of corrective information can ameliorate false beliefs. Schwind and colleagues (Schwind et al. 2012; Schwind and Buder 2012) suggest that explicitly recommending information that is inconsistent with prior preferences or beliefs (e.g., by highlighting the results in an online search) can be effective in reducing or correcting those misapprehensions.

Confirmation bias may also play a role in situations where efforts to deliver corrective information "backfire." *Worldview backfire effects* occur when corrections that challenge people's worldviews bolster their belief in the original misinformation (Ecker et al. 2022; Wittenberg and Berinsky 2020). This phenomenon was originally identified by Nyhan and Reifler (2010), who found that corrections that contradicted individuals' political beliefs increased their prior misperceptions.

Worldview backfire effects have been conceived of as a product of directionally motivated reasoning (Holman and Lay 2019; Wittenberg and Berinsky 2020). In short, individuals may be motivated to arrive at either an accurate conclusion or a particular, directional conclusion (Kunda 1990), and "worldview backfire effects transpire when directional motivations take precedence over accuracy goals" (Wittenberg and Berinsky 2020, 170). These effects are believed to be tied to both *confirmation bias* and *disconfirmation bias*, the latter of which involves calling to mind opposing arguments, or "counterarguing," when faced with ideologically dissonant information (Wittenberg and Berinsky 2020). In essence, when a correction challenges false beliefs that are central to one's identity, they may discount the accurate information and generate counterarguments to it (Ecker et al. 2022; Wittenberg and Berinsky 2020). For instance, someone identifying as an antivaxxer might perceive information proving that the risks of a vaccine do not outweigh the risks of a disease to be an identity threat, and subsequently ignore this worldview-inconsistent evidence while selectively focusing on evidence that supports their own position (Ecker et al. 2022).

To avoid worldview backfire effects, corrections "should be tailored to their target audience: the subset of people for whom these corrections would feel the most threatening" (Wittenberg and Berinsky 2020, 171). Specifically, corrections should be framed to be consonant with their target audience's values and worldviews (Lewandowsky et al. 2012; Wittenberg and Berinsky 2020). When attempting to convince someone with an "ecocentric" outlook that nanotechnology is safe, for example, the target may be more receptive to evidence of the technology's safety if its use is portrayed as part of an effort to protect the environment; here, potential benefits and opportunities are highlighted rather than risks and threats (Lewandowsky et al. 2012). Corrections that threaten a target's worldview can also be made more palatable by pairing them with a self-affirmation or identity affirmation, which involves a message or task that affirms one's basic values (Lewandowsky et al. 2012) or highlights sources of self-worth (Ecker et al. 2022).

The Continued Influence Effect (CIE)

The fact that misinformation may continue to influence people's thinking – even after a retraction has been acknowledged and recalled – is known as the *continued influence effect* (CIE) (Johnson and Seifert 1994; Seifert 2002). The CIE has been explained by *dual-process theory* and *mental model theory* (Wittenberg and Berinsky 2020). Dual-process theory assumes that memory retrieval can be automatic (fast and unconscious) or strategic (deliberate and effortful) (Wittenberg and Berinsky 2020). While automatic processing is "relatively acontextual, distilling information down to only its most essential properties," strategic processing is needed "to retrieve specific details about a piece of information" (Wittenberg and Berinsky 2020, 174). From this perspective, it is possible that individuals could automatically retrieve a piece of misinformation and fail to recall relevant features, such as its source or perceived accuracy (Wittenberg and Berinsky 2020).

On the other hand, mental model theory suggests that people "construct models of external events in their heads, which they continuously update as new information becomes available" (Wittenberg and Berinsky 2020, 174). Retractions may create a gap in an individual's model of an event, motivating them to continue to invoke the original misinformation (Ecker et al. 2022; Lewandowsky et al. 2012; Wittenberg and Berinsky 2020). This effect can be reduced by providing an alternative causal explanation for an event that fills the gap left behind by the retracted misinformation (Ecker et al. 2010; Ecker et al. 2022; Lewandowsky et al. 2012; Seifert 2002). For instance, to correct the misperception that a fire was caused by negligence, a causal explanation ("there is evidence for arson") is more effective than a plain retraction ("there was no negligence") (Ecker et al. 2022, 21). One study found that a specific warning describing the CIE reduced, but did not eliminate, participants' continued reliance on outdated information (Ecker et al. 2010). When the warning was combined with a plausible alternative explanation for the retracted information, the CIE was further reduced – though still not fully eliminated (Ecker et al. 2010).

Moving Beyond Panaceas

At the outset of this section, we asked the question of whether information is a "panacea" for the misinformed. Not surprisingly, the answer is a resounding "no." Complex problems such as the issue of mis/disinformation require complex, multifaceted, and multitiered solutions (Ostrom and Cox 2010), and are resistant to simple, single-strategy approaches. Addressing the pressing issue of false beliefs – and false information – will require interventions with information consumers, information providers and producers, and information hosting platforms, and regulatory bodies including governments at the very least. Moreover, and critically, while interventions at the individual level including training and appropriate information design will be essential to reducing the impact of mis/disinformation, interventions with hosting platforms including regulation by government and other bodies will be critical to an effective and appropriate governance response.

CONCLUSION

To be misinformed is to hold false beliefs, and the most obvious source of false beliefs is misinformation. In some cases, however, false beliefs arise even when the information received and processed is entirely accurate – and delivering accurate information does not always or necessarily correct prior misconceptions.

In this chapter, we have demonstrated how information *use* – rather than the accuracy of information itself – can cause people to become and remain misinformed. In our discussion, we have focused on one specific context – that of COVID-19 – and within that context we have documented particular situations in which cognitive biases have (or could have) a meaningful effect on reasoning and decision making. Throughout the chapter, we have referred to many other circumstances in which cognitive biases can influence decisions, and information professionals should be aware of and responsive to these potential influences whenever there is a situation of judgment or decision making under uncertainty (Kahneman, Slovic, and Tversky 1982), including, for example, elections (and campaign materials), healthcare decisions, or decisions about resource allocation (e.g., to climate change initiatives). The list of cognitive biases we have highlighted in this discussion are illustrative rather than exhaustive, and there are many resources identifying additional biases and heuristics (e.g., Kahneman, Slovic, and Tversky 1982), including discussions of the use of these techniques to promote particular viewpoints or decisions (e.g., Thaler, Sunstein, and Balz 2013).

Cognitive biases and heuristics affect information processing, with impact on attitudes, beliefs, and choices. We rely on them because they *work* – making "good enough" reasoning possible in the context of cognitive limitations. But "good enough" is far from perfect, and from time to time – as discussed in this chapter – these biases and heuristics lead us astray. The *bad* news is that these biases are persistent and unconscious; the *good* news is that they can be to some extent mitigated by careful information design and education. Rather than provide an exhaustive review, our aim has been to highlight examples of biases in information processing that contribute to the formation and maintenance of false beliefs, and to identify possible strategies that reduce their impact. These strategies, among others, can help information professionals ensure that the accurate information they provide supports equally accurate beliefs. These strategies cannot replace other aspects of governance solutions, including platform regulation (Ostrom and Cox 2010), but they form part of an effective multi-sectoral response to the problem of misinformation.

REFERENCES

Advocacy Advance. 2021. "Bicycle Injury and Fatality Statistics during the Pandemic." May 27. www.advocacyadvance.org/2021/05/bicycle-injury-and-fatality-statistics-during-the-pandemic/.

Alicke, Mark, Mary Lou Klotz, David L. Breitenbecher, Tricia J. Yurak, and Debbie S. Vredenburg. 1995. "Personal Contact, Individuation, and the Better-Than-Average Effect." *Journal of Personality and Social Psychology* 68 (5): 804–825. https://doi.org/10.1037/0022-3514.68.5.804.

Allyn, Bobby, and Barbara Sprunt. 2020. "Poll: As Coronavirus Spreads, Fewer Americans See Pandemic as a Real Threat." *NPR*, March 17. www.npr.org/2020/03/17/816501871/poll-as-coronavirus-spreads-fewer-americans-see-pandemic-as-a-real-threat.

Azarpanah, Hossein, Mohsen Farhadloo, Rustam M. Vahidov, and Louise Pilote. 2021. "Vaccine Hesitancy: Evidence from an Adverse Events Following Immunization Database, and the Role of Cognitive Biases." *BMC Public Health* 21 (1686): 1–13. https://doi.org/10.1186/s12889–021-11745-1.

Bak-Coleman, Joseph B., Ian Kennedy, and Jevin D. West et al. 2022. "Combining Interventions to Reduce the Spread of Viral Misinformation." *Nature Human Behaviour*. https://doi.org/10.1038/s41562-022-01388-6.

Banerjee, Ritwik, Joydeep Bhattacharya, and Priyama Majumdar. 2021. "Exponential-Growth Prediction Bias and Compliance with Safety Measures Related to COVID-

19." *Social Science and Medicine* 268 (January): 1–9. https://doi.org/10.1016/j.socscimed.2020.113473.

Banerjee, Ritwik, and Priyama Majumdar. 2020. "Exponential Growth Bias in the Prediction of Covid-19 Spread and Economic Expectation." IZA Discussion Paper No. 13664. http://dx.doi.org/10.2139/ssrn.3687141.

Beach, Lee Roy, and Richard G. Swensson. 1967. "Instructions about Randomness and Run Dependency in Two-Choice Learning." *Journal of Experimental Psychology* 75 (2): 279–282. https://doi.org/10.1037/h0024979.

Benedictus, Leo. 2021. "Most People Admitted to Hospital with Covid-19 Are Vaccinated." *Full Fact*, October 21. https://fullfact.org/health/economist-vaccination-status/.

Berenbaum, May R. 2021. "On COVID-19, Cognitive Bias, and Open Access." *Proceedings of the National Academy of Sciences* 118 (2): 1–3. https://doi.org/10.107/pnas.2026319118.

Best US Casinos. 2022. "Mysterious Slot Lions Share Hits for $2.4M at MGM." Accessed October 21. www.bestuscasinos.org/vegas/lions-share-jackpot-won/.

Birch, Jonathan. 2021. "Animals, Humans and Pandemics: What Needs to Change?" *The London School of Economics and Political Science*, March 9. www.lse.ac.uk/philosophy/blog/2021/03/09/animals-humans-and-pandemics-what-needs-to-change/.

Bishop, D. V. M., Jackie Thompson, and Adam J. Parker. 2022. "Can We Shift Belief in the 'Law of Small Numbers'?" *Royal Society Open Science* 9 (3): 1–27. https://doi.org/10.1098/rsos.211028.

Bode, Leticia, and Emily K. Vraga. 2015. "In Related News, That Was Wrong: The Correction of Misinformation through Related Stories Functionality in Social Media." *Journal of Communication* 65 (4): 619–638. https://doi.org/10.1111/jcom.12166.

Bottemanne, Hugo, Orphée Morlaàs, Philippe Fossati, and Liane Schmidt. 2020. "Does the Coronavirus Epidemic Take Advantage of Human Optimism Bias?" *Frontiers in Psychology* 11: 1–5. https://doi.org/10.3389/fpsyg.2020.02001.

Braun, Barbara L., Jinnet B. Fowles, Leif Solberg, Elizabeth Kind, Margaret Healey, and Renner Anderson. 2000. "Patient Beliefs about the Characteristics, Causes, and Care of the Common Cold." *Journal of Family Practice* 49 (2): 153–156.

Brezis, Mayer, Daphna Halpern-Reichert, and Mitchell J. Schwaber. 2004. "Mass Media-Induced Availability Bias in the Clinical Suspicion of West Nile Fever." *Annals of Internal Medicine* 140 (3): 234–235. https://doi.org/10.7326/0003-4819-140-3-200402030-00024.

Bryden, Joan. 2021. "Canadians Far More Wary of AstraZeneca than Other COVID-19 Vaccines: Poll." *CP24*, March 30. www.cp24.com/news/canadians-far-more-wary-of-astrazeneca-than-other-covid-19-vaccines-poll-1.5367801?cache = yes%3FautoPlay%3Dtrue%3FclipId%3D89950.

Burkell, Jacquelyn. 2004. "What Are the Chances? Evaluating Risk and Benefit Information in Consumer Health Materials." *Journal of the Medical Library Association* 92 (2): 200–208.

Cai, Jane. 2020. "Wuhan Quarantine: Shutting Down a City Five Times the Size of London." *South China Morning Post*, January 23. www.scmp.com/news/china/politics/article/3047365/wuhan-quarantine-shutting-down-city-five-times-size-london.

Centers for Disease Control and Prevention. 2022. "COVID-19 Vaccines Work." June 28. www.cdc.gov/coronavirus/2019-ncov/vaccines/effectiveness/work.html.

Chen, Daniel L., Tobias J. Moskowitz, and Kelly Shue. 2016. "Decision Making under the Gambler's Fallacy: Evidence from Asylum Judges, Loan Officers, and Baseball Umpires." *Quarterly Journal of Economics* 131 (3): 1181–1242.

Christensen, Love. 2021. "Optimal Persuasion under Confirmation Bias: Theory and Evidence from a Registered Report." *Journal of Experimental Political Science* 10 (1): 4–20. https://doi.org/10.1017/XPS.2021.21.

CNN Wire. 2014. "Couple Wins $2.4 Million Jackpot at Vegas Slot Machine that Hasn't Paid Out in 20 Years." *Fox 8*, August 25. https://myfox8.com/news/couple-wins-2-4-million-jackpot-at-vegas-slot-machine-that-hasnt-paid-out-in-20-years/.

Cook, John, Ullrich Ecker, and Stephan Lewandowsky. 2015. "Misinformation and How to Correct It." In *Emerging Trends in the Social and Behavioral Sciences: An Interdisciplinary, Searchable, and Linkable Resource*, ed. Robert A. Scott, and Stephen Michael Kosslyn. Hoboken, NJ: Wiley, 1–17.

Demi. 1997. *One Grain of Rice: A Mathematical Folktale*. New York: Scholastic,

Devis, Deborah. 2021. "Why Are There So Many Vaccinated People in Hospital?" *Cosmos*, September 20. https://cosmosmagazine.com/health/covid/why-are-there-so-many-vaccinated-people-in-hospital/.

Dickson, David. 2005. "The Case for a 'Deficit Model' of Science Communication." *SciDevNet* 27: 1–6. www.scidev.net/global/editorials/the-case-for-a-deficit-model-of-science-communic/.

Diemand-Yauman, Connor, Daniel M. Oppenheimer, and Erikka B. Vaughan. 2011. "Fortune Favors the Bold (and the Italicized): Effects of Disfluency on Educational Outcomes." *Cognition* 118 (1): 111–115. https://doi.org/10.1016/j.cognition.2010.09.012.

Dolinski, Dariusz, Barbara Dolinska, Barbara Zmaczynska-Witek, Maciej Banach, and Wojciech Kulesza. 2020. "Unrealistic Optimism in the Time of Coronavirus Pandemic: May It Help to Kill, if So—Whom: Disease or the Person?" *Journal of Clinical Medicine* 9 (5): 1–9. https://doi.org/10.3390/jcm9051464.

Dolinski, Dariusz, Wojciech Kulesza, Paweł Muniak, Barbara Dolinska, Rafał Węgrzyn, and Kamil Izydorczak. 2022. "Media Intervention Program for Reducing Unrealistic Optimism Bias: The Link between Unrealistic Optimism, Well-Being, and Health." *Applied Psychology: Health and Well-Being* 14 (2): 499–518.

Ducharme, Jamie. 2022. "Almost 60% of Americans Have Had COVID-19, CDC Says." *Time*, April 26. https://time.com/6170735/how-many-people-have-had-covid-19/.

Ecker, Ullrich K. H., Stephan Lewandowsky, and Michelle A. Amazeen et al. 2022. "The Psychological Drivers of Misinformation Belief and Its Resistance to Correction." *Nature Reviews Psychology* 1 (1): 13–29. https://doi.org/10.1038/s44159–021-00006-y.

Ecker, Ullrich K. H., Stephan Lewandowsky, Briony Swire, and Darren Chang. 2011. "Correcting False Information in Memory: Manipulating the Strength of Misinformation Encoding and Its Retraction." *Psychonomic Bulletin and Review* 18 (3): 570–578.

Ecker, Ullrich K. H., Stephan Lewandowsky, and David T. W. Tang. 2010. "Explicit Warnings Reduce but Do Not Eliminate the Continued Influence of Misinformation." *Memory and Cognition* 38 (8): 1087–1100. https://doi.org/10.3758/mc.38.8.1087.

Egger, Sam, and Garry Egger. 2022. "The Vaccinated Proportion of People with COVID-19 Needs Context." *The Lancet* 399 (10325): 627. https://doi.org/10.1016/S0140–6736(21)02837-3.

Ehrenstein, Boris P., Frank Hanses, Stefan Blaas, Falitsa Mandraka, Franz Audebert, and Bernd Salzberger. 2010. "Perceived Risks of Adverse Effects and Influenza Vaccination: A Survey of Hospital Employees." *European Journal of Public Health* 20 (5): 4959499.

European Medicines Agency. 2021. "COVID-19 Vaccine AstraZeneca: Benefits Still Outweigh the Risks Despite Possible Link to Rare Blood Clots with Low Blood

Platelets." March 18. https://www.ema.europa.eu/en/news/covid-19-vaccine-astrazeneca-benefits-still-outweigh-risks-despite-possible-link-rare-blood-clots-low-blood-platelets.

The Exposé. 2022. "Italy's National Institute of Health Reveals the Fully Vaccinated Now Account for 7 in Every 10 Covid-19 Deaths." April 9. https://expose-news.com/2022/04/09/italy-7-in-10-covid-deaths-vaccinated/.

Fallis, Don. 2020. "The Epistemic Threat of Deepfakes." *Philosophy and Technology* 34 (4): 623–643.

Felgendreff, Lisa, Lars Korn, Philipp Sprengholz, Sarah Eitze, Regina Siegers, and Cornelia Betsch. 2021. "Risk Information Alone Is Not Sufficient to Reduce Optimistic Bias." *Research in Social and Administrative Pharmacy* 17 (5): 1026–1027. https://doi.org/10.1016/j.sapharm.2021.01.010.

Ferreira, Jennifer. 2022. "Making Sense of the Numbers: Greater Proportion of Unvaccinated Are Being Hospitalized." *CTV News*, February 8. www.ctvnews.ca/health/coronavirus/most-covid-19-hospitalizations-in-provinces-are-among-the-vaccinated-here-s-why-1.5770226.

Festinger, Leon. 1957. *A Theory of Cognitive Dissonance*. Vol. 2. Palo Alto: Stanford University Press,

Fischhoff, Baruch, Paul Slovic, and Sarah Lichtenstein. 1977. "Knowing with Certainty: The Appropriateness of Extreme Confidence." *Journal of Experimental Psychology: Human Perception and Performance* 3 (4): 552–564.

Fox, Christopher John. 1983. *Information and Misinformation: An Investigation of the Notions of Information, Misinformation, Informing, and Misinforming*. Westport, CT: Greenwood.

Fragkaki, Iro, Dominique F. Maciejewski, Esther L. Weijman, Jonas Feltes, and Maaike Cima. 2021. "Human Responses to Covid-19: The Role of Optimism Bias, Perceived Severity, and Anxiety." *Personality and Individual Differences* 176: 1–7. https://doi.org/10.1016/j.paid.2021.110781.

Frischmann, Brett M., Michael J. Madison, and Katherine J. Strandburg. 2014. "Governing Knowledge Commons." In *Governing Knowledge Commons*, ed. Brett M. Frischmann, Michael J. Madison, and Katherine J. Strandburg, 1–43. Oxford University Press.

Gates, Bill. 2020. "Responding to Covid-19: A Once-in-a-Century Pandemic?" *New England Journal of Medicine* 382 (18): 1677–1679. www.nejm.org/doi/full/10.1056/nejmp2003762.

Gerhold, Lars. 2020. "COVID-19: Risk Perception and Coping Strategies." https://doi.org/10.31234/osf.io/xmpk4.

Gronholt-Pedersen, Jacob. 2021. "Woman Who Died after AstraZeneca Shot Had 'Highly Unusual' Symptoms, Officials Say." *Global News*, March 16. https://globalnews.ca/news/7696802/astrazeneca-denmark-blood-clot-suspension/.

Guterres, António. 2022. "All Hands On Deck to Fight a Once-in-a-Lifetime Pandemic." *United Nations*. Accessed October 21. www.un.org/en/un-coronavirus-communications-team/all-hands-deck-fight-once-lifetime-pandemic.

Hayes, Kimani. 2021. "Gottlieb Says Vaccines Still Effective at Preventing Serious Illness from Delta Variant." *CBS News*, August 2. www.cbsnews.com/news/covid-19-vaccines-delta-variant-effectiveness-gottlieb/.

Henry, David, and Paul Glasziou. 2021. "How Well Do COVID Vaccines Work in the Real World?" *The Conversation*, July 1. https://theconversation.com/how-well-do-covid-vaccines-work-in-the-real-world-162926.

Hernandez, Ivan, and Jesse Lee Preston. 2013. "Disfluency Disrupts the Confirmation Bias." *Journal of Experimental Social Psychology* 49 (1): 178–182.

Holman, Mirya R., and J. Celeste Lay. 2019. "They See Dead People (Voting): Correcting Misperceptions about Voter Fraud in the 2016 US Presidential Election." *Journal of Political Marketing* 18 (1–2): 31–68. https://doi.org/10.1080/15377857.2018.1478656.

Johnson, Hollyn M., and Colleen M. Seifert. 1994. "Sources of the Continued Influence Effect: When Misinformation in Memory Affects Later Inferences." *Journal of Experimental Psychology: Learning, Memory, and Cognition* 20 (6): 1420–1436.

Jones, Martin, and Robert Sugden. 2001. "Positive Confirmation Bias in the Acquisition of Information." *Theory and Decision* 50 (1): 59–99.

Kahneman, Daniel, Paul Slovic, and Amos Tversky. 1982. *Judgment under Uncertainty: Heuristics and Biases*. Cambridge: Cambridge University Press.

Kampf, Günter. 2021. "COVID-19: Stigmatising the Unvaccinated Is Not Justified." *The Lancet* 398 (10314): 1871. https://doi.org/10.1016/S0140–6736(21)02243-1.

Kerr, John R., Alexandra L. J. Freeman, Theresa M. Marteau, and Sander van der Linden. 2021. "Effect of Information about COVID-19 Vaccine Effectiveness and Side Effects on Behavioural Intentions: Two Online Experiments." *Vaccines* 9 (4): 1–22. https://doi.org/10.3390/vaccines904379.

Kim, Jooyeon, Behzad Tabibian, Alice Oh, Bernhard Schölkopf, and Manuel Gomez-Rodriguez. 2018. "Leveraging the Crowd to Detect and Reduce the Spread of Fake News and Misinformation." In *Proceedings of the Eleventh ACM International Conference on Web Search and Data Mining*, 324–332). https://doi.org/10.1145/3159652.3159734.

King, Wendy C., Max Rubinstein, Alex Reinhart, and Robin Mejia. 2021. "COVID-19 Vaccine Hesitancy January–May 2021 Among 18–64 Year Old US Adults by Employment and Occupation." *Preventive Medicine Reports* 24: 1–9. https://doi.org/10.1016/j.pmedr.2021.101569.

Kirzinger, Ashley, Grace Sparks, and Mollyann Brodie et al. 2021. "KFF COVID-19 Vaccine Monitor: July." *KFF*, August 21. www.kff.org/coronavirus-covid-19/poll-finding/kff-covid-19-vaccine-monitor-july-2021/.

Kunda, Ziva. 1990. "The Case for Motivated Reasoning." *Psychological Bulletin* 10 (3): 480–498. https://doi.org/10.1037/0033-2909.108.3.480.

Lammers, Joris, Jan Crusius, and Anne Gast. 2020. "Correcting Misperceptions of Exponential Coronavirus Growth Increases Support for Social Distancing." *Proceedings of the National Academy of Sciences* 117 (28): 16264–16266. https://doi.org/10.1073/pnas.2006048117.

Lewandowsky, Stephan, Ullrich K. H. Ecker, Colleen M. Seifert, Norbert Schwarz, and John Cook. 2012. "Misinformation and Its Correction: Continued Influence and Successful Debiasing." *Psychological Science in the Public Interest* 13 (3): 106–131. https://doi.org/10.1177%2F1529100612451018.

Lewandowsky, Stephan, and Paul W. Smith. 1983. "The Effect of Increasing the Memorability of Category Instances on Estimates of Category Size." *Memory and Cognition* 11 (4): 347–350.

Ling, Rich. 2020. "Confirmation Bias in the Era of Mobile News Consumption: The Social and Psychological Dimensions." *Digital Journalism* 8 (5): 596–604. https://doi.org/10.1080/21670811.2020.1766987.

Mamede, Sílvia, Marco Antonio de Carvalho-Filho, and Henk G. Schmidt et al. 2020. "'Immunising' Physicians against Availability Bias in Diagnostic Reasoning: A Randomised Controlled Experiment." *BMJ Quality and Safety* 29 (7): 550–559.

Mazerolle, John. 2021. "Great COVID-19 Bicycle Boom Expected to Keep Bike Industry On Its Toes for Years To Come." *CBC News*, March 21. www.cbc.ca/news/business/bicycle-boom-industry-turmoil-covid-19-1.5956400.

Meppelink, Corine S., Edith G. Smit, Marieke L. Fransen, and Nicola Diviani. 2019. "'I Was Right about Vaccination': Confirmation Bias and Health Literacy in Online Health

Information Seeking." *Journal of Health Communication* 24 (2): 129–140. https://doi.org/10.1080/10810730.2019.1583701.

Modgil, Sachin, Rohit Kumar Singh, Shivam Gupta, and Denis Dennehy. 2021. "A Confirmation Bias View on Social Media Induced Polarisation during Covid-19." *Information Systems Frontiers* 26 (2): 417–441.

Monte, Lindsay M. 2021. "Household Pulse Survey Shows Many Don't Trust COVID Vaccine, Worry about Side Effects." *United States Census Bureau*, December 28. www.census.gov/library/stories/2021/12/who-are-the-adults-not-vaccinated-against-covid.html.

Nyhan, Brendan, and Jason Reifler. 2010. "When Corrections Fail: The Persistence of Political Misperceptions." *Political Behavior* 32 (2): 303–330. https://doi.org/10.1007/s11109–010-9112-2.

Ostrom, Elinor. 2005. "Doing Institutional Analysis: Digging Deeper than Markets and Hierarchies." In *Handbook of New Institutional Economics*, ed. Claude Menard and Mary M. Shirley, 819–848. Boston: Springer.

2009. *Understanding Institutional Diversity*. Princeton, NJ: Princeton University Press,

Ostrom, Elinor, and Michael Cox. 2010. "Moving Beyond Panaceas: A Multi-tiered Diagnostic Approach for Social-Ecological Analysis." *Environmental Conservation* 37 (4): 451–463.

Panthagani, Kristen. 2022. "Q: If 50% of COVID Hospitalizations Are among the Vaccinated, Does that Mean the Vaccines Aren't Working?" *Dear Pandemic*, February 3. https://dearpandemic.org/base-rate-fallacy/.

Pascual-Leone, Alvaro, Gabriele Cattaneo, Dídac Macià, Javier Solana, José M. Tormos, and David xBartrés-Faz. 2021. "Beware of Optimism Bias in the Context of the COVID-19 Pandemic." *Annals of Neurology* 89 (3): 423–425. https://doi.org/10.1002/ana.26001.

Prakash, Shivesh, Ruth M. Sladek, and Lambert Schuwirth. 2019. "Interventions to Improve Diagnostic Decision Making: A Systematic Review and Meta-analysis on Reflective Strategies." *Medicsal Teacher* 41 (5): 517–524. https://doi.org/10.1080/0142159X.2018.1497786.

Quon, Alexander. 2022. "Why Experts Say Premier Moe's Claim Vaccines No Longer Protect against COVID-19 Is Based on Misunderstanding." *CBC News*, February 3. www.cbc.ca/news/canada/saskatchewan/covid-19-transmission-scott-moe-1.6336479.

Rajsic, Jason, Daryl E. Wilson, and Jay Pratt. 2018. "The Price of Information: Increased Inspection Costs Reduce the Confirmation Bias in Visual Search." *Quarterly Journal of Experimental Psychology* 71 (4): 832–849.

Raude, Jocelyn, Marion Debin, and Isabelle Bonmarin et al. 2020. "Are People Excessively Pessimistic about the Risk of Coronavirus Infection?" https://psyarxiv.com/364qj/.

Reuters staff. 2020. "Impact of Coronavirus Will be Felt for Decades to Come, WHO Says." *Reuters*, July 31. www.reuters.com/article/us-health-coronavirus-who-idUSKCN24W27L.

2021. "Austria Suspends AstraZeneca COVID-19 Vaccine Batch after Death." *Reuters*, March 7. www.reuters.com/article/health-coronavirus-austria-nurse-idUSL8N2L506P.

Roney, Christopher J. R., and Lana M. Trick. 2003. "Grouping and Gambling: A Gestalt Approach to Understanding the Gambler's Fallacy." *Canadian Journal of Experimental Psychology* 57 (2): 69–75.

Roozenbeek, Jon, Claudia R. Schneider, and Sander Van Der Linden et al. 2020. "Susceptibility to Misinformation about COVID-19 around the World." *Royal Society Open Science* 7 (10): 1–15.

Rosenberg, J. 2001. "Young People in the United States Are Often Misinformed about the Proper Use of Condoms." *Perspectives on Sexual and Reproductive Health* 33 (5): 235.

Rubin, Victoria L., Chris Brogly, Nadia Conroy, Yimin Chen, Sarah E. Cornwell, and Toluwase V. Asubiaro. 2019. "A News Verification Browser for the Detection of

Clickbait, Satire, and Falsified News." *Journal of Open Source Software* 4 (35): 1–3. https://doi.org/10.21105/joss.01208.

Rumilly, Marc (@MarcRummy). 2021. "'X% of Hospitalized Patients Are Fully Vaccinated' Doesn't Always Mean What You Think It Does. The Real Meaning Depends on the Base Rates. Without Knowing That, the Quote Can Mean Anything." Twitter post, July 23.

Savolainen, Reijo. 2009. "Information Use and Information Processing: Comparison of Conceptualizations." *Journal of Documentation* 65 (2): 187–207.

Schlager, Tobias, and Ashley V. Whillans. 2022. "People Underestimate the Probability of Contracting the Coronavirus from Friends." *Humanities and Social Sciences Communications* 9 (1): 1–11. http://dx.doi.org/10.1057/s41599-022-01052-4.

Schmidt, Henk G., Sílvia Mamede, Kees Van Den Berge, Tamara Van Gog, Jan L. C. M. Van Saase, and Remy M. J. P. Rikers. 2014. "Exposure to Media Information about a Disease Can Cause Doctors to Misdiagnose Similar-Looking Clinical Cases." *Academic Medicine* 89 (2): 285–291.

Schonger, Martin, and Daniela Sele. 2021. "Intuition and Exponential Growth: Bias and the Roles of Parameterization and Complexity." *Mathematische Semesterberichte* 68 (2): 221–235.

Schwind, Christina, and Jürgen Buder. 2012. "Reducing Confirmation Bias and Evaluation Bias: When Are Preference-Inconsistent Recommendations Effective – and When Not?" *Computers in Human Behavior* 28 (6): 2280–2290.

Schwind, Christina, Jürgen Buder, Ulrike Cress, and Friedrich W. Hesse. 2012. "Preference-Inconsistent Recommendations: An Effective Approach for Reducing Confirmation Bias and Stimulating Divergent Thinking?" *Computers and Education* 58 (2): 787–796.

Scherer, Laura D., and Gordon Pennycook. 2020. "Who Is Susceptible to Online Health Misinformation?" *American Journal of Public Health* 110 (S3): S276–S277.

Seifert, Colleen M. 2002. "The Continued Influence of Misinformation in Memory: What Makes a Correction Effective?" In *Psychology of Learning and Motivation*, Vol. 41, 265–292. Cambridge, MA: Academic Press,

Seneff, Stephanie, and Greg Nigh. 2021. "Worse than the Disease? Reviewing Some Possible Unintended Consequences of the mRNA Vaccines against COVID-19." *International Journal of Vaccine Theory, Practice, and Research* 2 (1): 38–79.

Sharon, Aviv J., and Ayelet Baram-Tsabari. 2020. "Can Science Literacy Help Individuals Identify Misinformation in Everyday Life?" *Science Education* 104 (5): 873–894. https://doi.org/10.1002/sce.21581.

Simis, Molly J., Haley Madden, Michael A. Cacciatore, and Sara K. Yeo. 2016. "The Lure of Rationality: Why Does the Deficit Model Persist in Science Communication?" *Public Understanding of Science* 25 (4): 400–414. https://doi.org/10.1177/0963662516629749.

Simonetti, Omar, Mariano Martini, and Emanuele Armocida. 2021. "COVID-19 and Spanish Flu-18: Review of Medical and Social Parallelisms between Two Global Pandemics." *Journal of Preventive Medicine and Hygiene* 62 (3): E613–E620. https://doi.org/10.15167/2421-4248/jpmh2021.62.3.2124.

Smitham, Eleni, and Amanda Glassman. 2021. "The Next Pandemic Could Come Soon and be Deadlier." *Center for Global Development*, August 25. www.cgdev.org/blog/the-next-pandemic-could-come-soon-and-be-deadlier.

Stapel, Diederik A., Stephen D. Reicher, and Russell Spears. 1995. "Contextual Determinants of Strategic Choice: Some Moderators of the Availability Bias." *European Journal of Social Psychology* 25 (2): 141–158.

Statista. 2022. "Number of Confirmed COVID-19 Deaths in Canada from December 14, 2020 to August 21, 2022, by Vaccination Status." www.statista.com/statistics/1257040/number-covid-deaths-canada-by-vaccination-status/.

Tasnim, Samia, Md Mahbub Hossain, and Hoimonty Mazumder. 2020. "Impact of Rumors and Misinformation on COVID-19 in Social Media." *Journal of Preventive Medicine and Public Health* 53 (3): 171–174.

Thaler, Richard H., Cass R. Sunstein, and John P. Balz. 2013. "Choice Architecture." In *The Behavioral Foundations of Public Policy*, ed. Eldar Shafir, Ch. 25. Princeton, NJ: Princeton University Press.

Thomas, Naomi, and Jason Hanna. 2021. "Moderna's Covid-19 Vaccine Shows 93% Efficacy through 6 Months, as Company Expects to Finish Application for Approval this Month." *CNN*, August 5. www.cnn.com/2021/08/05/health/moderna-coronavirus-vaccine-efficacy/index.html.

Tversky, Amos, and Daniel Kahneman. 1971. "Belief in the Law of Small Numbers." *Psychological Bulletin* 76 (2): 105–110.

1973. "Availability: A Heuristic for Judging Frequency and Probability." *Cognitive Psychology* 5 (2): 207–232.

1974. "Judgment under Uncertainty: Heuristics and Biases." *Science* 185 (4157): 1124–1131.

Villanova, Daniel. 2022. "Linear Biases and Pandemic Communications." *Medical Decision Making* 42 (6): 765–775.

Vraga, Emily K., and Leticia Bode. 2017. "Using Expert Sources to Correct Health Misinformation in Social Media." *Science Communication* 39 (5): 621–645. https://doi.org/10.1002/pra2.2015.145052010083.

Vraga, Emily K., Melissa Tully, and Leticia Bode. 2022. "Empowering Users to Respond to Misinformation about Covid-19." *Media and Communication (Lisboa)* 8 (2): 475–479. https://doi.org/10.17645/mac.v8i2.3200.

Wason, Peter C. 1960. "On the Failure to Eliminate Hypotheses in a Conceptual Task." *Quarterly Journal of Experimental Psychology* 12 (3): 129–140.

Weinstein, Neil D. 1983. "Reducing Unrealistic Optimism about Illness Susceptibility." *Health Psychology* 2 (1): 11.

Wheeler, Gregory. 2018. "Bounded Rationality." *Stanford Encyclopedia of Philosophy*. November 30. https://plato.stanford.edu/entries/bounded-rationality/.

Wise, Toby, Tomislav D. Zbozinek, Giorgia Michelini, Cindy C. Hagan, and Dean Mobbs. 2020. "Changes in Risk Perception and Self-Reported Protective Behaviour during the First Week of the COVID-19 Pandemic in the United States." *Royal Society Open Science* 7 (9): online 200742.

Wittenberg, Chloe, and Adam J. Berinsky. 2020. "Misinformation and Its Correction." In *Social Media and Democracy: The State of the Field, Prospects for Reform*, ed. Nathaniel Persily and Joshua A. Tucker, 163–198. Cambridge: Cambridge University Press.

Wogan, Tim. 2014. "Gambler's Fallacy Trips up Goalies." *Science*, July 31. www.science.org/content/article/gamblers-fallacy-trips-goalies.

Xu, Shan, Ioana A. Coman, Masahiro Yamamoto, and Christina Jimenez Najera. 2022. "Exposure Effects or Confirmation Bias? Examining Reciprocal Dynamics of Misinformation, Misperceptions, and Attitudes toward COVID-19 Vaccines." *Health Communication* 38 (10): 1–11.

Yahoo Finance. 2022. "61% of Americans Underestimate Their Odds of Contracting Long COVID-19." May 24. https://finance.yahoo.com/news/61-americans-underestimate-odds-contracting-140100108.html.

Zheng, Caifang, Weihao Shao, Xiaorui Chen, Bowen Zhang, Gaili Wang, and Weidong Zhang. 2022. "Real-World Effectiveness of COVID-19 Vaccines: A Literature Review and Meta-analysis." *International Journal of Infectious Diseases* 114: 252–260. https://doi.org/10.1016/j.ijid.2021.11.009.

4

Information Hazing

An Examination through Computer Science Education

Elizabeth Wickes and Melissa G. Ocepek

INFORMATION HAZING

Misinformation is experienced in many places and forms throughout a person's lifetime. Unfortunately, one common place where misinformation can have a long and lasting impact is within the classroom. As instructors working within the field of information science, we have seen students struggle with understanding the norms and rules of academia. Inadequate or incorrect guidance can cause negative impacts for students. These include missed opportunities, peer ostracism, anxiety, and lowered grades. These have long- term implications in terms of the students lived experience as well as larger societal representation in computing fields.

We define information hazing as the use of information to directly and indirectly harass and/or exclude newcomers. It is often used to create strong bonds of in and out groups in social environments that have shared values and social cohesion. It is used to demonstrate one's current membership within a group, or worthiness for gaining membership. This concept derives from the research literature on hazing in academic and work settings as well as research on bullying in higher education.

The concept of information hazing provides a new framework to consider how the norms of early computing education encourage malinformation and misinformation to spread and privilege students who are more familiar with computer science concepts, culture, and industries.

Most traditional examples of hazing showcase overt and direct behaviors between individuals. Information hazing is different in several distinct areas. First, it is often a covert approach, residing within policies or other governance structures of the social group. This content will often blend into surrounding policies, hiding behind other discussions of ethics and morality. Second, being embedded within governance and decision-making powers, it acts at scale on newcomers rather than individual actors. Third, these two factors intersect to obfuscate an individual person's ownership or connection to these governance structures. For example, the justification for a decision would come from the system rather than a person. Fourth, this

FIGURE 4.1 Visual themes from information hazing: an examination through computer science education.

independence allows the systems to operate outside the normal barriers of time. Policy language and precedents can last beyond the original author's lifespan or tenure within the group, with authorship credit often forgotten and attributed just to group norms.

These factors combine to make intersectional traps that further advance information hazing's ability to exist under the normal checks and balances designed to dismantle inequitable and unfair practices. While scholars have addressed many elements of computing education that have made it unwelcoming to many, no studies have considered how the governing documents of class including the syllabus, honor code, student code, and other academic integrity related policies begin the process of creating an unwelcoming environment when they are shared before or on the first day of class. These policies have been studied under the lenses of professional practice, detection, and prevention; there appears to be a gap of investigation into the more holistic impact of the policies on the student experience. In this chapter, we examine and consider the governing documents of first computing courses with an eye to information hazing. A form of hazing that we believe is all too prevalent in computing education and is a factor in the reason many nontraditional computer science students struggle with their grades, mental health, and long term prospects in the field.

Hazing

As long as higher education has existed so has hazing. Plato described what modern-day scholars consider hazing at his Academy in 387 BC (Nuwer 1999). Since then, public attitudes around hazing have shifted and currently forty-four states have laws on the books to criminalize and deter hazing (StopHazing 2022). Hazing, much like information, is a term with many definitions. Most contain three elements: a power in-balance, abuse, and initiation.

Hazing is the exercise of control or power, usually in a punitive or abusive manner, by persons who are members of a group or organization and directed at individuals seeking admission to the group. Hazing often is an initiation ritual. (Guynn and Aquila 2004, 1).

There are also distinctions in the literature between mental hazing, physical hazing (Nuwer 1999), sexual hazing (Guynn and Aquila 2004), and academic hazing (Costa Pinto et al. 2020). Academic hazing is the most similar to information as it is within the context of initiation and transition into higher education.

A few notable attributes of hazing in general, which also apply to information hazing include:

- it can happen regardless of malicious intent (Allan and Madden 2012);
- those who have been hazed believe it is their right or even responsibility to haze others, and that it is used to maintain the values, principles, and exclusivity of the group (Guynn and Aquila 2004).

Along with exclusivity, hazing emphasizes conformity, making mass initiations particularly difficult for members of a potential group who are different, including but not limited to members who do not belong to the dominate race, sexual orientation, gender, appearance, or ideology. Finally, much like the topics throughout this book, hazing is tied to misinformation, malinformation, and deceit.

> Nearly all acts of hazing involve deception. Hazers lie all the time to newcomers. First they lie about the severity of the hazing, which is intended to build fear in the initiate. Then as newcomers invest more and as the initiation process nears an end, hazing escalates, always remaining secret. Hazers also lie to one another, to adults, and to themselves to rationalize that the brutal practices build group unity. (Guynn and Aquila 2004, 26)

Applying this framework onto a critical study of early computing education allows for a more nuanced lens in which to consider the possible impacts of certain policies and procedures that create an unwelcome and exclusionary academic environment for many students.

GOVERNING KNOWLEDGE COMMONS

Governing Knowledge Commons (GKC) is a descriptive framework for analysis, defining knowledge commons, and further investigating how the people within that knowledge commons manage governance. The classroom space has been discussed by others as a knowledge commons (Madison, Frischmann, and Strandburg 2009). Classrooms operate under many agents of control, but the syllabus is one of the most direct and personal factors between the instructor and the student. Instructors often have a wide range of options for requirements and rules with little oversight or review. These factors make syllabi, as governance documents, extremely interesting for analysis.

We see that information hazing can be enacted by a knowledge commons (within the classroom) by means of governance (the syllabus). As a document directly related to control, power, and punishment, the syllabus uses policy and procedures to enact this control at scale. Control over behaviors, specifically around information access and use, aligns with the GKC framework of analysis while still creating a clear link between the framework and our novel concept of information hazing.

Previous works on commons governance have identified the university and its various units and departments as constructed cultural commons where knowledge is produced and disseminated throughout various levels of classrooms, research labs, and administrative level (Madison, Frischmann, and Strandburg 2009). We build on this work and its focus on the formal and informal institutional structures that influence a knowledge commons including policies, structures, and social norms. For our exploration of information hazing in top-tier computer science classrooms, we view the contextual space of higher education, research universities and small liberal arts colleges and universities, and the physical, digital, social, and structural spaces of the classroom. The classroom and its related digital spaces are our action arena. The actors in our exploration are the students, instructors, and teaching assistants who populate introductory computer science education spaces. The resources that move throughout this commons space include the informational and intellectual material that constitutes the content of the class as well as the information that makes up the policies and rules that constrain the learning environment.

Syllabus as Governance

Outside of direct observation, understanding the power dynamics of any classroom is difficult. Students' lived experiences within these spaces is an essential element of such a description. However, our focus is on examining the governance structure rather than field observation. The course syllabus defines the elements of power, punishment, and control over the classroom. They are presented to students as the governing documents for individual courses. Being used in partnership with any existing student codes, handbooks, or other policies, syllabi have a responsibility to inform students of instructor expectations and policies for that class. Faculty have indicated that their academic integrity policies were sourced from a variety of levels within their institution, often being "defined, redefined, or interpreted at multiple levels" (Simon et al. 2018, 116). The syllabus should then be the final operationalized product of these influences, representing the official rules for their particular space.

Hazing behaviors often appear within the initial onboarding or other introduction to a space. The notion of an introductory or other "first" programming (CS1) course and the resulting classroom (virtual and physical) fit these frameworks as a knowledge commons with formal governance structures, but also serve as a first or early experience within a professional community of programmers The syllabus and classroom connect as "introductory" at two levels: the syllabus as an introduction to the classroom at the micro level and the course itself being the introduction to the field at a macro level.

We examined introductory courses to better understand and describe the interplay between governance and hazing. Our working definition required each course to be a student's initial exposure to programming topics, without any substantial prior programming experiences. These tend to have similar topics across many institutions, allowing us to focus on any differences in governance and information.

A full course load for many undergraduates is between four to six classes running concurrently each semester, meaning that students often operate under four to six different sets of classroom policies that also exist under departmental or program policies. The academic term changeover means that expertise in specific policies is reset as new courses and policies come into effect. Expertise can be difficult to gain, particularly for rarer events like academic misconduct.

Close examination of syllabi within this educational space provides key evidence to observe governance of these knowledge commons, while maintaining our lens of information hazing on novices. Previous research (Simon et al. 2018) surveyed and interviewed faculty members on their academic integrity policies, which were gathered as part of data collection. These findings provide a baseline to validate our sample and observe differences. They found that faculty used informal explanations alongside official policy: "some noted that the policies are not uniformly enforced or, more commonly, that the policies are customized – officially or not – for a discipline, a course, or even an individual assessment" (Simon et al. 2018, 116). Thus, we can expect that statements included within our samples to be incomplete as compared to what students may be presented within the classroom, but provide a fuller picture of what a faculty member would be required to enforce.

Academic Integrity

Academic dishonesty policies exist to provide definitions and explain consequences, and encompass things like cheating on exams, using fabricated data, forgeries, and plagiarism. These governing documents are created with a variety of names, but students are expected to understand and abide by them. Smaller units, such as departments within a university, often create their own versions meant to work in conjunction with existing policies or provide contextual clarifications. These higher-level policies are broad and meant to cover issues across many courses. The lowest level of these policies are the ones for each course or section and are generally required for each course. As discussed, the syllabus stands as a governing document for the physical classroom space but also the learning space where any work is being completed, including outside the classroom.

The higher-level policies often exist within a system for processing accusations and sanctions. Some academic units may allow faculty members to resolve cases of academic integrity independently, but additional groups exist to process appeals and provide other kinds of oversight. These systems are generally very formal and can have names like the "Honor Council" or "Integrity Committee."

These are formal systems with specific governance charges to uphold. Yet the policies that touch students directly come from the syllabus, where the faculty attempt to remind students of the policies and provide contextual understanding. This project looked at syllabi from introductory computer science courses, so the faculty are expected to provide instruction on how the policies apply to writing computer code and introduce students to (very probably) a new way of handling attribution and citation.

Within Computer Science

Computing work generates a variety of written content, such as reports, documentation, presentations, and books. These items exist to support and explain the other items: source code, executable content, data, and more. Most domains have content needing some special handling, but the written content related to computing works well with traditional plagiarism, attribution, and crediting paradigms. Source code and other outputs challenge the traditional approaches of detection. New definitions are needed because, unlike with prose writing, novelty and originality appear differently within source code (Simon et al. 2013). Issues related to data (CODATA-ICSTI Task Group on Data Citation Standards and Practices 2013) and source code (Smith, Katz, and Niemeyer 2016) citation are current areas of research, particularly for those working in scientific reproducibility (Vicente-Saez and Martinez-Fuentes 2018). Source code citation and attribution within introductory computer science homework shares some of these similar concerns but is much more focused on issues of plagiarism and detection rather than reproducibility or citation style and object identifiers.

Existing policies around academic integrity for assessments that involve writing prose are insufficient to cover how much of this new content is being used. Altering a quotation from literature may make it no longer blatant plagiarism, the same is not the case for code. Content copied directly from another source almost always requires at least some minimal adaptation to slot into an existing program, even if your intent is to keep the core behavior the same (Wu et al. 2019). Providing attribution for reused code is considered important within computing, but methods for providing such attribution are highly dependent on context and commonly accepted standards (Simon et al. 2016), making the practice not obvious and not represented under existing general plagiarism guidance beyond simply asking students to provide attribution.

COLLECTION METHODS

Sample Design

Our goal was to review a broad collection of introductory computer science syllabi, hoping to observe many forms of information governance and document standard structural patterns.

We restricted our sample to United States higher education institutions and focused on two groups: "Top" ranked undergraduate computer science programs according to US News and World report (Table 4.2) and top programs within small liberal arts colleges (SLAC) (Table 4.1). The US News and World report rankings for 2022 were obtained and saved for later reference. The actual ranked value of each institution was not retained or considered during this investigation. These labels are not mutually exclusive and not intended to be used as such.

SLACs tend to lack representation within nationally ranked lists of colleges by major. Lists of top SLACs for computer science (CS) were obtained from four separate sources.[1] Institutions listed within these lists were combined, and the number of times each appeared was tallied. The final list of candidates was obtained

[1] www.collegetransitions.com/dataverse/best-small-colleges-for-computer-science. www.collegevine.com/schools/best-liberal-arts-colleges-for-computer-science. www.koppelmangroup.com/blog/2019/11/4/the-best-small-liberal-arts-colleges-for-computer-science. https://academicinfluence.com/rankings/discipline/best-computer-science-liberal-arts-colleges.

TABLE 4.1 *List of courses from SLACs in sample*

Institution	Course	Term	Title
Amherst College	CSSC 111	Spring 2020	Intro to computer science I
Bowdoin College	CSCI 1101	Fall 2019	Intro to computer science
Bowdoin College	CSCI 1103	Fall 2021	Programming with data
Carleton College	CS 111	Winter 2019	Intro to computer science
Colgate College	COSC 101	Spring 2022	Intro to computing I
Grinnell College	CS 151	Spring 2022	Functional problem solving
Harvey Mudd College	CS 5	Spring 2022	Intro to computer science
Mount Holyoke College	CS 100	Fall 2019	Intro to computer science
Mount Holyoke College	CS 151	Spring 2022	Introduction to computational problem solving
Swarthmore College	CS 21	Spring 2022	Intro to computer science
Vassar College	CMPU 101	Spring 2020	Problem solving and abstraction
Wellesley College	CS 111	Spring 2022	Computer programming and problem solving
Wellesley College	CS 115	Spring 2020	Computing for the sociotechno web
Wesleyan University	COMP 112	Fall 2017	Intro to programming
Williams College	CSCI 134	Spring 2022	Intro to computer science

TABLE 4.2 *List of courses from top-ranked computer science programs in sample*

Institution	Course	Term	Title
Carnegie Mellon Univ	15–112	Spring 2022	Fundamentals of programming
Cornell Univ	CS 1110	Spring 2022	Intro to computing using python
Cornell Univ	CS 1112	Spring 2022	Intro to Ccmputing using MATLAB
Georgia Institute of Tech	CS 1301	Fall 2017	Intro to computing
Princeton Univ	COS 126	Spring 2022	Computer science: an interdisciplinary approach
Stanford Univ	CS 101	Spring 2022	Intro to computing principles
Stanford Univ	CS 106A	Spring 2021	Programming methodologies
Univ California Berkeley	CS10	Spring 2022	The beauty and joy of computing
Univ California Berkeley	CS61A	Spring 2022	Structure and interpretation of computer programs
Univ of Illinois	CS 105	Spring 2022	Intro computing: nontech
Univ of Illinois	CS 101	Spring 2022	Intro to programming for engineers and scientists
Univ of Michigan	EECS 183	Fall 2020	Elem programming concepts
Univ of Texas Austin	CS 312	Fall 2021	Intro to programming
Univ of Texas Austin	CS 303e	Spring 2022	Elements of computers and programming
Univ of Washington	CSE 142	Winter 2022	Intro to computer programming I

by selecting all the institutions that appeared on at least two of these lists and were not present within the US News and World report ranking group.

The use of "highest ranked" programs within our search boundary has multiple reasons. Top programs across many types of institutions should provide an opportunity to find examples of creative nontraditional practices and possibly faculty who are engaging in currently understood best practices within the computing education field. More practically, we believed these programs had the highest likelihood to have syllabi readily available within their general online resources. We believe that these highly respected programs would serve as an informal standard that other institutions would look toward. With openly available syllabi, we not only gain access but other programs looking to adopt similar practices would as well.

Syllabus Collection

Computer science course descriptions were reviewed for each institution. Specific courses were selected according to the following rules:

- must be a general programming instruction course (versus general computer skills, business applications, etc.);
- may not be a hybrid course covering programming and statistics (e.g., data science courses);
- presume no prior personal or academic programming experience;
- not have any specialty hybrid content denoted in the title;
- multiple courses were allowed so long as the target audiences were fundamentally different.

Tables 4.1 and 4.2 describe the identified courses from these institutions.

Data Analysis

Once all the documents were collected, we used Atlas.ti qualitative coding software to analyze the documents. We applied thematic analysis (Braun and Clarke 2006) to identify policy language, descriptions of cheating, collaboration, and other attributes related to out-of-class learning. Thematic analysis provides a flexible framework for identifying, analyzing, and reporting themes in qualitative data, including documents. Thematic analysis, unlike many other qualitative analysis frameworks, is not tied to any larger theory or narrowly drawn process. We went through the six phases of thematic analysis beginning with familiarizing ourselves with the data, generating codes together through discussion based on the documents and our past familiarity with research literature and experience as instructors in higher education. We then considered the larger trends and themes in the codes, discussed them and their relationships to each other and the research literature. While we focused on open coding, we also found connections with our codes and the GKC, further supporting our findings that the classroom does represent a knowledge commons space and syllabi are useful artifacts to observe many of these structures. Finally, producing this chapter to where we hope to provide illustrative examples that highlight the opportunities and concerns, we see in this relevant cross-section the best of computing education in higher education across the US.

FINDINGS AND DISCUSSION

Policy presentation within the courses was quite varied, with many different section headers, placements, and so on. Most were introduced using a mix of labels like academic integrity, plagiarism policy, honor code, and collaboration policy. Some linked to separate pages dedicated to specific policies or longer documentation on their expectations for allowable group work, and so on. Many linked to, referenced, or directly quoted the honor code or student code from their department or institution. Most of these went to campus-level honor codes or similar policies, meant for general academic work, but a few had computer science or programming specific guidance.

Observed themes include:

- using behavioral guidance rather than definitions;
- framing peers as information sources to be feared;
- white, western ethics as the only moral compass.

Our findings highlight many of the strategies instructors take to explain these policies, including the variance of approaches and (natural) stumbling points of attempting to work through such a difficult task. Our lens remains focused on understanding the student impact that goes beyond simply understanding the policies and acceptable behaviors.

Advice, not Governance

Plagiarism policies within syllabi serve to provide instruction to students about producing original work and respecting the work of others. These normally remind students to focus on creating original content, use quotations to reference another author's words directly, paraphrase as needed, and so on. Defining these concepts is difficult and an ongoing area of discussion among instructors and researchers as new tools for content creation are adopted. Simon et al. (2018) investigated the language within plagiarism policies for computing classes and stressed that instructors should not depend on student understanding of these concepts even for prose writing, and that there is a strong need for explicit instruction and contextualization of these policies for code writing work.

Do Your Own ~~Work~~ Typing

Terms related to ownership were commonly used within these plagiarism policies, matching many traditional definitions of prose originality. A writer has their "own voice" and writes in their "own words," for example. Attribution, crediting, quotations, and other techniques are used to respect this ownership when an author's words are reused or for a work reference. Plagiarism can be considered theft of these ideas or fraudulently presenting another person's effort as your own. Most began their policies with some instruction to only submit assignments containing their own original content.

> Assume all work you hand in is to be yours and yours alone. (Wellseley CS 111)

> All the work that you submit must be yours. (Mt. Holyoke CS 100)

> Work you submit must be predominantly your own. (University of Washington, CSE 142)

Other policies focused on the creative actions taken to create the work, implying that the act itself generated the originality and ownership. These generally add some clarity by specifying the expected actions that the students should be doing alone, but some still fall back to referencing ownership.

> Each individual programming assignment must be coded by you. (Georgia Tech CS 1301)

> You must write your own solutions. (UC Berkeley 61a)

> The solutions you write up and submit must be your own. (UT Austin CS 312)

> Claiming or submitting as your own any work or code you did not fully author, explicitly or implicitly, no matter how small. (CMU 15–112)

The last example (CMU 15–112) is from a long list of actions that are considered academic integrity violations. Not only is "fully author" not defined, but "explicitly or implicitly" were confusing modifiers. Students may interpret this to say that each line of code they produce must be completely unique and original, no matter how small. Many core programming patterns contain very short lines of code that have little to no room for alterations.

Interestingly, many used the action of typing the content into the computer as a replacement for the more creative actions. This is an objective measure and provides an actionable direction to follow.

> You should type and debug all the code you submit. (Stanford CS 101)

> You have to be the one that types all of the work that you do in this class … You should be the one entering responses to the textbook, pre-lecture activities, and homework. (UIUC CS 105)

Behavioral guidance was commonly provided in lieu of a specific definition for original work. Originality is difficult to explain and gauge, even in the worlds of art, music, and writing, for example. This is compounded with the challenge of explaining it to undergraduate students who likely still struggle with fully understanding plagiarism requirements for prose writing. Formal definitions may not be easily understood by the students or explained by instructors. Behavioral examples may be a fall back for faculty struggling to provide contextual precision from general descriptions provided by their institutions but represent a very realistic workaround the issues within policy language.

The use of typing as a litmus test for original work was an interesting way to provide some objective measure for a deeply complex and subjective concept. This direction is relatively simple to explain to students and based on an everyday action that they would understand. The implicit meaning behind this rule accurately acknowledges that influence and inspiration for writing the code likely came from many sources. So long as the work was typed "by your own hand" it will be deemed acceptable.

Even when using specific examples, some held brief descriptions with little extra clarity. Verbal guidance might be given informally during class time, but the syllabus remains the primary governing document and is usually treated as the final word when disagreements occur. Similar short policies were found by (Simon et al. 2018) and were described as "snippets of advice" (118).

When documented governance rules are unclear, those that prohibit a strong knowledge commons from forming prevent essential knowledge sharing among peers. Students may attempt to read between the lines and fill in gaps with representations of their own risk tolerance, policies from prior courses, or their knowledge of common industry practices. Requiring students to use their own sense of risk when interpreting the meaning of a policy will likely result in very different interpretations. Students from over-policed groups will be viewing academic policies through the lens of their lived experiences, likely resulting in very different interpretations than peers from under-policed groups.

Biased Threads

Faculty members usually have a wide range of freedom to contextualize policies and requirements as needed for their courses, while also utilizing the course management tools purchased by the wider university. Changes may stem from using different types of assignments or teaching techniques but may also be based on personal preference. Policies on late penalties, for example, can be a large area of difference. Some faculty see being late to class or late submitting assignments as a form of professional disrespect while others see one or both as a less of a matter to be punished, and more a byproduct of life. These differences are often due to personal experiences or professional expectations within their discipline. Personal preferences like these often end up encoded within policies.

These differing preferences among faculty are seen across the syllabi in our sample, even within the same institution. Some suggested that solutions and related code were completely forbidden and tainted any original work the student attempted to produce, while another directed students to use the solutions to complete the assignments.

> Looking at solutions from other students or any other source (including the web), or collaborating to write solutions to individual work, is considered a violation of the honor code. (Mt. Holyoke CS 100)

> Homeworks are an opportunity to engage with the material and practice your programming skills. To support you, solutions will be available on moodle. It is your responsibility to engage with them in a way that strategically moves your learning forward. (Mt. Holyoke CS 151)

These students will have dramatically different experiences and expectations for future programming courses. One student may leave feeling like they missed out on the important reflective nature of working with open solutions, while another is not used to the isolation and struggle of closed material problem solving.

Faculty customizing policy choices allows them the freedom to quickly adapt to students and situations, try things out, and find the right fit for their content. Our own course policies differ based on our preferences and the kind of content that we teach. Differences in belief about collaboration versus independence is a specific pedagogical choice, but other areas of difference may allow for misconceptions or bias to appear in policy.

Source plagiarism detection systems are commonly used to compare submitted code to other submissions and other known solutions. While there are some differences in how each system processes and calculates similarity, the core action is comparing the text patterns within files. These tools appeared to strongly influence the policy guidance given for each course. Many policies mentioned using a detection system and described them as systems to detect cheating or plagiarism. Rarely were they mentioned by name or other details on how violations were measured.

Viewing source code and prose writing as the same type of creative activity is common and generally unproblematic. However, treating these two things as the same when attempting to define attribution and plagiarism creates problems. Plagiarism in source code is nothing new, and has been discussed long before these detection tools existed. As discussed by Gibson (2009), providing attribution for reused or referenced source code requires adaptation depending on context. He named his guidance a "code of practice" because he wanted it to be seen as a style guide where source code authors could make informed decisions rather than a strict set of rules. When used in a normal context of a software project or company, programmers have the freedom to make these contextual choices.

Some policies directed students to existing academic writing policies for their programming assignments, even directly comparing writing code to writing essays. These materials remain valuable for prose writing but usually lack guidance for source code attribution if they even discuss source code at all.

> Don't cheat. Read *Originality and Attribution: A guide for student writers at Vassar College*. Copying someone else's code is plagiarism. School policy dictates

> instructors must report all suspected incidents of cheating to their department chair. Please don't put yourself or your professor in that position. When in doubt, ask your professor before seeking help from another source. (Vassar CS 101)

> One of the key things to understand about programming, and computer science in general, is that it is a writing-heavy discipline. When you create a computer program, you are writing a document, just like you write documents in an English class (or any class that involves essays). Therefore, many of the same rules that apply to writing essays also apply to computer programs, particularly regarding plagiarism. (Cornell CS 1110)

This approach appears to be an honest but common misconception that goes directly against known and documented limitations of this approach. Given their lack of awareness, experience, and guidance, novice programmers could easily violate these policies without knowing it (Simon et al. 2014). Riedesel et al. (2012) stressed the need for institutions to have a well-understood policy for source code plagiarism, while Simon et al. (2018) found that instructors will still amend it with their own preferences and contextual needs. Discussion about this issue has grown recently, making it unlikely that a faculty member would discover it otherwise unless specifically searching for it.

The Vassar (CS 101) example below is notable because this quotation is the entirety of the academic integrity policy section. Additionally, the book mentioned appears to be from 1980 and only accessible from an administrator's office. Students are told to read a writing handbook for guidance about source code attribution, when the handbook contains no reference to programming, so they end up with no actual guidance. Additionally, these students may believe that it should be there, but they lack the skill to find it.

> Plagiarism is relatively easy to detect in a programming class. Do not take shortcuts. Always do your own work. Note that we encourage discussion on course content. (UIUC CS 101)

Some policies gave direct instructions on how to adjust their study behaviors and conversations to better fit with how these systems work. Code or notes resulting from student discussions were often restricted in addition to conversations. Many explained that written notes allowed code from each student to look too similar and could cause it to be flagged as plagiarism by detection tools.

> Students sometimes unwittingly get "carried away" and start writing code discussing a solution to an assignment, only to turn in suspiciously similar code that is flagged by our plagiarism detection system, which is then subject to academic dishonesty policies. (UT Austin CS 312)

One policy even went as far as suggesting that students wait thirty minutes after any interaction to write any code as a method to prevent accidental replication between parties.

> A good rule of thumb to ensuring your collaboration is allowed is not take written notes, photographs, or other records during your discussion and wait at least 30 minutes after completing the discussion before returning to your own work. (You could use this time to relax, watch TV, listen to a podcast, or do work for another class.) For most students, this will result in you only bringing the high-level concepts of the collaboration back to your work, and ensuring that you reconstruct the ideas on your own. (U Washington ESE 142)

Remembering elements of code from conversations or other indirect connections were not covered within working definitions of source code plagiarism (see Novak, Joy, and Kermek 2019). Yet producing solutions that were too similar due to this was explicitly labeled as cheating because the detection system flags these similarities. The detection systems are literally defining the forms of cheating within these courses.

Policies based on misconceptions can be more than just misinformation. Directions instruct students seeking to follow the rules that help exist within certain documents. Finding no actual reference to their problem has several consequences. In the short term, the student may not meet expectations due to lack of guidance and question their own understanding or seeking out help again. Longer term, the frustration of being directed to use nonexistent information may deter the student from progressing with a subject area.

While the comparison between source code and prose may be simple misinformation, telling students to seek help within policies containing no guidance on source code attribution is malinformation at best and gaslighting at worst.

Certain Uncertainty

Concerns about communication and access to prohibited information were common. The previous section covered the actions faculty required and suggested students perform to ensure originality. Discussions of communication came up in elements related to the plagiarism detection systems, but this next section looks deeper into issues of communication. Student conversations were strongly policed by many policies and faculty used policy space to add their own moral judgments and threats about cheating and plagiarism.

Code (Of/In) Silence

Faculty used words like "collaboration," "conversations," "getting help," and "discussions" within policies to describe interpersonal interaction between students about the homework assignments. These descriptions took up most of the policies and provided rules for interactions: specifying the modality, content, and even language. Student collaboration to complete a group assignment was usually differentiated from more generally talking about the assignment content, with the informal conversations being the instructor's largest concern.

This policy area had some of the broadest set of rules and requirements for students, ranging from effectively no restrictions to only speaking to instructional staff for the entire term. These policy sections had the most examples and detailed descriptions of expectations. Most policies allowed or encouraged some form of discussion between students and gave details on the range of acceptable topics for their interactions.

> Note that we encourage discussion on course content. (UIUC CS 101)

> Discussing ideas and approaches to problems with others on a general level is encouraged. (Swarthmore CS 21)

> We believe that collaboration fosters a healthy and enjoyable educational environment. For this reason, we encourage you to talk with other students about the course material and to form study groups. (Wellesley CS 115)

Programming assignments can involve substantial problem solving, research, trial and error, and resolving bugs within the code. Students struggle with these areas throughout CS1 courses, making them common conversation topics within a classroom and learning group. Specific areas like approaches for problem solving, programming concepts and patterns, and seeking help from peers were usually broken into clusters approved for considerations or any off-limits topics.

> In most cases, you may discuss concepts (algorithms, ideas, approaches, etc.) described in the readings, lab exercises, or during class with anyone. (Carleton CS 111)

> You are free to discuss ideas and approaches with other students, and then implement the solution yourself. (Stanford CS 101)

> We recognize that collaboration can help you master course material. In fact, there are certain ways in which we will encourage you to collaborate. These include: discussing course content at a high level, getting hints or debugging help, talking about problem-solving strategies, discussing ideas together. (Mt. Holyoke CS 151)

The modality was usually required to be in person with verbal communication. The language descriptions often had the modifiers of "natural" or "human" language as a method of restricting anything like a programming language to be spoken aloud.

> Such a description should be in English or another natural human language. (UW CSE 142)

> Verbal collaboration without code sharing ... (Bowdoin CS Collaboration Policy)

> You may collaborate with other students by talking about the problem or your solution in a natural language (e.g., English), but you may not use any formal language, and especially not Python. (Wellseley CS 111)

> Roughly speaking, it is okay to share ideas but it is not okay to share any artifacts (code, write-up, etc.). Here is a good way to think about it: you and a classmate can

> get together, discuss ideas, and even write some code. However, you are expected to leave that meeting with nothing – no notes and certainly no code – and write up your own solution. (Colgate COSC 101)

This final example (Colgate COSC 101) gives some insight into why there were these requirements. Code is plain text, making it very easy to share within text platforms. Students often have group chats or another text-based communication setup, so sharing portions of their code is quite easy and could possibly include a dozen students as recipients. Limiting the language spoken generally forces students to speak in broader, more conceptual terms, rather than specific elements of the assignment being discussed. These two specific behavioral requirements encompass many common concerns without lengthy behavioral lists.

Content and modality were not the only limited areas for interaction. Specific groups of people were also prohibited. A few mandated students to hold no homework-related interactions with anyone other than approved course staff. While very strict, this rule was directly stated and essentially eliminates collusion.

> Discussing any part of assignments or assessments with anyone else (besides 112 TA's and course faculty), no matter how briefly or casually, in person or via Discord/ WeChat/Zoom etc. unless the assignment explicitly allows it … all homework exercises are solo, meaning that you must not collaborate in any way. (CMU 15–112)

> Students may only collaborate with fellow students currently taking CS 1301, the TA's and the instructor. (Georgia Tech CS 1301)

Peers were commonly presented as a source of seeing, hearing, or receiving forbidden information and collaboration turning into collusion. These two concepts are intrinsically related behaviors, as they are referring to the same set of social actions that people may take toward a project (Fraser 2014). Collaboration describes all the possible social actions taken toward a project or goal, while collusion covers any of those actions that are not allowed. This means that the very definition of collusion depends on the rules and scope for collaboration to be defined.

> There is a difference between learning collaboratively and completing the work for someone else. This can be a subtle but important distinction. … The following types of collaboration are encouraged … The following types of collaboration are prohibited and may constitute academic misconduct. (UWashington CSE 142)

> You may not share information about your solution in such a manner that a student could reconstruct your solution in a meaningful way (such as by dictation, providing a detailed outline, or discussing specific aspects of the solution). (Swarthmore CS 21)

> All assignments must be done individually. No collaboration of any kind is allowed. Any cases of inappropriate collaboration (cheating) have to be reported to the department chair and the Academic Panel, and will be dealt with promptly. (Vassar CS 101)

Many attempted to define the level of detail students could include in their conversations, hoping to highlight the tipping point into collusion. While not formal definitions, providing examples of unacceptable behaviors is an appropriate way to approach this challenge. Unfortunately, many examples explicitly told students they should already know what is meant by these labels.

> Collaboration and learning from one another are encouraged. Copying and cheating are strictly forbidden. You are expected to be able to distinguish the two. If you're contemplating an action and you're not sure into which category it falls, you should consider whether what you intend to submit is the product of your own efforts and represents your own understanding of the ideas involved. If it is/does not, then you should not submit the work as your own. Wesleyan imposes an Honor Code (to be found in the Student Handbook). You are expected to abide by it in all of your courses, including this one. (Wesleyan COMP 112)

Encoding very fine lines of communication into clear and actionable definitions is not an easy task. Varied interpretations (Joy et al. 2013; Sheard et al. 2002; Simon et al. 2013) and operating definitions (Fraser 2014; Riedesel et al. 2012) about academic integrity exist within computing faculty and students. Faculty are often left solely responsible for writing contextualized governance. Lack of specificity in policies passed to the faculty seems to be a related issue, where faculty "find their role to be one of contextualization" (Simon et al. 2018) of the policies they are given. Several of their respondents mentioned using behavioral examples because policy language alone was insufficient for students to understand expectations, which was also found within our sample. Providing these examples opened other areas for misinterpretation.

> Looking at code related to an assignment together is considered crossing the line into cheating territory! (UT Austin CS 312)

This example (UT Austin CS 312) came from the middle of a large paragraph (about 430 words) that represents just a portion of the academic integrity policy content of the syllabus. The paragraph gives numerous examples of cheating behavior, including the example above. The phrasing "code related to an assignment" is likely intended to mean looking at another student's code for an assignment. This opened a reasonably descriptive policy to potentially include any class sources related to the assignment. Students carefully reading the policy and taking this statement very literally could naturally include things like the professor's notes, textbook, or even their class notes. Each of those elements likely contain elements for the homework.

Peers were often presented as a source of forbidden knowledge and just asking for help could potentially result in harm to both parties. This harm ranged from formal sanctions due to violating policies to informal where learning opportunities were missed.

> Please don't put your friends at risk by asking them for unauthorized help. (Cornell CS 1110)

> You rob yourself and others of learning how to approach difficult programming problems, an essential skill for future classes. (UC Berkeley 10a)

> Providing help beyond what is allowed is as much of an infraction of the honor code as receiving help. (Bowdoin CS Collaboration Policy)

As several syllabi reported, once a student sees an example of the solution, their memory and ability to perform "original" work has been forever tainted. Students must complete the problems without seeing external sources and are instructed within the policies to maintain the isolation, protect their work, and to avoid situations where they may inadvertently expose themselves or others. While some policies allow for students to cite what they have looked for or other sources of help, some disallow nearly all forms of communication. These environments may differ between instructors and institutions such that not all grades and assessment outcomes may be considered equal. While one student has the benefit of a study group or partner to discuss concepts and concerns with, another may only access help during office hours or email with course staff.

Students were also warned to protect their work from peers who may maliciously view and copy content. An unknowing victim would receive the same severity of consequences as the student who took the direct action to copy the work.

> You are expected to take reasonable measures to protect your work from unauthorized access by others, including your electronic files, print-outs, and written work. (UT Austin CS 312)

This presentation of peers as being sources of constant potential risk isolates students and creates an oppositional existence instead of collegial. However, this may only be the case for certain groups of newcomers. Students with higher risk tolerance and a better understanding of higher education practices may not feel the same way about their peers and are able to more freely engage in group studying and social support for learning. Their lowered risk awareness also makes them a riskier person to socialize with for students afraid of forbidden information exposure, further expanding their isolation.

With collaboration and originality presented as mutually exclusive, compounded with the individual academic assessment justification, peers tended to be described as objects or sources to be feared. They are potential sources of exposure to external ideas and code, but also active threats of observation. Several policies instruct students to actively maintain a two-way barrier from viewing and being viewed. This includes incidents where another student deliberately uses someone else's laptop to see the answer. The laptop's owner is deemed just as guilty for not locking down their computer.

Normal plagiarism policies for academic writing refer to overt actions taken by students: the inclusion of text from another work without proper attribution, purposefully copying text, and others. A purposeful choice, even if not understood to be an infraction, can be pointed to as the problem. Treating passive infractions as overt and deliberate choices to cheat requires students to be constantly diligent of the actions and intents of their peers. The result may look the same: Code submitted by students could be tagged as possible plagiarism by a detection tool, but the reason is left unclear without investigation. Students and faculty end up navigating this confusing combination of understanding the tool, the policies, and behaviors.

Importing Morality

The origin story of any syllabus involves reuse, boilerplate requirements, and educated guesswork. Faculty may inherit policies from previous instructors of a course, use a template with a campus or department boilerplate, and then add their own content and personalization. Once created, review services may not be offered or required before the syllabus is used for a class. Errors or problematic language easily fall through the cracks.

Although not generally required, many faculty elected to include justification and reasoning behind their policies. These sections had some of the most informal and emotionally changed content out of all the content reviewed for this project. Many intended to defend the policy choices, explained why it was worth following, and highlighted the severity of the sanctions. Some went further and presented moral values or other ethical incentives for abiding by their governing rules. These justifications generally fell into two categories: ethics and respect or learning goals.

Justification claims under ethics invoked a variety of concepts in their persuasive arguments: honor, respect, fairness, honesty, personal responsibility, ethics, moral codes, and so on. The objects these concepts acted on were often a larger social group, such as the institution, learning community, and peers. Positive policy examples set a tone of pride for students to adopt and maintain.

> For your sake and the sake of the Claremont-Colleges community, please conduct yourself with the highest level of academic integrity. (Harvey Mudd CS 5)

> Simply put, academic integrity is about respecting yourself and respecting others. You respect yourself by submitting work completed through your own effort; you respect others by acknowledging contributions from others when such external contribution is allowed. (Cornell CS 1112)

The implication of these arguments was left unsaid: that violating this respect can cause harm to yourself; but it also threatens to withhold trust or approval of the students should they break these rules. Negative examples focused more on this consequence.

> Sadly, we encounter such violations almost every semester, and the rulings of the Honor Code Council tend to be harsh. For your sake and ours, please don't be tempted to cheat; you are likely to be much better off getting a poor grade on an problem set or exam than you are if you are found guilty of cheating! (Wellesley CS 111)

> We begin by choosing to trust each of you individually. Do not be one of the few who loses that trust. If you cheat, expect to be caught, and expect significant consequences. Use common sense and understand these rules. (CMU 15–112)

These may lack overt emotionally charged words, but describing these acts as being forbidden, learning shortcuts, and plain cheating set a judgmental and threatening tone. Many of the longer policies contained emotional discussions of the morality around cheating, yet the amount of actionable information about cheating was quite low.

> Many students begin every assignment by immediately going to Google, trying to find something that might keep them from having to solve the problem for themselves. *This is an incredibly stupid thing to do* ... you're starting down a moral slippery slope that's liable to send you over a cliff ... a complete solution that some idiot has posted on GitHub; it will be too tempting not to use it ... *Don't do it!* If you are caught, you will deeply regret it. And even if you're not caught, you're still a cheating low-life. (UT Austin CS 303E)

> It's not that people can be divided into cheaters and noncheaters in some preordained way, though that is an easy way to think about life. It's more than the stress and bad decision making of a particular *situation* make a cheater of someone. (Stanford CS 106a)

Others maintained a more pragmatic framing around fairness and legitimacy. These focused on the learning assessment process and the importance that grades accurately represent their understanding of the content. Cheating or subverting the expected values of the policy would make it unfair for the other students following the morally correct path or unfair to your own learning.

> ... does not contribute toward developing mastery. In addition, this deprives you of the ability to receive feedback and support from the course staff in addressing the areas in which you are struggling. (U Washington CSE 142)

> Cheating not only robs you of an opportunity to learn, it also devalues your peers' hard work. Because of this, we take cheating very seriously in this course. (UIUC CS 101)

Some added computing industry practices to frame the purpose of their requirements. These references were uncommon, but almost all did so to explain why they promoted collaboration.

> We believe that collaboration fosters a healthy and enjoyable educational environment. For this reason, we encourage you to talk with other students about the course material and to form study groups. Programming assignments in this course can be challenging. Also teamwork is the norm in the CS industry. Given the above, some of the assignment work is required to be done with a partner, while some is required to be done individually. (Wellesley CS 115)

> Note: details differ from place to place, but the shared-software-development approach described above is very common in professional practice. Sharing the

> building and understanding of software is the characteristic computing skill of our era! (Harvey Mudd CS5)

Authentic assessment is the concept of a course using assessments that closely match authentic actions within the practice. As discussed by Simon and Sheard (2015), these are valuable for learning but often require higher levels of interaction with the instruction team and are harder to grade at scale and involve collaboration. These factors put adoption of authentic assessment at odds with the focus on individual learning and assessment needed for awarding individual grades.

These justifications also highlight the two general approaches taken toward academic integrity: consequences and virtuousness. The former seeks to leverage punitive measures to ensure that students do not act on the desire to cheat, while the latter attempts to fill in educational and mindset gaps to prevent the desire in the first place.

Introductory students within these restrictive and highly individualistic classes get an inauthentic and overly policed experience within computer science. At a micro level, individual students work under a punitive and unclear policy. At a meso level, looking at the collection of peers operating under these policies, experienced students are put at the advantage both in permissible communication and deepening peer connections. Students with novice level experience work in isolation while being instructed to fear betrayal by peers.

In considering the macro level in all these policies, many students are being prepared to operate in an environment that mischaracterizes the essential operations within this profession. Code is commonly discussed, shared, borrowed, and inherited within teams. The errors and omissions described here may seem like honest mistakes from an instructor navigating a vague space, but we argue that it is precisely this imprecision that is the point. We argue that the hazing element comes from how the policies describe information that should be accessed or restricted. These policies may be unintentionally problematic at an individual level, but when situated among their peers, they replicate those in power who seek to promote same or similar group learners to achieve success and value placing additional checks against marginalized groups.

Encoding White Spaces

Academic integrity does not have a single universal definition. Global and historical traditions intersect as well as discipline specific interpretations. As discussed by (Macfarlane, Zhang, and Pun 2014), discussions and literature in this area can include academic integrity, honor codes, and professional ethics. Synonyms abound and most works spend a nontrivial amount of time clarifying their viewpoint and scope for their research. Growing education globalization, information access, and technologies for text similarity detection have reinforced the need for institutions to have a shared understanding of academic integrity issues and active training on the topics.

Modern tools to detect plagiarism and other academic dishonesty issues may be presented as objective measures, but each require review and interpretation. This means that the gap between being flagged and being accused is once again expected to be filled by "gut" feelings. Faculty must understand these tools but also navigate their own policies to determine if further action is needed. Vague and open-ended policy description leave decision making susceptible to unconscious biases and subjective opinions. These tools presented as objective measures remain dependent on evaluators to make objective judgments.

Unconscious bias and suspicion put black students at higher risk of being reported by peers for perceived infractions. This is directly discussed by participants in Feagin and Sikes (1995). Participants reported being accused of plagiarism by instructors solely because they did not often speak up in class and the professor simply did not believe their work was original. Those with the power to assess student work may easily be swayed by unconscious biases about the innate abilities of students of color.

Historically white colonial traditions can be found within modern expectations of academic behavior. Bonilla-Silva and Peoples (2022) outlined these acts and their result of perpetuating erasure of nonwhite traditions. Higher education curricula not only stress white content, but also values and traditions. While not directly discussed by Bonilla-Silva and Peoples, the elements of whiteness held up within these institutions can be extended to the policies of the classroom and therefore the syllabus.

The long history of nonwhite students being subjected to the varied feelings and perceptions of faculty members continues to this day and certainly within computer science classrooms. Policies being written and enforced with technical review tools to support them provide a sense of authority and likely make faculty feel comfortable that they are working within a "color-blind" system, but do very little to provide safeguards for students who have historically been subjected to higher levels of scrutiny and mistrust. Without clear process documentation and clear policies to be accused of violating, the power here rests solely within the accusing faculty's perception of the truth and internal definitions of the rules.

Disproportionate impact cannot be measured as a byproduct of the intent of the faculty member writing the policy. Students will act according to the policies as they are given and interpreted through their understanding of rights and risk. This means that each student will have a unique interpretation and thus experience different impacts on their behavior. This alone will always be the expectation, but becomes problematic when certain groups of students will be systematically impacted in different ways along the lines of race, socioeconomic status, and other social factors.

New tools and techniques used in response to these challenges of student growth are constantly created and adopted. Some may have existed a decade ago but were not used at the scale they are at the time of this writing. Creating the code database mentioned in the example below was not truly possible until recently. Faculty writing these syllabi likely learned how to program under very different systems; most importantly, their work was less likely to be monitored at such a degree.

> Plagiarism detection. We use sophisticated software tools to detect plagiarism. For example, we compare each submission (including interim submissions) against a code database, which includes all submissions from this offering and previous offerings of the course. While we take no pleasure in bringing cases to the Committee on Discipline; it is our duty to keep the playing field level for the overwhelming majority of students who work very hard in this course and follow the rules. (Princeton COS 126)

While these policies are attempting to dismantle a knowledge commons forming within the classroom, it is actually dismantling it only for these risk-averse students. Students with in-group connections can be told or mentored through how to subvert or avoid detection, preserving the knowledge commons just for them. Out-group students must remain isolated and fearful, providing evidence to the in-group members that any outsiders who were able to complete courses like this must be exceptional and worthy of academic respect.

> Utilizing discussion with peers to facilitate your learning is a critical skill for success in computer science. However, at the same time, you must be aware that getting stuck and pushing through challenging problems is essential for robust learning. (Grinnel CSC 151)

The struggle through these introductory programming courses is often seen and framed as a rite of passage, one that the instructor survived and thus so must they. Their survival is framed as tribute to the industry. Yet the struggle and isolation created under some of these policies is a lie. Those working in industry borrow and build on external code constantly. They also work together in teams. While pair programming is a common tool for code review and mentorship, large projects are split up among members of a team who will work on independent elements, but each with the support and guidance of other team members.

Conclusions

Much of the research has been on what people consider misconduct, and whether students committing misconduct are aware they are doing so. We seek to flip this question, and ask: How far might a student go to avoid misconduct out of a fear of committing an infraction? Are the forms of these policies doing a disservice for students in fully recognizing their own behavior as misconduct; but do the policies also open up the possibility for students to incorrectly limit their own behaviors?

Overly restrictive policies also create stressors via isolation and fear. Students are isolated from peers in their work with the content, with the potential to reinforce prior feelings of otherness and being an impostor. As defined by many of the policies, communication and sharing resources with peers is something to be feared. All of this creates an environment rife for misinformation to spread.

Hazing also continues after the first programming class. These strict policies and foundational fears are cemented during job interviews. Whiteboard coding questions are the epitome of fictional practices treated as a norm within computer science. Presenting handwritten syntactically perfect code on a whiteboard with no tools or resources, under time constraints and observation, creates a high stress environment that represents very little of what the actual day to day job is like. The growing cost of higher education and the student loan crisis make these interviews a high-stakes finale to the overall system of hazing within the computing industry. However, many will complete these efforts because they have been prepared by prior hazing rituals to expect the initiation costs to be high. The other element of the lie that makes this hazing is that all these tasks are meant to be a measure of your worth within a giant meritocracy, when networking, bias, alumni networks, and programming interview cram schools are all leveraged by others to bypass these checks of merit.

Employers are outsourcing the measurement and evaluation of a candidate's skill in the field to the grades assigned during academic evaluation (Simon and Sheard 2015). By outsourcing this evaluation work, they measure elements of hireability as those who are successful within the university grading system. Obviously not the sole factor used by many companies, grade point average (GPA) and other exam ranks do play an important role in determining candidates and finalists for internships and full-time work. Students who have more access to support or more freedom to subvert assessment strategies have a stronger chance of ranking higher and receiving these positions.

Academic dishonesty policies are contributors to creating an environment where students expect to be creating source code in isolated, high pressure, and competitive environments. Many startup companies expect staff to show their dedication to the project by working extreme hours within similarly high-pressure situations. A student graduating out of such a program and landing a role within a company may have little experience or basis to push back on these unhealthy expectations.

Hazing is more than simple gatekeeping, but an effort to ensure that all who pass through come to believe that those filtering structures are valid and important. The goal is not just to reenact them when they later come into their own house of power but to have those in power genuinely believe in those structures and the reasoning behind them. Information hazing, therefore, is setting these structures off and making the filters based on how information is used and accessed by those newcomers. Rather than overt acts upon them or specific demands for action, this hazing is more subtle. It circumvents many of our normal triggers to view something with suspicion or questioning.

The information hazing observed within this study hides behind western ethics and neoliberal morality frameworks. After all, questioning something that is supposed to uphold honor and integrity must make that questioning be in opposition to those factors. Those not interested in upholding the virtues of honesty must therefore be against it. However, holding the rigid moralistic jargon up to a lens of information use and access allows for objections that are based in tangible ways: that many of these statements are problematic due to wide ambiguity of or a misrepresentation of the natural process of computing work.

In closing, we want to note that although we have highlighted many aspects of information hazing that we believe are likely to cause harm in the classroom and beyond, we do not write with this to blame the instructors who have created, curated, and crafted the syllabi in our sample. We know how hard teaching is and how complicated plagiarism and academic integrity are to teach, enforce, and navigate. We created this work with the goal of identifying problems in order to create positive change, and we hope that this chapter helps us to see the things that we have ignored or thought were the last best option.

REFERENCES

Allan, Elizabeth J., and Mary Madden. 2012. "The Nature and Extent of College Student Hazing." *International Journal of Adolescent Medicine and Health* 24 (1): 83–90. https://doi.org/10.1515/ijamh.2012.012.

Bonilla-Silva, Eduardo, and Crystal E. Peoples. 2022. "Historically White Colleges and Universities: The Unbearable Whiteness of (Most) Colleges and Universities in America." *American Behavioral Scientist* 66 (11): 1490–1504. https://doi.org/10.1177/00027642211066047.

Braun, Virginia, and Victoria Clarke. 2006. "Using Thematic Analysis in Psychology." *Qualitative Research in Psychology* 3 (2): 77–101. https://doi.org/10.1191/1478088706qp063oa.

CODATA-ICSTI Task Group on Data Citation Standards and Practices. 2013. "Out of Cite, Out of Mind: The Current State of Practice, Policy, and Technology for the Citation of Data." *Data Science Journal* 12: CIDCR1–CIDCR7. https://doi.org/10.2481/dsj.OSOM13-043.

Costa Pinto, Ligia M., Carla Sá, Nuno Soares, Sílvia Sousa, and Marieta Valente. 2020. "The Case for Academic Hazing as a Rational Choice: An Economic Approach." *Economic Analysis and Policy* 66 (June): 51–62. https://doi.org/10.1016/j.eap.2020.02.004.

Feagin, Joe R., and Melvin P. Sikes. 1995. "How Black Students Cope with Racism on White Campuses." *Journal of Blacks in Higher Education* 8: 91–97. https://doi.org/10.2307/2963064.

Fraser, Robert. 2014. "Collaboration, Collusion and Plagiarism in Computer Science Coursework." *Informatics in Education* 13 (2): 179–195. https://doi.org/10.15388/infedu.2014.10.

Gibson, J. Paul. 2003. "Software Reuse in Final Year Projects: A Code of Practice." Report NUIM-CS-2003-TR12, Maynooth: National University of Ireland, November.

2009. "Software Reuse and Plagiarism: A Code of Practice." *ACM SIGCSE Bulletin* 41 (3): 55–59. https://doi.org/10.1145/1595496.1562900.

Guynn, Kevin L., and Frank D. Aquila. 2004. *Hazing in High Schools: Causes and Consequences*. Bloomington, IN: Phi Delta Kappa Educational Foundation.

Joy, Mike S., Jane E. Sinclair, Russell Boyatt, Jane Yin-Kim Yau, and Georgina Cosma. 2013. "Student Perspectives on Source-Code Plagiarism." *International Journal for Educational Integrity* 9 (1): 3–19. https://doi.org/10.21913/IJEI.v9i1.844.

Macfarlane, Bruce, Jingjing Zhang, and Annie Pun. 2014. "Academic Integrity: A Review of the Literature." *Studies in Higher Education* 39 (2): 339–358. https://doi.org/10.1080/03075079.2012.709495.

Madison, Michael J., Brett M. Frischmann, and Katherine J. Strandburg. 2009. "The University as Constructed Cultural Commons." *Journal of Law and Policy* 30: 365. https://doi.org/10.31228/osf.io/z8x7m.

Novak, Matija, Mike Joy, and Dragutin Kermek. 2019. "Source-Code Similarity Detection and Detection Tools Used in Academia: A Systematic Review." *ACM Transactions on Computing Education* 19 (3): 1–37. https://doi.org/10.1145/3313290.

Nuwer, Hank. 1999. *Wrongs of Passage: Fraternities, Sororities, Hazing, and Binge Drinking.* Bloomington: Indiana University Press.

Riedesel, Charles P., Alison L. Clear, Gerry W. Cross, Janet M. Hughes, Simon, and Henry M. Walker. 2012. "Academic Integrity Policies in a Computing Education Context." In *Proceedings of the Final Reports on Innovation and Technology in Computer Science Education 2012 Working Groups*, 1–15. ACM, Haifa. https://doi.org/10.1145/2426636.2426638.

Sheard, Judy, Martin Dick, Selby Markham, Ian Macdonald, and Meaghan Walsh. 2002. "Cheating and Plagiarism: Perceptions and Practices of First Year IT Students." In *Proceedings of the 7th Annual Conference on Innovation and Technology in Computer Science Education*, 183–187. ACM, Aarhus. https://doi.org/10.1145/544414.544468.

Simon, and Judy Sheard. 2015. "Academic Integrity and Professional Integrity in Computing Education." In *Proceedings of the 2015 ACM Conference on Innovation and Technology in Computer Science Education*, 237–141. ITiCSE 2015. Association for Computing Machinery, New York. https://doi.org/10.1145/2729094.2742633.

Sheard, Simon, Beth Cook, Judy Sheard, Angela Carbone, and Chris Johnson. 2013. "Academic Integrity: Differences between Computing Assessments and Essays." In *Proceedings of the 13th Koli Calling International Conference on Computing Education Research*, 23–32. ACM, Koli. https://doi.org/10.1145/2526968.2526971.

2014. "Student Perceptions of the Acceptability of Various Code-Writing Practices." In *Proceedings of the 2014 Conference on Innovation and Technology in Computer Science Education*, 105–110. ITiCSE 2014. Association for Computing Machinery, New York. https://doi.org/10.1145/2591708.2591755.

Simon, Judy Sheard, Michael Morgan, Andrew Petersen, Amber Settle, and Jane Sinclair. 2018. "Informing Students about Academic Integrity in Programming." In *Proceedings of the 20th Australasian Computing Education Conference*, 113–122. ACE 2018. Association for Computing Machinery, New York. https://doi.org/10.1145/3160489.3160502.

Simon, Judy Sheard, Michael Morgan, and Charles Riedesel et al. 2016. "Negotiating the Maze of Academic Integrity in Computing Education." In *Proceedings of the 2016 ITiCSE Working Group Reports*, 57–80. ITiCSE 2016. Association for Computing Machinery, New York. https://doi.org/10.1145/3024906.3024910.

Smith, Arfon M., Daniel S. Katz, and Kyle E. Niemeyer. 2016. "Software Citation Principles." *PeerJ Computer Science* 2 (September): e86. https://doi.org/10.7717/peerj-cs.86.

StopHazing. 2022. "States with Anti-Hazing Laws." Stophazing.Org. 2022. https://stophazing.org/policy/state-laws/.

Vicente-Saez, Ruben, and Clara Martinez-Fuentes. 2018. "Open Science Now: A Systematic Literature Review for an Integrated Definition." *Journal of Business Research* 88 (July): 428–36. https://doi.org/10.1016/j.jbusres.2017.12.043.

Wu, Yuhao, Shaowei Wang, Cor-Paul Bezemer, and Katsuro Inoue. 2019. "How Do Developers Utilize Source Code from Stack Overflow?" *Empirical Software Engineering* 24 (2): 637–673. https://doi.org/10.1007/s10664-018-9634-5.

5

Common Nonsense about Password Security and the Expert–Layperson Knowledge Gap

Brett M. Frischmann and Alexandria Johnson

INTRODUCTION

Creating and managing passwords is familiar to most of us. We use passwords on computing devices every day. To access social media accounts, check the balance on online banking apps, or send an email, individuals must authenticate themselves, often by logging into a computer system with a username and password. Password creation and management practices require knowledge that may seem commonsensical. Yet common sense about passwords and security is often misleading nonsense. Many people hold incorrect beliefs about what makes a password secure, the nature and origin of security threats, and what practices compromise or conversely strengthen security. Substantial research has shown that user-chosen passwords are highly predictable and follow similar patterns; individuals create weak passwords based on memorability rather than on secureness, reuse passwords, and often use personal information in passwords that is easily discovered or guessed. Yet, less research has been done to understand the origins and persistence of common nonsense about password security. Several research questions arise: What beliefs do users have about creating secure passwords, security threats, and best practices? Where do these beliefs come from? Do users realize that they are selecting easily guessable passwords, and if so, is this intentional? Do users believe that the passwords they create are secure? Are users more focused on convenience and memorability than security? Are individuals aware of password security risks? Do individuals

Charles Widger Endowed University Professor in Law, Business and Economics, Villanova University, Charles Widger School of Law; Affiliated Faculty, The Vincent and Elinor Ostrom Workshop in Political Theory and Policy Analysis, Indiana University Bloomington; Affiliate Scholar, Center for Internet and Society, Stanford Law School; Trustee, Nexa Center for Internet and Society, Politecnico di Torino. J.D. Georgetown University Law Center; M.S., Columbia University; B.A., Columbia University.

J.D. Candidate Duke University School of Law; B.A. Political Science, Villanova University; B.A., Criminology, Villanova University.

FIGURE 5.1 Visual themes from common nonsense about password security and the expert–layperson knowledge gap.

not understand these risks or do individuals understand the risks and create weak passwords anyway?

While the gap between lay and expert knowledge about password security is well documented, less well understood are why the gap persists and what are the origins of incorrect beliefs and misguided security practices. In this paper, we examine the gap and identify knowledge governance dilemmas that help explain its persistence. We use the Governing Knowledge Commons (GKC) framework to structure our study and frame the sets of conceptual and empirical questions we investigate. The security knowledge ecosystem is complex. We focus on password security.

We initially framed our study in terms of two knowledge communities – lay persons and security experts, but we learned during interviews with experts that a better structure would differentiate and explore the interactions between three communities – lay persons, professionals/practitioners, and experts. Notably, it may be a stretch to describe lay persons and professionals/practitioners as communities. Actors in these classifications are members of various communities within which password security knowledge is shared and acted upon, but unlike the password security expert community, these various communities are not organized around or otherwise focused on password security knowledge. More important than community definitions, at least for purposes of this study, are the relationships, interactions, knowledge flows, and governance dilemmas among lay persons, professionals/practitioners, and experts.

We employed two primary methods: (i) a systematic literature review to gather information about password security, password behaviors, expert knowledge on passwords, and lay knowledge about passwords; and (ii) semi-structured interviews of password security experts to supplement the findings from the literature review and gain first-hand information and perspectives. We briefly describe each and provide more detail and data in an Appendix available via the GKC repository (an open data portal).[1]

Our literature review involved five steps. First, we collected peer review publications, focusing on research articles and user studies. We searched major research databases, including Google Scholar, Science Direct, ACM, and Springer Link. Keywords included variations of password, user security perceptions, password creation behaviors, and password security awareness methods. Second, we screened research articles by assessing if the title, abstract, and full article matched the relevant subject matter. We also ensured that the research literature met five inclusion criteria.[2] Third, we identified additional publications by examining references of all the articles we had included and repeating the second step. Fourth, we

[1] https://hdl.handle.net/2142/117212.

[2] The article was (i) peer-reviewed, (ii) written in English, (iii) available full text, (iv) involved a user study, survey, questionnaire, and/or interviews, and (v) primarily focused on password security perceptions, behaviors, and practices.

identified additional articles through suggestions by experts during interviews and by drawing from the publication pages of experts we interviewed. The resulting sample size is ninety-seven articles. (N = 97). Fifth, we qualitatively analyzed the research literature using both the GCK framework and a basic set of research questions regarding password security beliefs and practices.[3]

After protocol review by the Villanova University IRB, we conducted eight semi-structured interviews with professionals directly involved in password security and/or usability.[4] We used the systematic literature review and the GKC framework to provide structure and generate interview questions.[5] We recorded, transcribed, analyzed, and summarized each of the interviews. Then, once all interviews were completed, we reviewed the transcripts again. We summarize the results of our qualitative analysis of the literature and interviews below.

BACKGROUND: PASSWORDS AND SECURITY

Basic Overview

Passwords have become a recurring aspect of everyday life in our digital networked world because of the many different devices and online accounts we use daily that require authentication. A 2020 study by Nordpass, a password manager website created by Nord Security, revealed that the average user has around 100 passwords, a 25 percent increase from 2019 (Bobba & Paruchuri 2022).

Passwords are also familiar to most people because of their offline use in other social contexts, for example, as a signal of membership in a club. Passwords take many forms and have a long history. People learn about passwords not only in practice but also in literature. One of our favorite examples, described by Martin Paul Eve (2016) in his book *Password*, is Daedalus's labyrinth, "designed as a spatial-control mechanism for determining the unique identity of a single individual based on knowledge of its topology. … For everyone but Daedalus, the labyrinth was supposed to be, quite literally, a death trap." Theseus escaped by circumventing the identification function and thus, according to Eve, "is one of the earliest species of geek that we now would call a hacker or cracker (3–4).

3 See Appendix https://hdl.handle.net/2142/117212.

4 We initially contacted interviewees with whom we have had professional contact in 2022, and then we added a few experts based on recommendations from the initial interviewees. We interviewed the following experts: Steven M. Bellovin, Zachary Bornstein, Lorrie Cranor, Cormac Herley, Arvind Narayanan, Bruce Schneier, David Thaw, and Rick Wash. We provide more detail about the interviewees in the Appendix. After interviews were completed, we consulted Xavier de Carné de Carnavalet, who provided written comments on our study.

5 We provide more detail, including interview questions and data in the Appendix.

Passwords are a knowledge-based means for authentication.[6] Knowledge (of a password) serves as a proxy for identification in a system designed to control access to resources or systems. Alternative means of authentication that do not rely on knowledge may be based on "something you have, such as a token or a particular mobile phone [or] something you are, that is, some form of biometric" (Bellovin 2016, 107). Steve Bellovin explains that one must understand the "*systems* nature of authentication," meaning: "The total environment – who will use it, how you deal with lost credentials, what the consequences are of lack of access or access by the wrong person, and more." Ultimately, he concludes: "The most important question of all is how people will actually use the authentication technology in the real world" (107).

So how do people use passwords? What do people believe about the passwords they use? Do they understand password strength, security threats, and relevant consequences? These are some of the basic background questions we sought to answer in our review of the literature.

Literature Review and Interviews

Our literature review aimed to answer questions about password security beliefs, where these beliefs originate, how these beliefs are transmitted more broadly, and how the beliefs shape everyday actions and practices. Many articles we reviewed used either surveys alone or surveys with lab experiments as their methodology.[7] We group our findings into two categories: lay knowledge and expert knowledge.

Lay Knowledge (Dilemmas) Concerning Password Security

We collected data on user beliefs about the nature and origins of security threats, what users believe makes a password secure, and what users believe compromise or strengthen security. Across the articles that focused on understanding users' perceptions, beliefs, and behaviors, some common practices and beliefs among users were as follows:

Practices:

- choose passwords from a limited set of alpha-numeric characters;
- use names, slang words, dictionary words, or consecutive digits as passwords;[8]

[6] Every security expert we interviewed responded to the preliminary questions – Why do we use passwords? What problem do passwords solve? – with authentication.

[7] Many studies recruited subjects attending colleges and universities, a demographic that may not be representative of the larger population.

[8] A 2021 Nordpass study, https://nordpass.com/most-common-passwords-list/, revealed the five most common passwords across fifty countries were:

1. 123456
2. 123456789
3. 12345
4. qwerty
5. password

- use personal information when creating passwords because it makes the passwords more memorable and easier to recall;
- reuse identical passwords across multiple sites.

Beliefs:

- adding numbers makes a password more secure than using only letters;
- adding an exclamation mark or other symbol at the end of a password makes it more secure;
- using a word that is difficult to spell as a password is more secure than an easy-to-spell word;
- a password is secure if the password is longer and contains uppercase letters, digits, and symbols;
- attacks on passwords are targeted at specific individuals;
- using personal information, such as a birthday, in a password is secure if such information is not on their social media accounts;
- password strength meters accurately measure the security of a password.

Misconceptions about password security persist despite growing public awareness of security threats. There are many reasons. We highlight those that surfaced most often and prominently in the literature and interviews.

First, lay people struggle to create and manage an ever-growing list of secure passwords.[9] In a (somewhat dated) large-scale study of password habits, Dinei Florencio and Cormac Herley (2006) found that on average, people tend to have twenty-five accounts that require passwords, actively use around seven distinct passwords, and reuse passwords across sites/accounts. The researchers concluded that their large-scale study confirmed conventional wisdom about the large number of passwords that users maintain, the poor quality of those passwords, and the frequency of password reuse. In our interview, Herley confirmed that these basic observations from over fifteen years ago remain salient. The number of passwords people create and manage has risen significantly, possibly by an order of magnitude, and people continue to reuse passwords. Herley emphasized that people sometimes exercise common sense effectively when creating, managing, and even reusing passwords. There is a rough and often implicit cost–benefit analysis that people must engage in, given the growing number of password-protected accounts. Contrary to security advice not to reuse passwords often given by experts, Herley told us, "I reuse passwords all the time. I do it without fear, without shame." Using his own experience as an example, he explained that for 200 or so password-

9 Most experts suggested that password managers are a useful solution to this problem. As Bruce Schneier explained: "Password Managers change everything. If I don't have to remember passwords, and if I don't have to repeatedly type them, I can choose strong random passwords. Even better, I can have my password manager choose them for me. [For example,] I just asked PasswordSafe for a password. It gave me 'k%wo=-x{Y_XTpwWz@L'."

protected accounts, 150 are low-value accounts for which weaker and even reused passwords might be justified. There are just too many accounts to create and keep track of strong, distinct passwords, and if a low-value account is hacked, there is less downside so long as it does not provide access to a high value account or enable another type of attack (e.g., phishing). Thus, one password security dilemma is knowing how to partition among low- and high-value (risk) accounts, and a second one is knowing how to respond when an account is compromised.

Second, lay people tend to have an incomplete understanding of the nature and variety of password security threats, which has changed substantially over the past few decades. The experts we interviewed all emphasized there are many different threats to consider when evaluating password security. Some threats involve "guessing a password," and these threats are often what people think about when considering what is necessary for a secure password.[10] Password guessing can be as simple as it sounds – a person trying to guess the password. That is a threat model most people can understand. But it can get much more complicated, and correspondingly more difficult for lay people to understand, when the person (guesser/attacker) uses different tools, ranging from surveillance tools (e.g., to collect information about likely passwords)[11] to computational tools (e.g., a password-guessing algorithm). Further, it can be difficult to appreciate the differences between attacks directed a specific target and undirected attacks, online versus offline attacks,[12] and other variations of guessing attacks.

To make matters more complicated, many threats to password-protected resources do not involve password guessing at all. For example, a person might look over your shoulder and observe you entering your password. Or a person might trick you into downloading malware onto your computer and that software may capture your keystrokes and thus your password(s). Or a person may hack the system for which a password is required and thereby obtain the password (along with others); if that password had been used for more than one password-protected account (password reuse), then those other accounts could be compromised. Notably, each of these threats occurs without any guessing. While it may be easy for someone to catch the person looking over one's shoulder and respond to that threat, the other threats require more knowledge for lay people to be able to manage their security. For

[10] According to Herley, "30 or 40 years ago, maybe the dominant threat against passwords was a password guessing attack ... that hasn't been the case for a long, long time, and the threats we have against password secure resources tend not to be guessing."

[11] Surveillance tools that enable the person to observe you entering your password obviate the need for guessing and thus fit into the next category we discuss.

[12] In his interview, Arvind Narayanan explained: "Online guessing attack is where somebody tries putting in different passwords into the login screen. An offline guessing attack is where somebody breaks into the server and steals the password database; usually those passwords might be hashed, which is a type of encryption, but it can be reversed." And so, one relevant but often overlooked password security consideration is "how well does your password resist password hashing?"

example, our literature review and interviews suggested that people tend to underappreciate the risk of password reuse because they fail to understand the external effects from a hacked system. (The hack of system 1 where password A is obtained compromises systems 2, 3, . . . N where A is also used as a password. Lay people who understand password security threats in terms of targeted attacks or guessing attacks may fail to appreciate fully the risk of password reuse.)

Misunderstanding the nature and variety of threats can strengthen incorrect beliefs and lead to poor security practices. For example, according to two studies (Ur et al. 2016a, 2016b) and (Ur et al. 2015) that directly examined user perceptions of password security threats, most users believed that attackers were strangers located far away from them; the users were concerned with attackers having access to and using their personal information. In these studies, users avoided using their own personal information such as birthdays and pet names, but some used the personal information of their family and friends instead to create passwords. The authors concluded that the users' misconceptions about password security directly related to their misunderstanding or lack of knowledge about automated password guessing attacks, which would be able to guess those passwords. Thus, another dilemma to consider is how user (mis)perceptions of security threats shape user beliefs about what constitutes a secure password or practice.

A third reason concerns how lay people learn about password security. Our literature review did not provide much insight on this issue. For the most part, the studies we reviewed aimed to uncover lay persons' beliefs and practices, but the studies did not examine learning processes or the origins of those beliefs. There are some studies that criticize security advice and suggest that lay people reject the advice because it is too difficult to follow given the ever-growing number of passwords to manage. (We discuss these studies below.) The experts we interviewed suggested that people learn through their everyday experiences with passwords (supporting, if not confirming, our hypothesis that password knowledge is often in the realm of everyday common sense knowledge). People engage with password creation regularly, encountering password composition rules and password security meters. In addition to their instrumental function (creating secure passwords), these tools serve an important, generally overlooked, pedagogical function, which is to teach users about password security (see Frischmann & Selinger 2018; Weizenbaum 1976). Unfortunately, these tools can easily mislead and teach the wrong lessons. Consider the following stylized example that we raised during interviews.[13]

Jeremy must create a new password. He encounters a typical password creation interface, a text box to enter characters, password composition rules displayed

[13] During interviews, we discussed slight variations: **Brett**, **Brett1**, and **Brett1!** The difference in names makes a slight difference because of the additional length (one extra character). We switched to Jeremy for the write-up only because the example better illustrates the potential shift in a strength meter assessment from weak to strong.

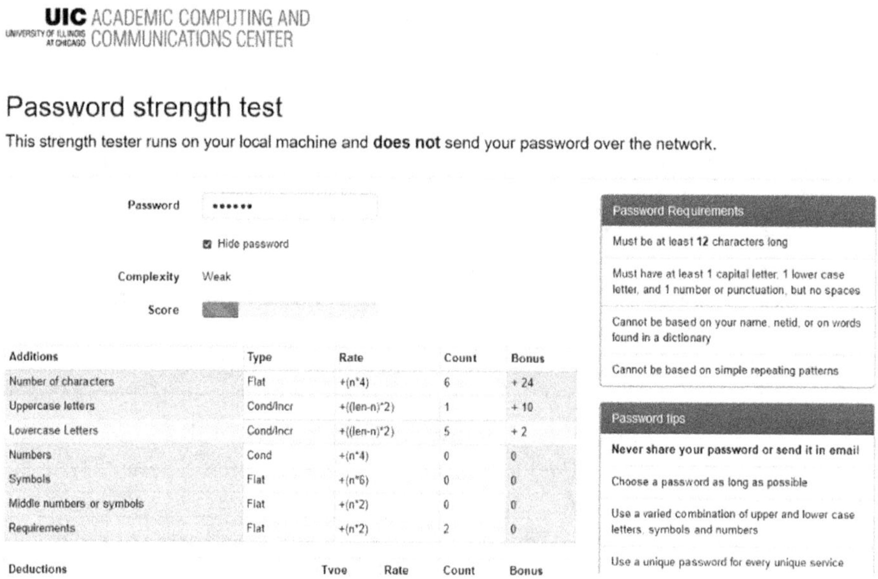

FIGURE 5.2 Password strength test.

prominently to one side, and a password strength meter that updates the color (red, yellow, green) with each new character displayed prominently on the other side.

> **Step 1.** Jeremy begins by typing the following characters: **Jeremy**
> The security meter remains *red.*
> **Step 2.** Jeremy continues by adding the following character: 1
> The security meter turns *yellow.*
> **Step 3.** Jeremy continues by adding the following character: !
> The security meter turns *green.*[14]

Figures 5.2 and 5.3 demonstrate the first and third steps using the Password Strength Test, available at the UIC Academic Computing and Communication Center.[15]

What might Jeremy learn from this experience? What have these tools taught him about password security? During interviews, the experts noted how there are various ways in which the tools could teach the wrong lessons (i.e., misinformation about password security).

First, the security meter implies marginal security improvements that may not be genuine or meaningful improvements. Adding a single number and/or symbol at the

[14] This is a stylized hypothetical. Most password security meters would not shift from red to yellow to green based on the changes we suggest. Our point is only to illustrate how password meters teach users about password security.

[15] See www.uic.edu/apps/strong-password/.

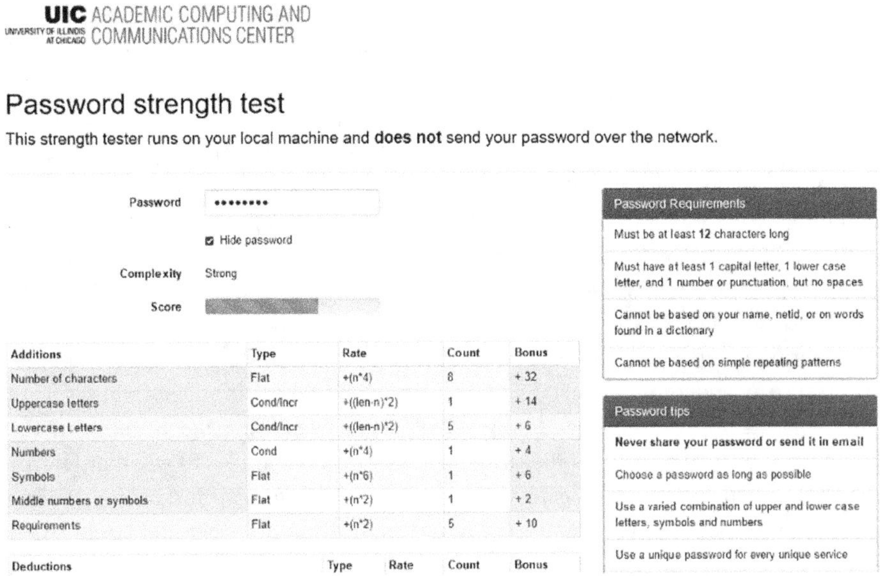

FIGURE 5.3 Password strength test.

end of a common name does not substantially improve the strength of the password. In the specific example, **Jeremy**, **Jeremy1**, and **Jeremy1!** are all relatively weak passwords. The marginal difference between the first two passwords is negligible; the third is an improvement, but it is not substantial. The differences among the three do not impose much of a burden on automated guessing attacks.[16]

Second, in more absolute terms, the security meter suggests **Jeremy1!** is a strong password when in fact, it is not. This is an incredibly important source of consequential misinformation. Jeremy, like most lay persons, is likely to trust the digital tool, rely on its "advice" and the weak password, and go no further to create an even stronger password or learn more about password security. How can this possibly occur? In practice, security meters are based on compliance with password composition rules and essentially that means strength depends on checking a series of boxes, such as including characters from different character classes (upper and lowercase letters, numbers, symbols), having enough characters (e.g., length of password is 8, 12, or possibly more characters), and not being on a blacklist of already known common passwords (Carnavalet & Mannan 2014). For most password strength meters, the first two examples in the hypothetical would remain in the red (weak) because they lacked a symbol and would be too short (password length under eight

[16] We checked a few online tools for checking password strength and estimated times for the password to be cracked. While the estimates varied, they all suggested that an automated password guessing program would take seconds to crack the first and second passwords and minutes or hours to crack the third password.

characters); the third example would pass some, however, because it checks those two boxes.[17] To be clear, the point we are illustrating with this hypothetical is not limited to the **Jeremy1!** example or even password strength meter tools. Below, we discuss other examples related to password security, but it is important to highlight how this examination is relevant to other digital networked technologies and social dilemmas concerning mis- and dis-information.[18]

Third, above and beyond the specific attempt to create a password, are the lessons learned about password security that Jeremy may carry into his next password creation experience or that he may relate to other people. We know very little about these types of effects. None of the studies we reviewed considered them.

Security meters could be designed to be better pedagogical tools. For example, when Jeremy added the numeral 1 to his name, the tool could suggest more secure alternatives using the same addition, for example by placing the number in the middle rather than the end of the string of characters. Regardless of whether Jeremy chose that option, he could learn why it was more secure with just a little explanation. Conversational prompts during password creation, for example, could teach while also nudging users toward stronger passwords (Khern-am-nuai 2022). As Arvind Narayanan suggested after discussing the stylized hypothetical: "Well-designed password meters can both help users create stronger passwords and teach them what makes a stronger password." Experts have found that password meters with a variety of visual appearances led users to create longer passwords, but significant increases in resistance to password-cracking algorithms were only achieved using meters that scored passwords stringently (Ur et al. 2012). These findings support the push for wide-scale deployment of more stringent password meters to train users to create stronger passwords more routinely. Yet, as we discuss below, this has not happened. Professionals and practitioners who design and deploy password creation tools typically balance security, usability, and market incentives to keep consumers using their services. Further, it is not clear that consumers want to learn about security when they are asked to create passwords, which is often when they are excited to try out a new service.[19] (Again, we must emphasize that these observations are not

[17] See, for example, the Password Strength Test, available at the UIC Academic Computing and Communication Center, www.uic.edu/apps/strong-password/ (weak rating for the first two, but strong rating for the third); Is My Password Vulnerable?, available at https://nordpass.com/secure-password/ (weak rating for all three).

[18] See generally Frischmann and Selinger (2018), *Re-Engineering Humanity*. See also Haenschen, Frischmann, and Ellenbogen (2020) (manipulating the Facebook birthday notification tool, causing it to distribute fake birthday notifications, to examine the pedagogical function of the tool and how it shapes human thinking).

[19] As Bruce Schneier noted, "people are asked to create passwords at moments when they are not ready to do it. You know how it happens. You're signing up for some new service, all excited about trying it out. And then you're suddenly asked to create a password. What you want most is for that screen to go away. You don't want to think about security. So, of course you do a bad job."

necessarily limited to password creation tools; we strongly suspect that many other digital tools and interfaces could be designed to be better pedagogical tools but face similar dynamics.)

Another common experience lay people have with passwords is forgetting them. That is, people often forget a password and then engage with password recovery tools, for example, responding to a set of security questions (sometimes called "secret questions" or "challenge questions") that substitute for the password as a knowledge-based means of authentication (Bonneau et al. 2015a). Password recovery tools also have instrumental and pedagogical functions. People need to access password-protected resources, and the recovery tools serve that need. The security dilemma is two-fold: First, the password recovery tools are not always secure. The types of security questions a person must answer may be easier to guess than a password (Bonneau et al. 2015a). There are more secure password recovery tools (e.g., SMS and email-based recovery; Bonneau et al. 2015a). But again, businesses face tradeoffs between security, usability, and convenience (Bonneau et al. 2015a). Second, since password recovery tools are often easier to use than remembering a strong password, some people (learn to) adopt a strategy of relying on them rather than passwords as a means for authentication and obtaining access to resources. We can see the pedagogical function of password security tools in terms of how experience using the tools shapes everyday security beliefs and practices.

When comparing the behavior of experts to that of nonexperts, Iulia Ion, Rob Reeder, and Sunny Consolvo found that experts report installing software updates, using two-factor authentication, using unique passwords, and using passwords managers to remain safe online, while nonexperts report using antivirus software, using strong passwords, only using known websites, and changing their passwords frequently (Ion et al. 2015). As a result of their findings, Ion et al. suggest that better messaging along with systems and usability work are necessary to get nonexperts to follow the same security practices that experts do.

Expert Knowledge (Dilemmas) Concerning Password Security

In this section, we explain what we learned from the literature review and interviews about expert knowledge, how it has changed over the past few decades, and what are the "best practices" currently recognized by security experts. Our abbreviated history of expert knowledge about password security focuses on major themes related to knowledge dilemmas within the community, rather than specific actors or their contributions. While the literature review provides support for the findings we highlight, most of the insights are drawn from the interviews.

In the early days of computing and networking, password security experts mostly were academic researchers with expertise in computer science, mathematics, and engineering. These experts thought about password security as a mathematical problem. Security experts mostly worried about automated guessing attacks – that is, guessing attacks leveraging computational tools.[20] The offense and defense of password security was, at its core, a computational problem. As such, expert knowledge about password security was based heavily on the mathematical theory of computation. Expert knowledge thus led to certain prescriptions, namely that users should create sufficiently strong passwords to withstand an attack. Experts measured strength in terms of complexity (e.g., Shannon entropy), which usually involved a calculation based on the length of the password and the number of different character classes. For a reasonably accessible and thorough explanation, see *Appendix A: Estimating Entropy and Strength*, in the NIST Electronic Authentication Guideline (Burr et al. 2013b). Notably, as we discuss below, security experts now regard complexity/entropy as a *bad* proxy for guessability and instead prefer to evaluate password strength using an adversarial guessing approach (Lee et al. 2022).

Expert knowledge about password security shifted dramatically over the past few decades, and it continues to change. According to interviewees, during the 2000s, significant changes in expert knowledge emerged as knowledge grounded in theory failed to match reality (everyday life experience). Bonneau et al. discusses the evolution of passwords suggesting that estimates of password strength, models of user behavior, and policies related to password composition work in theory but can be unsupported in practice and even contradicted when observed empirically, possibly leading the research community to focus on the wrong threats (Bonneau et al. 2015b). We highlight two themes that surfaced repeatedly during interviews – first, changes in the technological environment (or reality on the ground), and second, corresponding changes in the expert community. We then turn to the role of professionals and practitioners in the password security ecosystem and explore how these actors mediate between the experts and lay people. We conclude with a brief discussion of the latest expert knowledge regarding best practices and knowledge dilemmas associated with translating such knowledge to actual practice by professionals, practitioners, and lay people.

CHANGES IN THE TECHNOLOGICAL ENVIRONMENT Over the past few decades, digital networked computing technologies evolved considerably and diffused widely. The number of users and the number of accounts each user must manage grew incredibly.[21] Password-protected resources became part of everyday life experience

[20] Password theft, whether by someone looking over your shoulder or more sophisticated forms of espionage, existed as a threat model, but it was less prominent as a threat for academic security researchers to focus on because there were fewer targets (opportunities for such attacks). As noted in the text, this changed with the diffusion of digital networked technologies and proliferation of password-protected accounts.

[21] "Password strength restrictions go back to about 1979 when a power user might have 3 logins on 3 different computers. And so very few computers, and ... almost nothing is a service login as opposed to a system login. Users had no local storage and no local computing, and the rules made a lot more sense then. The situation today is completely different. [E]veryone's got the local storage. My wristwatch has got more storage than the computers I was using in the late

for billions of people. This shift dramatically expanded the attack surface where the offense and defense of password security played out and, as a result, fundamentally altered the epistemic demands on password security experts. Not surprisingly, security threats multiplied and changed. Initially, security experts (mostly) agreed that strong passwords were an appropriate solution for the threat of guessing attacks. But as password security became an everyday life concern for lay people (at home, work, and everywhere else), this conventional expert wisdom began to fail for a few reasons.

First, despite reasonable consensus among experts on what constitutes a strong password (complexity/entropy, though that would change), people nonetheless often chose and still choose terribly weak passwords (see section "Expert Knowledge (Dilemmas) Concerning Password Security"). Leaked datasets of millions of actual passwords, for example the 2009 RockYou.com database leak, confirmed what many considered to be common sense: Lay people were unable to keep up with the computational arms race and ever-growing number of passwords,[22] and they often defaulted to weak but memorable passwords and reused such passwords across different accounts. (As Herley and colleagues pointed out, lay people often were exercising common sense, performing rough cost–benefit analyses, and using stronger passwords for more important accounts.) This realization in the mid-2000s led to a shift in focus among security researchers toward the needs and capabilities of actual human beings involved in authentication processes. *Human factors and usability matter!* This realization also focused attention on the increasingly important role of professionals and practitioners who designed, deployed, and managed digital networked technologies, including password creation tools (e.g., human–computer interfaces, password composition rules, strength meters).

Second, automated guessing attacks became more sophisticated and varied. Attackers could use an array of widely available tools, including dictionaries, large datasets of stolen passwords, and even the various password creation tools themselves, to improve their guessing algorithms. Attackers easily could learn about how people tend to respond to password composition rules, for example by adding **1!** at the end of a character string deemed weak by a strength meter or by replacing letters with similar symbols (**$** for **s**). This knowledge advantaged attackers and confounded the Shannon entropy calculation used to determine password strength.

For some threats, such as online guessing attacks made at the login interface, experts determined effective countermeasures, such as limiting the rate at which guesses can be made by injecting exponentially increasing time delays between each

seventies and more compute power. I have hundreds of logins. They're mostly service logins to a website rather than a system logins that would give me access to system commands, and I'm not unique in that. And there's many more things about the, so the environment has changed completely." (Steve Bellovin Interview)

[22] Arguably, there is a tragedy of the commons style dilemma lurking here. However, it is not one we explored during our research.

wrong guess.[23] Despite the mathematical beauty of this solution, such friction-in-design measures are not, however, universally implemented. Again, math theory falls victim to pragmatic considerations of everyday life; human factors and usability matter. Such delays are inconvenient and that can be off-putting for consumers of digital services, and so, as companies engage in a (rough) cost–benefit tradeoff, they may forsake security in the name of consumer convenience (profits). Of course, as David Thaw pointed out during our interview, this countermeasure is effectively used to secure bank ATMs from PIN guessing attacks and ought to be used much more widely. Bruce Schneier made a similar point, noting that "systems that lock you out after only a few incorrect guesses work great; it's why your ATM card password can be secure with only four digits." But he warned of "a denial-of-service trade-off: I can deliberately lock you out of any system that has that feature by using your username and guessing wrong."[24]

For offline guessing attacks (e.g., following a database leak), countermeasures often are the burden of professionals operating web services. Xavier de Carné de Carnavalet (2022) observed:

> Proper password hashing, for example, should be applied to protect plaintext passwords; otherwise, even seemingly hard-to-guess passwords would be readily available in the [leaked] database. Inappropriate hashing algorithms such as MD5 and SHA1 have been extensively used, and one can still find recent breaches with such algorithms used. Besides the lag between expert and professional knowledge explained later [in this chapter], this is also the result of outdated resources being promoted on the first page of Google search results [and] how young professionals learned about authentication at school.[25]

Third, beyond guessing attacks, the threats multiplied, became more varied, and hybridized. Some attacks targeted vulnerable weak links. For example, attackers recognized that password hosts and login commands could be subverted. Phishing attacks, compromised servers, and compromised client hosts make it easier for attackers to steal passwords (Bellovin 2016). If an attacker has accomplished one of

[23] These measures would not counter offline guessing attacks, which for the sake of brevity, we do not discuss extensively.

[24] Xavier de Carné de Carnavalet explained: "To be effective, delays should be applied to accounts rather than IP addresses. Therefore, as delays could be the result of an attacker trying to guess an account password, they may be denying service to the rightful owner of the account." Pre-publication comments on "Common nonsense about password security and the expert–layperson knowledge gap." 2022. On file with authors.

[25] See Naiakshina et al. (2020, 8) ("Some [participants] further suggested that there is a lot of outdated information on the Internet with respect to security practices (PJ6, PS8), making Internet research rather challenging."); Naiakshina et al. (2017) (exploring poor password storage practices among developers); Krombholz et al. (2017, 1347) (study of HTTPS configuration by professionals: "Our participants reported that they came across outdated or simply wrong information in online tutorials."). Of course, the perpetuation of misinformation, including outdated information about security practices, on the internet is a more general problem.

these attacks, then a strong password does not provide any defense (Florencio et al. 2007); the attacker doesn't care about special characters, or any other suggested characteristics users are told will make their passwords stronger and more secure. Thus, security experts realized that even if necessary, strong passwords are not sufficient protection for password-protected accounts and resources. David Thaw told us that strong passwords were counterproductive in the sense that (i) they could contribute to "security theatre" by providing a false sense of security to users so long as they comply with password composition rules, and (ii) draw attention away from other more important security threats and countermeasures.[26] Lee et al. (2022, 572) similarly offer as a hypothesis to help explain "the disconnect between industry and the academic community" the idea that "Password policy is security theater: measures such as character-class PCPs, even if ineffective, may give users a false sense of security, and websites use them for this reason."

Initial thinking about security measures to counter the threat of an organizational attack also was grounded in the math theory mindset. Conventional wisdom among experts suggested that one way to minimize the downside risk of a data breach at a company where a password file (database) has been compromised is to require people change their passwords periodically; then in the event of a breach, there is a good chance that by the time someone tries to exploit the leaked data, it has already been replaced with a new password. Bellovin explains why this wisdom fails. Users dislike frequent requests to change their passwords and tend to use similar patterns when rotating passwords (e.g., adding a number or increasing an existing number by one), making the replacement passwords even easier to guess. Again, sophisticated attackers who pay attention to human factors and behavioral patterns easily learn to predict what people will do when forced to periodically change their passwords. Armed with such knowledge, attackers can more easily guess passwords using a hacked dataset as a baseline (input) for their guessing algorithm. The conventional wisdom thus backfires. We return to this example below as a mini-case study that highlights relevant knowledge dilemmas between experts, professionals/practitioners, and lay people.

The emergence of new threats and their successful exploitation forced the expert community to confront the notion that password security is not just a mathematical problem solved with strong passwords. Instead, it is multifaceted and necessarily embroiled with the many fallibilities of humans and organizations.

CHANGES IN THE PASSWORD SECURITY EXPERT COMMUNITY Interviewees mostly described the password security expert community as academic researchers who

[26] Thaw further noted: "'There are some attacks on [password authentication systems] about which scientists reasonably disagree as to whether or not the system provides sufficient defense on its own … so I respect that there can be reasonable disagreement, [but] I'm on the side [that] the system can provide reasonable defense and password complexity doesn't meaningfully add to that, but there's fair reasonable disagreement."

published research in peer review journals and attended academic conferences. Some industry researchers, often at research labs, who participated in these same activities also were considered members of the expert community.[27] Here we highlight changes in community membership (types of experts included) and corresponding changes in the types of research and knowledge developed within the expert community. One of our interviewees described the transformative shift in the expert community that occurred roughly over the course of a decade (late 2000s to 2010s) as the triumph of empirical observation over math theory, a demonstration that human factors matter and that usability needed to be considered by security experts. Essentially, changes in the technological environment described above drove changes in the expert community as the community expanded to integrate different expertise. In the first few decades of computing and networking, the security expert community included mostly computer scientists and engineers who approached password security more or less as a security optimization problem. For example, where guessing attacks were the threat to password-based authentications systems, password strength needed to be optimized considering the attacker's resource constraints; thus, the research question was how strong a password must be to withstand an automated guessing attack, taking into account expected computational resources of attackers. A similar mathematical, risk-based framing occurred for the threat of data breaches and stolen password files, and this framing led to security prescriptions against password reuse and in favor of organizational policies requiring periodic password changes.

In the 2000s to early 2010s, research from experts outside of the traditional computer security community highlighted how other sets of real-world constraints needed to be considered. Specifically, human factors and usability constitute real-world constraints on the effectiveness of strong passwords as knowledge-based authentication mechanisms. Lorrie Cranor stated that until this new research from usability researchers surfaced and gained traction within the security community, "there was basically no knowledge about usability of passwords other than just intuition." She further explained:

> Experts knew what was easier or difficult for them personally, but beyond that there was basically no knowledge. Most of the knowledge about security was based on math. We can calculate how many possible passwords exist if we limit ourselves to characters and keys on the keyboard. You can do the math and figure out the password space. So people were looking at that sort of thing, but what they weren't looking at is the fact that there are a lot of combinations of characters that no human would ever come up with on their own. The space of realistic passwords is

[27] Interviewees explained that communications within this community mostly occurs in traditional academic forums such as conferences, academic publications, and interpersonal communications (email, conversations, sharing drafts and research ideas). In addition, experts share knowledge using online media, such as blogs and social media, as well as traditional media, such as newspapers and television appearances.

> much smaller than the space of all possible passwords unless you have a computer randomly generating them for you. Over the past 15 years, we've learned a lot about what kinds of passwords humans *actually* create when left to their own devices. And we've also learned about how to nudge humans towards creating stronger passwords.[28]

The expert community gradually broadened beyond conventional computer security experts to include experts in usability, information technology, and other related scientific and engineering disciplines as well as experts from adjacent social sciences such as psychology, economics, and cybercrime. The community remained an academic one, in the sense that these experts participated in academic conferences and published research in peer review publications, regardless of whether they were affiliated formally with academic institutions. This led to the development of new cross-disciplinary conferences and specialties, such as usable security.

When asked how expert knowledge had changed over the past two decades, Arvind Narayanan replied:

> Computer security [is] a field that has often fetishized mathematical knowledge over human subjects experiments, for various cultural reasons, historical accidents, etc., and so . . . a lot of the earlier experts' wisdom and decrees regarding password security often involved burdening users, and we never really tested to see how effective [those burdens] are, and so a lot of the things that experts know better today are the opposite of things they might have said 20 years ago. So what are some of those things?
>
> Never write down your password. That's one piece of bad password wisdom. The problem with that is because our memories are finite, and not writing down passwords means that people are going to reuse passwords, and that's a much bigger problem today. Most experts (today) will say it's OK to write down passwords, but think about the physical safety of where you write them down.
>
> A second one today is the importance of multifactor authentication; passwords are never going to be your only line of defense. It doesn't matter how clever your password is. It's probably going to get breached at some point.
>
> A third one is that a lot of the password-related advice these days would be directed at companies as opposed to users.
>
> . . .
>
> I think the security expert community is just a little bit more humane now than twenty years ago. Rather than treating users as the problem, [experts] recogniz[e] that people are overburdened as they go about their everyday lives, and designers have the primary responsibility for security.[29]

[28] Author interview with Arvind Narayanan, Professor of Computer Science, Princeton University. 2022.

[29] Author interview with Arvind Narayanan, Professor of Computer Science, Princeton University. 2022.

These observations resonated with some articles we reviewed from that transitional period that noted how experts sometimes misunderstood users. Herley (2009) noted that some experts may suggest that users are lazy, choose weak passwords, and ignore expert security advice; but he argued that these actions may be entirely justifiable. Users' rejection of security advice often is rational, Herley suggested. According to Herley, much of the public security advice for users is complex, outdated, and does little to address actual threats that users face, offering a poor cost–benefit tradeoff, and thus users reject the advice. Dinei Florencio, Cormac Herley, and Paul C. van Oorschot (2014) explored how users should manage large numbers of accounts and found that mandating that users only use strong passwords and not reuse passwords leaves users with an impossible task as the number of accounts they have has grown.

INTERMEDIATING ROLE OF SECURITY PROFESSIONALS AND PRACTITIONERS Changes in the technological environment were driven by and at the same time drove growth and expansion of an intermediate layer of computer and network security professionals and practitioners (hereinafter, "professionals" for brevity).[30] The expert security community came to recognize the importance of these actors and engaged with them more directly. Unfortunately, we were unable to fully explore this engagement during our research. It surfaced during interviews in a few different ways. Some interviewees noted how some industry security researchers actively participated in the academic security research community, whether by attending conferences or publishing in peer review journals. Others noted how IT, security, and other professionals often learned about password security while at university and then went on to careers where security was often one of many responsibilities. This intermediate layer of professionals played an increasingly important role in determining password security practices, by setting organizational policies and deploying technologies and services that required users to create and manage passwords.

While we were unable to fully map and explore the knowledge flows between these different communities (experts, professionals, lay people), we highlight observations drawn from interviews to help frame future research.

[30] The professional/practitioner community includes both security and IT professionals, such as Chief Information Security Officers (CISO), Chief Security Officers (CSO), security engineers, architects and analysts at companies, and even ethical hackers. These types of jobs entail implementing and testing security programs across an organization, overseeing the operations of the IT department, overseeing the physical and cybersecurity of a company, protecting company assets from threats, implementing quality control, conducting internal and external security audits, and various other related activities. Members of this community often join professional organizations such as Association for Computing Machinery (ACM), IEEE Computer Society, Information Systems Audit and Control Association (ISACA), Society for Information Management (SIM), and User Experience Professionals Association (UXPA). These types of professionals tend not to publish academic articles; instead, they may give talks and training sessions to other members of the community and lay people.

One theme that surfaced repeatedly in our interviews is the idea of a disconnect between expert knowledge and professional knowledge and practice, which, in turn, seems to trickle down and potentially generate lay person misunderstanding. This disconnect is, at least in part, the result of a lag in the knowledge flows between experts and professionals. The flow of knowledge from experts to professionals can be slow and delay the ability of professionals to implement new standards and translate that information into forms accessible to lay people. Reasons for this lag include the rapidly changing technological environment and different, and sometimes competing, incentives between the expert community and the professional community. According to David Thaw, it is easy to propagate information about security and best practices through the professional community because it is not held to evidence-based standards (at least, not the same as academics). Professionals have access to academic research, but that does not mean they keep up to date with it. They have mixed responsibilities and priorities and may view security primarily in terms of compliance.

Three notable concerns arise. First, professionals responsible for password security, business practices, and interface design may perceive security as one of many different competing considerations (business, technical, etc.). Second, a compliance mindset may lead professionals to ignore academic security research and instead rely on security standards set by insurance companies, auditors, other industry actors, or government. Finally, professionals may be overconfident and inaccurately see themselves as security experts or as being sufficiently up-to-date on security.

Expert consensus on what actions constitute best practices also struggles to keep up with the rapidly changing technological environment. As a result, what was accepted as best practices five to ten years ago can become outdated. Experts must correct their views first and then translate that knowledge to the broader public, including professionals and lay people. Knowledge transfer between experts as well as between experts and professionals may happen too slowly; technological changes may occur faster than expert knowledge can update and percolate. As a result, lay users may be the last community to become aware of new password security developments, and their security behaviors may be at odds with the behaviors experts characterize as best practices. Based on this lag-induced knowledge gap, lay users may unknowingly become accustomed to outdated security practices believing that they are increasing their security when they are in fact doing the opposite.

Experts and professionals also may have competing incentives, directly impacting how information is shared, which information is regarded as important, and how best practices and security advice get implemented. According to various interviewees, academic incentives prioritize publishing articles to increase prestige within the community. Arvind Narayanan emphasized his view that academic researchers have a responsibility to translate their findings to the public, including professionals and lay people, since most academics receive funding from taxpayers. He suggested that most fail to do this work and that universities should consider crediting such efforts in order to realign academic incentives with the public interest.

Interviewees suggested that the professional community has a compliance mindset. As a result, the professional community tends to evaluate security based on process rather than outcomes. One of the ways professionals evaluate security is through security checklists from organizations such as NIST or companies such as Deloitte. These checklists are sometimes forced upon professionals by their auditors and often are not updated to reflect the latest expert knowledge. Thus, a question arises as to whether professionals believe they are following the best security practices when they comply with these checklists. A related concern is that many professionals have been in the industry for years, but they are not getting retrained or refreshing their knowledge to reflect evolving expert knowledge. As a result, updated best practices and expert knowledge may not reach professionals and trickle down to lay people. While the lack of consistent training and refresher courses may be out of the control of professionals and more in the hands of the companies they work for, Rick Wash suggests that many professionals care more about checklists and not getting blamed for any problems than staying educated about security itself. Bruce Schneier similarly suggested that organizational inertia coupled with these compliance incentives do not allow for expert consensus to be integrated successfully in the professional community. Finally, Xavier de Carné de Carnavalet (2022) ties the observations of Wash and Schneier back to the concept of security theatre:

> bothering users with ineffective and sometimes counterproductive measures gives them a sense that a system is "secure", i.e., it tries to prevent attacks by requesting additional steps. … This could also explain why security practitioners are slow to adopt the latest NIST standard that deprecates PCPs. Those were at least some visible security steps, and it seems that improving security should be synonymous [with] adding constraints, not removing them.[31]

(Some) Best Practices

The conventional wisdom (consensus views) among security experts about best practices have changed over time as we have already described. Today, there are some identifiable best practices that security experts mostly agree on. We do not provide a comprehensive analysis of current best practices.[32] Instead, we focus on the password creation action arena[33] and highlight a recent study by Lee et al. 2022, which sought to empirically study whether 120 of the most popular websites followed

[31] Xavier de Carné de Carnavalet (2022) further noted: "Finally, recommending the opposite of what professionals have been praising for years may not be easy to do while keeping their face. This would be admitting they were wrong."

[32] A comprehensive study would require further empirical work.

[33] Refer to Figure 1.1: "An action arena 'refers to the social space where participants with diverse preferences interact, exchange goods and services, solve problems, dominate one another, or fight (among the many things that individuals do in action arenas)' (Ostrom2005, 14) – in other words, the place at which the exogenous variables combine in particular instances, leading over time to observed patterns of interactions and outcomes. A particular action arena involves

best practices. To do this, the authors identified the following "established" best practices based on prior research in security and usability:

1. **Blocklists**: (i) Check user passwords against blocklists that include leaked and easily guessed passwords. (ii) Block user passwords that appear on such lists. (iii) Prompt user to create a different password (Lee et al. 2022). (We might add: Inform user about the reason.)
2. **Strength meters and minimum strength requirement**: (i) Provide accurate, real-time strength estimates and feedback. (ii) Set minimum strength requirements based on an appropriate measure of password strength, namely guessability rather than complexity, Shannon entropy, or compliance with composition rules (Lee et al. 2022).
3. **Composition policies**. (i) Do not require specific character classes. (ii) Set minimum password length of eight characters (Lee et al. 2022).

Lee et al. (2022) explain how these practices are supported by recent research in security and usability. While precluding users from creating passwords found on blocklists is a low-cost security measure, Tan et al. (2020) tested different password security requirements in two experiments and found security–usability tradeoffs among different blocklist configurations. Also, in their study, strength meter and text feedback informed users about the reason for disallowing a password. Other research has shown that such feedback can nudge users to create stronger passwords.

Lee et al. (2022, 563) explain that minimum strength "requirements and strength meters are both effective and user-friendly." However, the preferred means for evaluating strength has changed from complexity to guessability, which is the "number of guesses needed to crack a password" (563). Determining strength thus involves testing the password against an adversarial neural network rather than calculating complexity/entropy. These developments led to a corresponding shift away from character-class requirements (e.g., password composition rules that require characters drawn from specific character classes) and toward minimum strength requirements of "at least 10^6 [guesses] to prevent online guessing attacks" (564).

Remarkably, Lee et al. (2022) found that out of 120 popular websites, only 13 percent followed the established best practices noted above. More than half of the websites did not check user passwords and allowed users to use the most common, leaked, and easy-to-guess passwords (e.g., "12345678"). Almost a decade ago, Carnavalet and Mannan (2014) published a damning empirical study of password meters that showed how poorly password meters fared in terms of evaluating password strength and guiding users to create strong passwords, in part because many of the meters measured strength based on entropy rather than guessability.

specific action situations and specific actors, along with those actors' identities and roles" (Frischmann, Madison, and Strandburg 2014, 14).

That study also showed inaccurate and inconsistent results across password meters, such that weak passwords were sometimes rated as strong. Yet Lee et al. (2022) found that little had changed. Surprisingly, most sites did not use password meters, despite plenty of research in the intervening period showing their value when properly deployed. The websites that did include password strength meters (i) misused the meters to nudge users towards including certain characters (i.e., to satisfy outdated character-class requirements) instead of serving the preferred best practice of encouraging users to freely construct strong passwords and (ii) did not include an evaluation of guessability (Lee et al. 2022). One positive note from Lee et al. (2022) was the observation that more than half of the websites have an eight-character minimum length requirement. The authors surmised, "Perhaps this is a result of updated guidance from NIST in 2017, which now recommends an 8-character minimum length for passwords, up from its previous recommendation of 6 characters" (Lee et al. 2022, 569).

Based on their findings, Lee et al. (2022) concluded that websites should review the best practices established by academic experts to adjust to their password policies. Moreover, the researchers concluded that because there seems to be a disconnect between academic password security experts and industry, future research should directly engage with system administrators to address the disconnect.

We conclude this section by noting another shift in best practice advice that surfaced during interviews.[34] One conventional wisdom dispelled by experts during interviews is the idea that people should not write down their passwords in a notebook or on a piece of paper; a few experts mentioned this example. They explained that experts now generally agree that writing down passwords is a good security practice so long as the physical password list is kept in a secure location because it helps people keep track of passwords and thus avoid (i) creating overly simple passwords to compensate for memorization difficulties and (ii) engaging with password recovery tools. The experts emphasized that the shift in advice reflects a better understanding of human factors and usability as well as prioritization of the security threats ordinary people face.

GOVERNANCE CHALLENGES

Three (meso-level) action arenas concerning everyday knowledge about password security surfaced in our study: The password creation action arena, the password recovery action arena, and the password expiration and replacement action arena. Each concerns a set of repeat interactions among stakeholders, where professional community stakeholders (e.g., employees of a company that design and operate the

[34] Other best practices noted by experts included using a password manager, two- or multifactor authentication, and even reusing passwords but only for low-value accounts.

password creation or password recovery tools or that implement a password expiration and replacement policy) determine the rules-in-use, informed to some degree by expert knowledge, and lay people take actions to create, recover, or replace passwords, possibly learning from their experiences. These action arenas present a series of knowledge dilemmas and governance challenges, which we discussed previously. We leave further analysis of them for future work and now look "upstream" from the password expiration and replacement action arena to examine the role of the National Institute of Standards and Technology (NIST) in shaping this action arena.

NIST on Password Expiration and Replacement: A Case Study

NIST, founded in 1901, serves as a nonregulatory agency funded by the United States Department of Commerce. The mission of NIST is to "promote U.S. innovation and industrial competitiveness by advancing measurement science, standards, and technology in ways that enhance economic security and improve our quality of life." Most important for our chapter is NIST's role as a standard-setting body. NIST standards generally establish best practices. Government agencies such as the FBI, NSA, and USDA must adhere to NIST guidelines, and in many sectors of industry, including cybersecurity, private companies tend to adopt NIST standards, although this can get complicated and be delayed as the specific case study below demonstrates.

To examine how NIST's capacity to set regulations, standards, and best practices in the password security industry can lead to governance challenges, we explore NIST's decision to change password expiration standards and the results in industry that followed. In 2003, NIST released "NIST Special Publication 800-63. Appendix A," which advised users to protect and secure their accounts by creating passwords with random characters, capital letters, and numbers, and to change their passwords regularly (McMillan 2017). Federal agencies, large corporations, and universities followed the advice. However, Bill Burr, who authored the publication, has since stated that the advice was largely wrong. Burr wanted to use real password data as a foundation of his advice, but there wasn't much of this data available to use as evidence for recommendations. Instead, security experts and professionals, including Burr, relied heavily on a paper written by Robert Morris and Ken Thompson (1979).

Recognizing the flaws and limitations of the 2003 publication, in June of 2017 NIST published "800-63-B: Digital Identity Guidelines: Authentication and Lifecycle Management." In this publication, NIST established new guidelines for password security to replace widespread practices and policies that made authentication mechanisms weak. One of the recommendations given by NIST was for organizations to no longer require password expiration, stipulating that users should only change their passwords if there is evidence of compromise (Grassi et al. 2017). The exact wording of the recommendation is as follows "Verifiers SHOULD NOT require memorized secrets to be changed arbitrarily (e.g., periodically). However, verifiers SHALL force a change if there is evidence of compromise of the authenticator." (Grassi et al. 2017).

The change in the guidelines resulted from academic research and expert consensus that password expiration is counterproductive to good password security. As we described extensively above, decades ago, the field of password security focused heavily on mathematical knowledge and theory and not on human factors and usability. Since password security ideas were based on math, rarely were expert recommendations tested empirically with users to measure their effectiveness. The modern practice of empirically testing password recommendations is relatively new. Today, security experts collect data on how often common attack types actually happen, and this data combined with empirically testing password recommendations against human usability has enabled experts to identify the following problems associated with password expiration: Lay users, when forced to comply with password expiration policies, often change their passwords in a trivial way, such as adding a 1 at the end of an already established password, making the password easy to guess; this doesn't solve the problem (attack type) as this type of behavior is predictable from an attacker perspective. Academic researchers conducted research, collected data, and published results (Chiasson & van Oorschot 2015; Cranor 2016; Habib et al. 2018; Spafford 2006; Yinqian Zhang et al. 2010; Zhang-Kennedy et al. 2016), and it became obvious to them that periodic password changes needed to be eliminated as a supposed best practice. Highlighting the impactful nature of such academic expert research, when creating their updated password guidelines, NIST cited academic research that reported on the ineffectiveness of password expiration policies (Habib et al. 2017; Herley & van Oorschot, 2012; Komanduri et al. 2011).

While NIST took into account up-to-date research findings and changed its standards, it is not clear whether the professional community has followed suit. Periodic password expiration and replacement appears to remain a widespread practice. Compliance with NIST's new standard seems slow, at best. Frankly, we asked interviewees about it, and while most speculated that industry has not yet adopted the new standard, no one knew of an empirical study of the issue. When asked about the reasons for this lag, our interviewees stressed the observations we noted in the previous section concerning the compliance mindset, conflicting incentives, and insufficient training and keeping up-to-date with expert knowledge.

CONCLUSION

Common nonsense about password security is a type of misinformation and digital illiteracy problem. In some ways, it is distinguishable from fake news and propaganda types that depend on trolls, bots, or others who deliberately pollute knowledge networks. Yet there are similarities, particularly with respect to how people learn about and from the digital tools they regularly use as well as the perpetuation of outdated information. This study examined different epistemic dynamics and dilemmas that may require different forms of governance. Unfortunately, we did not fully explore governance institutions within and between the different communities. That is an important topic for follow-on research. Our study shed light on the roles of experts and professionals, social demand for expert knowledge transfer via better and more widely accessible communication and education channels, and the societal risk of widening the lay–professional–expert knowledge gap. The GKC framework usefully structured our research methods and analysis, and it helped us identify different action arenas deserving of further study.

REFERENCES

'About NIST'. *NIST*, 10 July 2009. www.nist.gov/about-nist.

Bellovin, Steven M. 2016. *Thinking Security: Stopping Next Year's Hackers*. New York: Addison-Wesley.

Bobba, Sriram, and Vamsi Paruchuri. 2020. "Single Sign-on Using Contactless Smart Cards and Fingerprint Authentication." *Lecture Notes in Networks and Systems* (October): 158–166. https://doi.org/10.1007/978-3-030-90072-4_16.

Bonneau, Joseph, Elie Bursztein, Ilan Caron, Rob Jackson, and Mike Williamson. 2015a. "Secrets, Lies, and Account Recovery: Lessons from the Use of Personal Knowledge Questions at Google." In *Proceedings of the 24th International Conference on World Wide Web (WWW '15)*. International World Wide Web Conferences Steering Committee, Republic and Canton of Geneva, CHE, 141–150. https://doi.org/10.1145/2736277.2741691

Bonneau, Joseph, Cormac Herley, Paul C. van Oorschot, and Frank Stajano. 2015b. "Passwords and the Evolution of Imperfect Authentication." *Communications of ACM* (June). www.microsoft.com/en-us/research/publication/passwords-and-the-evolution-of-imperfect-authentication/.

Burr, William E., Donna F. Dodson, and Ebad A. Nabbus et al. 2013a "Electronic Authentication Guideline." National Institute of Standards and Technology (November). https://doi.org/10.6028/NIST.SP.800-63-2.

Burr, William, Donna F. Dodson, and W. Timothy Polk (eds.) 2013b. "Electronic Authentication Guideline." NIST Special Pub 800-63 Version 1.0, June 2004. (Later versions include Burr et al., NIST SP - 800-63-2, August.)

Chiasson, Sonia, and Paul C. van Oorschot. 2015. "Quantifying the Security Advantage of Password Expiration Policies." *Designs, Codes and Cryptography* 77 (2–3) (December): 401–408. https://doi.org/10.1007/s10623–015-0071-9.

Cranor, Lorrie. 2016. "Time to Rethink Mandatory Password Changes." Federal Trade Commission, 2 (March). www.ftc.gov/policy/advocacy-research/tech-at-ftc/2016/03/time-rethink-mandatory-password-changes.

de Carné de Carnavalet, Xavier, and Mohammad Mannan. 2014. "From Very Weak to Very Strong: Analyzing Password-Strength Meters – NDSS Symposium." (February). www.ndss-symposium.org/ndss2014/programme/very-weak-very-strong-analyzing-password-strength-meters/.

Egelman, Serge, Andreas Sotirakopoulos, Ildar Muslukhov, Konstantin Beznosov, and Cormac Herley. 2013. "Does My Password Go up to Eleven? The Impact of Password Meters on Password Selection." (January). www.microsoft.com/en-us/research/publica

tion/does-my-password-go-up-to-eleven-the-impact-of-password-meters-on-password-selection/.

Florencio, Dinei, and Cormac Herley. 2006. "A Large Scale Study of Web Password Habits." 1 (November). www.microsoft.com/en-us/research/publication/a-large-scale-study-of-web-password-habits/.

Florencio, Dinei, Cormac Herley, and Baris Coskun. 2007. "Do Strong Web Passwords Accomplish Anything?" USENIX (June). www.microsoft.com/en-us/research/publication/do-strong-web-passwords-accomplish-anything/.

Florencio, Dinei, Cormac Herley, and Paul C. van Oorschot. 2014. "Password Portfolios and the Finite-Effort User: Sustainably Managing Large Numbers of Accounts." (August). www.microsoft.com/en-us/research/publication/password-portfolios-and-the-finite-effort-user-sustainably-managing-large-numbers-of-accounts/.

Frischmann, Brett, Michael Madison, and Katherine Strandburg. 2014. *Governing Knowledge Commons*. Oxford University Press.

Frischmann, Brett, and Evan Selinger. 2018. *Re-Engineering Humanity*. Cambridge: Cambridge University Press (April). https://doi.org/10.1017/9781316544846.

Grassi, Paul A., James L. Fenton, and Justin P. Richer et al. 2017. *Digital Identity Guidelines: Authentication and Lifecycle Management*. Gaithersburg, MD: National Institute of Standards and Technology (June). https://doi.org/10.6028/NIST.SP.800-63b.

Habib, Hana, Jessica Colnago, and William Melicher et al. 2017. "Password Creation in the Presence of Blacklists." www.ndsssymposium.org/wpcontent/uploads/2017/09/usec2017_01_3_Habib_paper.pdf.

Habib, Hana, Pardis Emami Naeini, and Lorrie Faith Cranor et al. 2018. "User Behaviors and Attitudes under Password Expiration Policies." (August): 13–30. www.usenix.org/conference/soups2018/presentation/habib-password.

Haenschen, Katherine, Brett Frischmann, and Paul Ellenbogen. 2021. "Manipulating Facebook's Notification System to Provide Evidence of Techno-Social Engineering." *Social Science Computer Review* 40 (6): 1–18.

Herley, Cormac. 2009. "So Long, and No Thanks for the Externalities: The Rational Rejection of Security Advice by Users." (April). www.microsoft.com/en-us/research/publication/so-long-and-no-thanks-for-the-externalities-the-rational-rejection-of-security-advice-by-users/.

Herley, Cormac, and Paul van Oorschot. 2012. "A Research Agenda Acknowledging the Persistence of Passwords." *IEEE Security & Privacy Magazine* (January). http://research.microsoft.com/apps/pubs/default.aspx?id = 154077.

"How Secure Is My Password?" Accessed October 30, 2022. https://nordpass.com/secure-password/.

Ion, Iulia, Rob Reeder, and Sunny Consolvo. 2015. "('No) One Can Hack My (Mind'): Comparing Expert and (Non-Expert) Security Practices." (July): 327–346. www.usenix.org/conference/soups2015/proceedings/presentation/ion.

Khern-am-nuai, Warut, Matthew J. Hashim, Alain Pinsonneault, Weining Yang, and Ninghui Li. 2022. "Augmenting Password Strength Meter Design Using the Elaboration Likelihood Model: Evidence from Randomized Experiments." *Information Systems Research* (March). https://doi.org/10.1287/isre.2022.1125.

Komanduri, Saranga, Richard Shay, and Serge Egelman et al. 2011. "Of Passwords and People: Measuring the Effect of Password-Composition Policies." In *Proceedings of the SIGCHI Conference on Human Factors in Computing Systems*: 2595–2604. ACM. www.ece.cmu.edu/~lbauer/papers/2011/chi2011-passwords.pdf.

Krombholz, Katharina, Wilfried Mayer, Martin Schmiedecker, and Edgar Weippl. 2017. "'I Have No Idea What I'm Doing": On the Usability of Deploying HTTPS." In *Proceedings of the 26th USENIX Conference on Security Symposium (SEC.17)*. USENIX Association, USA: 1339–1356.

Lee, Kevin, Sten Sjöberg, and Arvind Narayanan. 2022. "Password Policies of Most Top Websites Fail to Follow Best Practices." (August): 561–580. www.usenix.org/conference/soups2022/presentation/lee.

Mazurek, Michelle L., Saranga Komanduri, and Blase Ur et al. 2013. "Measuring Password Guessability for an Entire University." In *Proceedings of the 2013 ACM SIGSAC Conference on Computer & Communications Security*, (November 2013): 173–186. CCS '13. Association for Computing Machinery, New York. https://doi.org/10.1145/2508859.2516726.

McMillan, Robert. 2017. "The Man Who Wrote Those Password Rules Has a New Tip: N3v$r M1^d!" *Wall Street Journal*, 7 August, sec. Page One. www.wsj.com/articles/the-man-who-wrote-those-password-rules-has-a-new-tip-n3v-r-m1-d-1502124118.

Morris, Robert, and Ken Thompson. 1979. "Password Security: A Case History." *Communications of the ACM* 22 (11) (1 November): 594–597. https://doi.org/10.1145/359168.359172.

Naiakshina, Alena, Anastasia Danilova, Eva Gerlitz, and Matthew Smith. 2020. "On Conducting Security Developer Studies with CS Students: Examining a Password-Storage Study with CS Students, Freelancers, and Company Developers." In *Proceedings of the 2020 CHI Conference on Human Factors in Computing Systems (CHI '20)*. Association for Computing Machinery, New York, 1–13. https://doi.org/10.1145/3313831.3376791.

Naiakshina, Alena, Anastasia Danilova, Christian Tiefenau, Marco Herzog, Sergej Dechand, and Matthew Smith. 2017. "Why Do Developers Get Password Storage Wrong? A Qualitative Usability Study." In *Proceedings of the 2017 ACM SIGSAC Conference on Computer and Communications Security (CCS '7)*. Association for Computing Machinery, New York. 311–328. https://doi.org/10.1145/3133956.3134082.

Ostrom, Elinor. (2005). *Understanding Institutional Diversity*. Princeton, NJ: Princeton University Press.

"Password Meter – A Visual Assessment of Password Strengths and Weaknesses." Accessed October 30, 2022. www.uic.edu/apps/strong-password/.

Paul Eve, Martin. 2016. *Password*. 1st ed. Object Lessons. Bloomsbury Publishing. www.bloomsbury.com/us/password-9781501314889/.

SearchSecurity. 2022. "What Is Cybersecurity? Everything You Need to Know | TechTarget." Accessed October 30, 2022. www.techtarget.com/searchsecurity/definition/cybersecurity.

Spafford, Gene. 2006. "Security Myths and Passwords." April 19. www.cerias.purdue.edu/.

Tan, Joshua, Lujo Bauer, Nicolas Christin, and Lorrie Faith Cranor. 2020. "Practical Recommendations for Stronger, More Usable Passwords Combining Minimum-Strength, Minimum-Length, and Blocklist Requirements." In *Proceedings of the 2020 ACM SIGSAC Conference on Computer and Communications Security* (October):1407–1426. Association for Computing Machinery, New York. https://doi.org/10.1145/3372297.3417882.

""Top 200 Most Common Password List 2021." Accessed October 30, 2022. https://nordpass.com/most-common-passwords-list/.

Ur, Blase, Jonathan Bees, Sean M. Segreti, Fumiko Noma, and Jonathan Bees. 2016a. "Do Users' Perceptions of Password Security Match Reality?" *Proceedings of the 2016 CHI Conference on Human Factors in Computing Systems*. ACM: 3748–3760. *doi.org (Crossref)*, https://doi.org/10.1145/2858036.2858546.

Ur, Blase, Fumiko Noma, and Jonathan Bees. 2016b. "I Added '!' at the End to Make It Secure: Observing Password Creation in the Lab." *Proceedings of the Eleventh*

USENIX Conference on Usable Privacy and Security, USENIX Association. (May): 123–140.

Ur, Blase, Patrick Gage Kelley, and Timothy Passaro et al. 2012. "How Does Your Password Measure Up? The Effect of Strength Meters on Password Creation." 65–80. www.usenix.org/conference/usenixsecurity12/technical-sessions/presentation/ur.

Weizenbaum, Joseph. 1976. *Computer Power and Human Reason: From Judgment to Calculation*. New York: W. H. Freeman.

Zhang, Yinqian, Fabian Monrose, and Michael K. Reiter. 2010. "The Security of Modern Password Expiration: An Algorithmic Framework and Empirical Analysis." In *Proceedings of the 17th ACM Conference on Computer and Communications Security* (October): 176–186. CCS '10. Association for Computing Machinery, New York. https://doi.org/10.1145/1866307.1866328.

Zhang-Kennedy, Leah, Sonia Chiasson, and Paul van Oorschot. 2016. "Revisiting Password Rules: Facilitating Human Management of Passwords." In *2016 APWG Symposium on Electronic Crime Research (ECrime)* (June): 1–10. https://doi.org/10.1109/ECRIME.2016.7487945.

6

Hacks, Fakes, and Hot Takes

Moderating "Bad Actors" on Google Maps Local Guides Platform

Rebecca Noone and Aparajita Bhandari

In 2019, Google Maps had a "business problem" (Lee 2018). Rob Copeland and Katherine Bindley (2019) of *The Wall Street Journal* reported millions of "fake businesses" appearing on Google Maps and filtering into local searches. According to their exposé, Google Maps users were being duped when enlisting the services of fraudulent businesses appearing on the map. Moreover, Google Maps was in part to blame since it had effectively brokered these scams. These businesses appeared real by virtue of being on the map, a map that, according to Copeland and Bindley "shapes what's real and what isn't for more than two billion monthly users" (Copeland & Bindley 2019).

Google's promise to organize the world's information and make it universally understandable and accessible (Google nd.a, 2022), a promise that includes spatial information through Google Maps (Graham and Dittus 2022) has been challenged by one of its main selling points: a democratic map where individuals can be part of mapping processes (McQuire 2019). Google Maps has long allowed for crowd-sourced contributions to its maps from systems like the now-defunct MapMaker, to GoogleMyBusiness, to Google MyMaps, to its most recent crowdsourcing endeavor, the Local Guides Platform. Not only did Google Maps know where you were, placing you always at the center of its global map (that blue pulsing orb of location awareness), but it also allowed for you to annotate its information, suggesting places to add, making sure your business profile was visible on the map, giving you a chance to create your own personalized map, and prompting you to rate places and add reviews to the map. Through these features, Google Maps presents as a democratic intervention into age old practices of cartography, but who was vetting all this information being added to the map?

Soon after the *Wall Street Journal*'s reporting gained traction, Google responded with a declaration of its commitment to "fighting fake business profiles and

Rebecca Noone, Lecturer, Information Studies, School of Humanities, University of Glasgow; Aparajita Bhandari, Assistant Professor, Department of English Language and Literature, University of Waterloo (Canada).

FIGURE 6.1 Visual themes from hacks, fakes, and hot takes: moderating “bad actors” on Google Maps Local Guides Platform.

fraudsters on the map" (Russell 2019) shared through their official blog *The Keyword*. Here Google outlined their processes of content moderation on the mapping platform. As is now standard in social media platform moderation, Google's approach to fighting "fake business profiles and fraudsters on the map" includes a combination of machine learning and human monitoring practices. Machine learning by way of a filtering software is what Ian Leader (2022), Google's Project Manager for User Content calls, "the first line of defense." Then, overseeing and augmenting the technical side of the monitoring platform are manual moderation practices that draw on both "skilled" and "voluntary" human labor.

Google Maps' manual process has two tiers. First, there are the "highly skilled human moderators" who, according to Leader, work "around the clock" to check for policy violators. Scholars like Sarah T. Roberts (2019) and Mary Gray and Siddharth Suri (2019) have shown that the working conditions of these content moderators are often precarious, under-supported, and reliant on the global economy of low-wage outsourcing. Second, Google Maps enlists contributors from their Local Guides Platform to help with the content monitoring processes. Through the Local Guides Platform, Google calls upon a volunteer group of "local experts" that moderate maps content including checking business information, adding missing places to the map, and monitoring reviews.[1] This chapter focuses on the second tier – the complicated process of monitoring crowdsourced content through crowdsourced moderation.

Google's platformization of mapping, as noted by Scott McQuire (2019), is a process by which Google Maps becomes "embedded" as the "foundational resource" for spatial information in a digital context. Powerful mapping platforms like Google Maps enact participatory protocols of locating and reporting misinformation to protect their claims to "authentic and reliable" information (wording from Leader 2022). In this chapter we present the results of a critical discourse analysis (Fairclough 2013) into the platform's product information, promotional materials, interface design, and corporate development strategies used by the Local Guide's Platform. We argue that this volunteer labor plays an important role beyond monitoring and managing mapped information – Local Guides Platform produces and enforces a narrative that misinformation on the map is the result of outsiders or "bad actors" who infiltrate an otherwise congenial "information ecosystem" of helpful place-based information (Kuo and Marwick 2021), thereby exonerating Google and Google Maps from their own mapping errors.

LOCAL GUIDES: THE WATCHFUL EYES OF GOOGLE MAPS

Participatory forms of content moderation have long been part of online communities such as reddit groups (Squirrell 2019), tumblr discourse communities

[1] To read more on Google Maps reviews from the perspective of Google, see Ian Leader (2022), Group Product Manager, User Generated Content, article "How Reviews on Google Maps Work: https://blog.google/products/maps/how-google-maps-reviews-work/.

FIGURE 6.2 A screenshot of Google Maps desktop with a focus on the UK-based department store Marks & Spencer.
*See the average rating after five stars under the Marks & Spencer title in the left-hand column and the hyperlink to click on the 1258 reviews.

(Tiidenberg, Hendry and Abidin, 2021), and more recently Discord (Kiene et al. 2019). Local Guides presents an interesting case study for content moderation due to its design and the function it serves as an appendaged community for the Google Maps Platform. This community not only regulates each other, and the place-based content Local Guides upload and share, but also regulates Google Maps content broadly – information that has become entwined in everyday navigation.

You may have come across Local Guides when using Google Maps. Their reviews are embedded in the sidebar of information that pops up when looking up a restaurant or a store (see Figure 6.2). You'll see their username and profile picture beside a rating, out of five stars, and their 200-word reviews (see Figure 6.3). Local Guides is a participatory program run through the Google Maps Platform. The Local Guides Platform enables users to interact with Google Maps through actions such as adding and editing reviews. Users are content producers and content moderators. While the Local Guides Platform can be accessed on Google Maps directly, to both read and add content, Local Guides also functions as a separate application or webpage where Local Guides users can add information to their personal Local Guides profiles and connect with other Local Guides through "following" them and reading their reviews.

Local Guides is a platform through which Google Maps users can engage with and annotate the location-based information they find on Google Maps. While Local Guides represent between only 5 and 10 percent of Google Maps users – as of

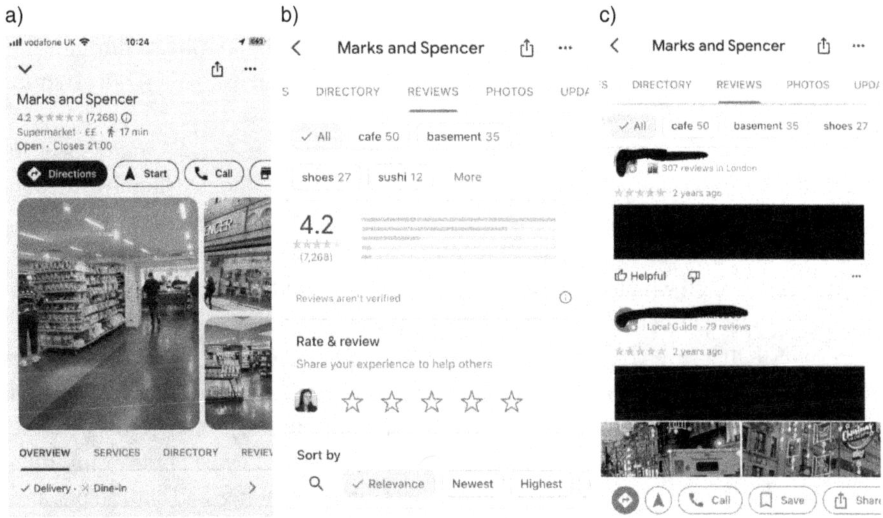

FIGURE 6.3 Three mobile screen shots of Local Guides Platform.
*From left to right, the main review interface with the average rating (4.2 out of 5) and the number of reviews (7,268), then scrolling down there is the types of topics covered by reviews (cafe, basement sushi), then the invitation to leave a review, then there are specific reviews by Local Guides and contributors.

2021 there were 150 million Local Guides in 24,000 cities, and collectively – they have contributed approximately 70 percent of reviews, photos, and other types of user-generated content on Google Maps (Reece 2021). Not all reviews on the map come from official Local Guides (presumably the other 30 percent of reviews); the platform provides the basis of review infrastructure, where even without an account, people can leave reviews of local places which are then visible on the maps.

Google promotes the Local Guides platform as an arena of conviviality where Local Guides play a supportive role on and off the map for their community (Bhandari and Noone, 2023) while at the same time monitoring and maintaining that which is happening locally. According to Google, Local Guide reviews "are a treasure trove of local knowledge that can point you to the places and businesses you'll enjoy most – whether it's a bakery with the best gluten-free cupcake or a nearby restaurant with live music" (Leader 2022). In this role, Local Guides are positioned as on the inside, proximate to Google Maps' grand vision of making the whole world mappable, or what Scott McQuire (2019) muses as "one map to rule them all." (For more on Google's mapping ambitions see Graham and Dittus 2022; McQuire 2019; Noone 2024; Plantin 2018; Wilkin 2019.) Google Maps, as such, is committed to adding more and more data to its maps through processes of data extraction, data processing, and crowdsourced data creation (McQuire 2019). Local Guides Platform thus enacts both a performative and a structural role in Google's spatial information database as maintainers of the map while setting the parameters of who is part of the Google Maps community and who is on the outside (Plantin 2018). Mapping is itself a power-laden process (Massey 1994) amplified through the Google Maps lens of what is important about space and what can be left out (Luque-Ayala and Neves Maia 2019).

We investigate the role crowdsourcing platforms such as Local Guides play in managing everyday online misinformation as it appears on Google Maps, while also playing a role in conditioning what misinformation on Google Maps looks like. Platforms like Google Maps also often leverage its crowdsourcing infrastructures – the systems through which users can add, share, modify, and comment on content – as testimony to its democratic form of information sharing. The hallmark of this online democracy is that "everyone" has a voice. But this supposed openness of the platform is not without its challenges, namely its vulnerability to misinformation that can be added by anyone wanting to do so (remember the fake businesses?). As a result of these challenges, platforms develop a content moderation system to weed out misinformation. Bringing these platform logic together – open crowdsourcing, bad faith actors, and moderation of content – this chapter calls attention to how popular conceptions of "misinformation" are shaped by this platform dynamic which ultimately serve to insulate "misinformation" from the structural question of *whose misinformation*?

Importantly, some of the most pervasive forms of errors on Google Maps are not the product of crowdsourcing but from Google's own top-down interpretation of spatial data. For example, Google Maps has been accused of leaving out and erasing local information often at the expense of marginalized communities. For example, the residents of the Fruit Belt Neighborhood in Buffalo, New York – a redlined neighborhood with a predominantly racialized population – has been fighting Google since 2008 because of the erasure of their neighborhood name on Google Maps, replaced by the name of the proximate The Medical Park development project in Buffalo (Dewey 2019; Noone 2024). Similarly, Google Maps removed the place names of Rincon Hill and South Beach in tech-industry dominated San Francisco, CA, replacing them with The East Cut, effectively erasing a historically low-income area from the map in favor of wealthy property development (Nicas 2018). The erasure of The Fruit Belt, Rincon Hill, and South Beach illustrates the larger structural dynamics of power perpetuated by Google's mapping apparatus and illustrates Google Maps priorities of commercial spaces and investment (Luque-Ayala and Neves Maia 2019). Moreover, when we zoom out on Google Maps, we see how the platform unevenly distributes the effects of errors, often targeting low-income and racialized populations like the Fruit Belt or Rincon Hill residents as erasable, and middle-class white suburban people who have mistakenly hired a fraudulent plumber (such as in the *Wall Street Journal* reporting), as the target population that needs protecting. These distinctive stakes of misinformation on Google Maps illustrate the importance and urgency of parsing what misinformation means in an online context in order to avoid perpetuating a status quo understanding of an information commons.

We argue that Google Maps situates misinformation as a glitch (building on the work of Benjamin 2019; Kuo and Marwick 2021; Nakamura 2013) or a technical problem that can be fixed with a mix of software and manual information coding. As such, Google tasks everyday users with the role of keeping "the information on Google relevant and accurate" and "keeping reviews authentic and reliable" (Leader 2022). The threat of misinformation is always framed through the lens of external bad actors while the Local Guide insiders keep this misinformation out. A global mapping project like Google Maps which leverages "objectivity" and "truth" must contend with the fact that forms of representation and knowledge production are always political (e.g., Graham and Dittus 2022; Massey 1994). Google's conception of misinformation on the map perpetuates a division of "with us or against us" in the process of map-making that prioritizes the commercial and the commoditizable (Noble 2018; Plantin 2018; Wilken 2019).

PLATFORM GOVERNANCE, CONTENT MODERATION, AND MISINFORMATION

Our case study of Local Guides brings together literature on content moderation (Gerrard and Thornham 2020; Gillespie 2018; Roberts 2019; Squirrel 2019) and literature on critical misinformation studies (Kuo and Marwick 2021; Nakamura 2013) to think about how constructions of insider authenticity vs. outsider bad actors are made and remade, enforced and reinforced through participatory efforts of content moderation. Then we considered the role misinformation plays in the construction of online places. Through an analysis of the governance documents, both regarding platform governance and community governance, we look at how Google Maps not only frames work and participation but also how they frame misinformation and accurate portrayals of place.

Beyond the digital, Rianka Singh (2018) reminds us that platforms have always been political technologies that negotiate spatial relationships to power, surfacing and elevating the views of some over the views of others, which continues into the digital context (Gillespie 2018). Critical analysis of social platform moderation processes has focused on biases of both the software-driven moderation practices, sometimes called machine learning solutions (Gerrard and Thornham 2020) as well as the moderations practices that rely on human labor (Carmi 2019), often operating in poor working conditions behind the scenes (Roberts 2019).

Platforms like Google Maps, and by extension Local Guides Platform, are an assemblage of governance tactics including laws, industry priorities, and the "owners' vision" of the purpose this platform serves. The structures and practices of platform governance help to shape what, how, and why certain types of information are included and supported by a particular platform while other forms are not (Gerrard and Thornham 2020). Processes of platform governance are not static but are iterative processes (Bucher 2018), often adapting to changing norms, new standards, and new contexts online (Gillespie 2015, 2018). Platform governance can be the terms and conditions which regulate the rules of the platform (Gillespie 2017; Gorwa 2019). Drawing from Katrin Tiidenberg, Natalie Ann Hendry, and Crystal Abidin's (2021), platform governance takes place through terms of service and explicit rules, but also through the design and functionality of the platform. These include the default settings of a platform, the functionality of interface, and the onboarding and registering processes (Bucher 2018; Tiidenberg, Hendry, and Abidin 2021). Therefore, platform governance is multifaceted – more than the formalized rules, platforms guide and condition participation implicitly and explicitly.

Applying the Governing Knowledge Commons Framework (Frischmann, Madison, and Strandburg 2014), we consider how platform governance structures come to be constructed and we question whose interests they may serve. Online platforms such as Google Maps coopt the language of the commons based on its facilitation of a network of community management and governance. And while a commons may emerge because of the construction of community, there is still a pull of top-down governance structure that serves the interests of the platform, in this case Google Maps broadly. This is in stark contrast to a governing structure grounded in the commons that serves the interests of the community, however fairly or unfairly that is distributed (Frischmann, Madison, and Strandburg 2014).

Gerrard and Thornham (2020) suggest Community Guidelines are an effective starting place to understand the platform's priorities since these are the public-facing and community-centered rules that structure participation and engagement. These are differentiated from the terms and conditions which are often crafted through legalese and more often set conditions on the platform's liability and positions their ownership. Community Guidelines, as Gerrard and Thornham identify, "lay out a given platforms' codes of conduct and nod to the political, economic, and social considerations of a given social media company" (pp. 1268). Moreover, community guidelines are more explicit than the platform's default settings of the platform or the interface designs that are often normalized through use. Community Guidelines in the context of use-based moderation practices (Squirrell 2019) provide a place of reference for users to turn to when looking to flag content.

But more than simply acting as a guide, Community Guidelines reveal the biases, politics, and normativities (Gerrard and Thornham 2020; Gillespie 2017) of the platforms. For example, what is classified as misinformation to begin with, what passes review as truth and what is flagged as false is based on dominant paradigms of knowledge production. Roberts (2019) argues that such guidelines regarding content moderation are developed "in the specific and rarefied sociocultural context of educated, economically elite, politically libertarian, and racially monochromatic Silicon Valley, USA" (93–94). Therefore, these values influence what passes review and what gets flagged but can often work abstrusely for users especially if these values are not clearly indicated in the guidelines (Gillespie 2017). Furthermore, research has shown that these opaque systems result in a type of "folk theories" of platform moderation in action to give personal meaning to how moderation decisions are made and acted on (Myers-West 2018). In other words, these are the reasons users give to why and how their content is removed from a platform, often blaming human intervention, like "a friend got mad at me" or the fear of "being targeted" (Myers-West 2018).

Platforms like Local Guides fit into the conception of a knowledge commons, as they actively draw from its collective sensibility but ultimately enforce terms and conditions that are not communal or democratic – using imaginaries of the knowledge commons for their own benefit. For example, Duguay, Burgess, and Suzor (2020) have examined the "patchwork platform governance" that is unevenly deployed in the context of queer content. Often this results in informal care and moderation practices in relation to people on social media platforms who are targeted and harassed, disproportionately affecting racialized and trans users (Schoenebeck, Haimson, and Nakamura 2021). A previous study on Local Guide reviews in Toronto found that racialized workers of chain stores are disproportionately named, targeted, and vilified in Local Guides reviews (Bhandari and Noone forthcoming), finding that challenge Local Guide's claims on communities (are these workers not part of the community?). Thus, the community guidelines and their implicit value systems construct a polarity of insiders and outsiders based on what and who is protected by platform regulation and who is not.

METHODOLOGY

While participatory processes online, including community forums, have always been moderated (see Gerrard and Thornham 2020), some platform governance practices like Local Guides structure what counts as "community" while drawing on the affective labor of users in the name of caring for the platform (Tarr and Alvarez León 2019). Rather than neutral processes, we look at how Google applies the language of "authenticity," "reliability," and "accuracy" as core values, while simultaneously calling on Local Guides to participate in maintaining these values. As such, we approach authenticity, reliability, and accuracy not as stable concepts but take a social constructivist approach to thinking how Google's language shapes these concepts and as such the role of Local Guides. This chapter offers a critical discourse analysis (Fairclough 2013) of these registers of Google Maps' promises as they relate to the contributions of Google Maps. We examine the formal and informal governance structures of the platform including the terms of use, the community terms of service, the press releases about the platform, and the content shared through the Local Guides connect platform. Using Kuo and Marwick's (2021) critique of the dominant disinformation discourses – as upholding systems of dominant knowledge – we consider how operationalizing location and place as that which can be knowable, consumable, transactional, and claimable already determines what relationships to space are possible and forecloses other relations to space. We apply Kuo and

Marwick's intervention to the disinformation discourse to our analysis of Google Maps and Local Guides Platforms to identify how dominant misinformation paradigms do not serve everyday users of the platform but rather serve to maintain a platform's authority. The findings are presented as a conceptual analysis of Local Guides grounded in a critical evaluation of the platform's discourse.

Additionally, Plantin's political economic analysis of Google Maps' crowdsourcing infrastructures sheds light on the seemingly contradictory simultaneous processes of decentralization and recentralization on the Google Maps platform. Plantin argues that the information collected through Google Maps' process of decentralization (in this case through crowdsourcing place-based information through Local Guides) are designed to service Google's market interest of dominating the highly competitive geospatial market. Therefore, Google's very construction of a Google Maps' common through the Local Guides platform recentralizes crowdsourced geospatial information in the name of serving Google's interests and not the commons (Plantin 2018). We identify through this recentralization process that Google Maps' platform regulates entry into the Local Guides community through a construction of "insider" vs. "outsider" knowledge (Kuo and Marwick 2021; McMillan Cottom 2020; Nakamura 2013; Washington and Kuo 2020). While the platform is an assemblage of users, the community guidelines enforce enclosures (Plantin 2018) that set limits on participation within the platform. Echoing Kate McDowell's chapter on Storytelling as Misinformation, we see how such enclosures are not insulated from the world but actively work to maintain the status quo.

Framing our analysis through this critique of the dominant misinformation paradigm – one championed by Kuo and Marwick (2021), McMillan Cottom (2020), Schoenebeck, Haimson, and Nakamura (2021), and Nakamura (2013), among others, who all in their work have shown how the dominant framework of misinformation is often framed on hegemonic ideals of truth and authenticity in systems designed to serve patriarchal and imperial systems – we argue that review and moderation practices on Local Guides further standardize constructions of misinformation as the product of a small group of outliers or "bad actors" in an otherwise convivial information ecosystem. Instead, the platform's governance of crowdsourced moderation, paired with the Google Maps' project of creating "one map to rule them all" (McQuire 2019) help to homogenize narratives that all territory is neutral and unpolitical (Massey 1994) that then reify the limited scope of Google's misinformation paradigm. Consequently, as Google Maps commoditizes relations to space (see Graham and Dittus 2022; McQuire 2019), it also forecloses what is considered trustworthy and useful content for the imagined Google Maps users, set against the fake or fraudulent information that deceives people. Thus, the question of misinformation is a question of not just strategies of keeping out information, but also a question of who gets to decide what misinformation about place is and how is this not only managed through the platform but becomes how we participate in knowledge creation on the platform.

LOCAL GUIDES AS DEMARCATIONS OF "THE INSIDER"

Google Maps' promotional material reveals a deliberate creation of a "community" on Google Maps through the Local Guides Platform that emphasizes sharing information about one's local space for the benefit of Google Maps users. This space is highly scaffolded and governed by the design of the platform evidenced through how the platform appears on Google Maps, the Community Policy about input, the onboarding processes of becoming a Local Guide, the incentive structure to maintain Local Guide participation, and the networking of review through the message board. These elements of design and communication effectively enclose what participation looks like on the platform (Plantin 2018) and condition participation as always in relation to Google Maps' project of platforming space (Graham and Dittus 2022). The platform thus structures what community membership looks like and what a Google Maps "insider" is – someone who participates in this convivial space of sharing location-based information.

Design and Guidelines

Local Guides Platform's Community Policy discursively frames the role Local Guides serves for Google Maps. Rather than punitive, this policy defines the function of Local Guides through celebratory language, praising the Local Guides platform as an online space that "brings Google Maps to life" where Local Guides "make it easier, tastier, and more fun to navigate the world" and "toward a better understanding of the places around them" (Google ndb, 2022). The Local Guides image is one of a user fastidiously "gathering facts" to bring their "on-the-ground expertise and a commitment to sharing everyday experiences that inform real-time decisions across the globe." To achieve this vision of what Local Guides is imagined to be, the policy recommends Local Guides generate "consistent contributions that express these values" and to respect a "community that takes the mission seriously." The Community Guidelines project a convivial online community who are there to both help the map and to help the map user.

Onboarding and Retention

During the onboarding process of joining Local Guides, which includes registration with the platform and links to relevant resources, Google shares the skills associated with reviewing places including photography skills, writing skills, and what to flag and edit. These are in the name of helping Local Guides produce "authentic" content. This includes instructions on how to take a photo of food to show it off in the best light, to how to write a review, to how to select a place to review, and how to take a "shelfie" of a store's interiors. These are offered in a YouTube video series titled "Quick Start Guide," the instructions embedded in the platform itself, the

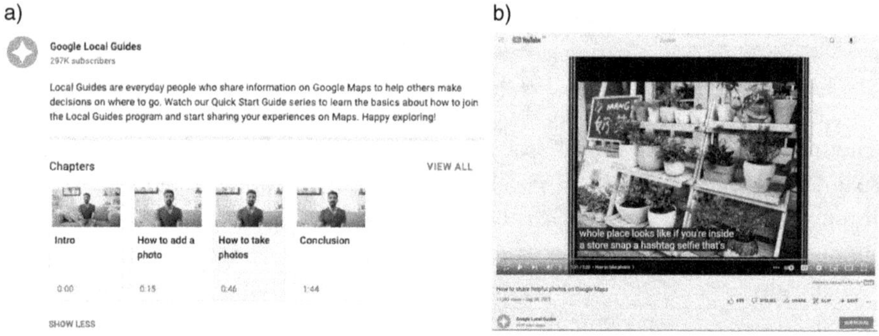

FIGURE 6.4 Screenshots of Local Guides YouTube channel, with the video* how to add a photo.
* Chapters of this instructive video include "how to add a photo" and "how to take a photo" and provide specific advice such as include the "hashtag shelfie."

blog posts by Google staff members found in The Keyword, and crowdsourced information sharing on Local Guides Connect. While engaging with these resources is not a mandatory part of engaging with the platform, nonetheless these didactic elements of participation are woven through the onboarding process of joining Local Guides (Figure 6.4).

In one such blog post titled Seven kinds of Local Guides You "might spot on Google Maps," Local Guides Team members Max Kaplan and Alex Vallis (2018), break down this Local Guides skillset into seven Local Guides archetypes that typify "specific passions and ways of sharing" and signal the type of content that Google wants for its map. It begins with the prompt "what kind [of Local Guide] are you?" The seven specialist skills they spot are the "visualist" who adds photos, the "fact hunter" who edits reviews, the "trailblazer" who adds new places to the map, "the sage" who recommends the best places to go by way of 5-star reviews, the "multi-media guru" who adds video to the map, the "connector" who hosts meet-ups for Local Guides, and "the advocate" who adds information about accessibility (Kaplan and Vallis 2018). The archetypes include tips for the fact hunter such as "uncover missing info to see which places near you need expertise" and for the trailblazer, "check out restaurants and local shops opening this year so you can add their first photos and get those views" (Kaplan and Vallis 2018). These classifications prioritize skills including technical savvy with photo and video and delineate claims to specialized knowledge embedded in the act of adding new places.

The categories are revealing of the type of information they want in the map – the pictorial, the latest, and compliance with (as they frame it) wheelchair accessibility, and designate roles to members of the community with a description, their "loves," and a "tip" on how to become such an expert. So while one of the expert types of knowledge is the "advocate" who is part of marking which places do and don't have stair access (and how this is verified remains unclear), access to space and

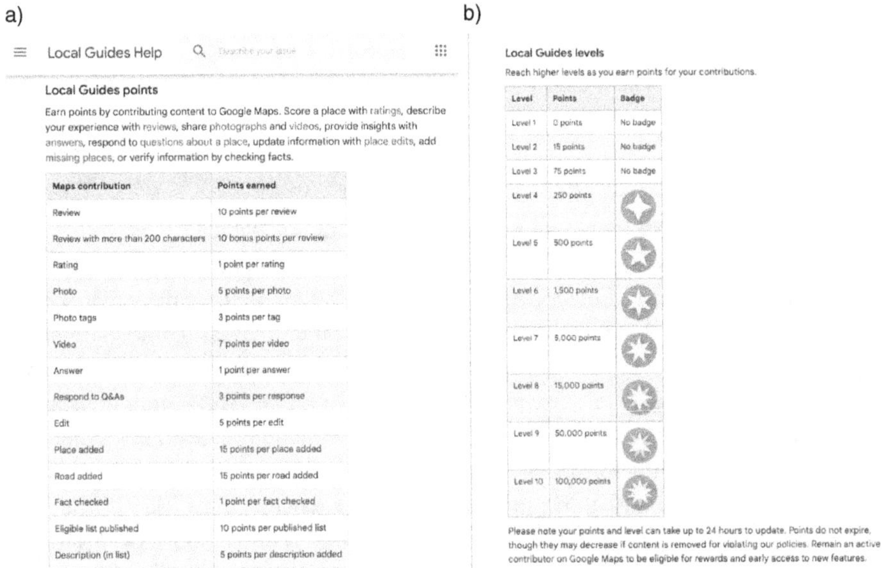

FIGURE 6.5 Google's official points system (a) and Local Guides levels (b).

claims on space are imagined as an experience of the "hottest restaurant," "where to find an ATM," or can help you "choose where to stay by the sea." The Local Guide's access to cultural capital (Bourdieu 1986), or cultural knowledge that awards Local Guides social status as "local expert" or "discerning food critic," is explicit, and uploading location-based information based on this cultural capital is a foundational part of what Local Guides "share." This cultural capital enables access to the Local Guides community.

As much as they are part of the Google Maps information infrastructure, Local Guides are not paid. Instead, their participation is incentivized through a points-based system that then allows them to access badges or levels. This system maintains platform membership after the onboarding process. There are ten levels or badges in total; the top level requires 100,000 points (see Figure 6.5). A review of 200 words receives 10 points. Rating a restaurant is 1 point. Adding a new place to the map is 15 points and fact checking and answering questions is 1 point (Google ndc, 2022). Upon reaching Level 4, Local Guides unlock their first "badge" (Google ndd, 2022), but when Local Guides was first developing, Level 4 Local Guides were rewarded 1 TB of Google storage (McCormick 2015). This has since been removed and replaced by a seemingly mercurial rewards system that is the subject of much speculation and information sharing among Local Guides on the Local Guides Connect Message Board (AriMar 2022). Local Guides Level 3 and above are encouraged to organize their own Local Guides meetup (Google nde, 2022). Top ranking Local Guides can become a Local Guides Connect moderator based on their contributions to the Local Guides connect platform (TraciC 2016). However, this proximity to Google is managed by Google. Community membership is measured through Local Guide's incentive program.

Networking and Connecting

Local Guides is also Google Maps' social infrastructure where users and Local Guides can follow each other, seeing all the places they have posted about around the world. In this sense, Local Guides combine locative media with social media logic. They have profiles, followers, follow other Local Guides, and upload content. Each profile contains a small blurb about the guide and lists the places they have reviewed and how many followers they have. Local Guides are a public persona – they can be followed and follow other Local Guides like the networking structures of other social media platforms. Like these platforms, Local Guides share how many people engage with the platform including how many people have viewed the post (Wells 2019). Therefore, the Local Guides Platform transforms Google Maps into a social media-style networking site where the content is location-based and funneled through the language of personal experiences but focused on leisure and mobility.

The feature to "follow" Local Guides launched in 2019 demonstrates an increased socialization of the platform (Wells 2019). These affordances offer visibility not just to those who use the map but also offer visibility and proximity to Google itself. Here we turn to a post by Scott Francis (2019) in DigitalMaas, a Local Marketing Platform that helps businesses manage their location information on sites such as Google, Facebook, and Apple, titled Become a Google Local Guide in 2020.

> Ever wanted to work for Google? It's the dream job for a lot of people. Sadly, for mere mortals like many of us, it's out of reach, generally reserved only for those with a special IQ and the ability to innovate and code up some life-changing tech. But there is a way to put that big G on your resume. Enter the Local Guide program. They're always hiring, and you're guaranteed to get the job! (Francis 2019)

Of course, this is just a single blog post, but this introduces us to the possibility of what else is going on in Google's often opaque infrastructure of information expertise. Google Maps extracts the labor of "authenticity" and realness and compensates in what Brooke Erin Duffy (2017) terms the "future reward system from present day productive activities" (7). Here the reward is proximity to Google achieved through community membership to Local Guides and visibility to Google through being a productive member of the Local Guides Platform. While this study did not cover the ways that Local Guides receive benefits from the places they review, previous work on Yelp (Luca and Zervas 2016) has shown that reviewers can be given special service, pay, or freebies, for their review. Moreover, business owners, and their family and friends can set up profiles that give the business glowing reviews. As such, we see how the internet review infrastructure is a fallible system that can be "gamed" or bribed.

One active part of this Local Guides community is their online message board, Local Guides Connect. This is an officially sanctioned message board designed by Google for Local reviewers, launched in 2016. Since then, Google reports that over 1 million Local Guides use Connect, collectively making hundreds of thousands of visits every month, making over 1.1 million comments, 5.6 million likes, and 22,000 comments marked as "solutions." Some Local Guides have risen the ranks to become message board moderators. While sharing similar conventions, Connect is different from other unofficial spaces such as the Local Guides Subreddit where Local Guides will seek guidance outside the Local Guides Connect page (albeit likely monitored by Google) and where redditors can use a different avatar when posing their questions. Here they ask questions about the reviewing process in the search of answers and responses that are not directly moderated by official Google Moderators and therefore not subject to Google's terms of service. For example, a Local Guide might ask why their review is not visible on Google Maps, or where their Local Guides tote bag is, the promised reward for reaching a certain level. Local Guides Connect on the other hand is intended to be a space to be seen by Google, to make one's work visible to the people who work at Google, in the hopes of being featured in Google press releases or invited to the annual Local Guides Connect conference at the Google Maps Headquarters in Mountain View California (Reece 2021).

In celebration of Local Guides Connect's five-year anniversary in July 2022, Google Moderators shared why Connect is important for them (TraciC 2022). Googler TraciC's post to the message board reflecting on the five-year history of Connect prompted other Local Guides to share their "favorite Connect memories." These included responses from Local Guides about how appreciative they are of the community, with comments such as: "for me, Connect is like a melting pot. It's not just improving Google Maps or contributing to the community, it's like making family away from your family. Connect helped me learn and improve myself in many ways and I made a wonderful family out here who are more than just friends to me." This Connect Moderator calls attention to ideas of family whilst doing the work of "improving Google Maps." Another moderator comments that this is a group that is "passionate about using and improving Maps" as well as a "colorful and vibrant community of friendly, helpful Local Guides from all over the globe." These moderators asked Local Guides "what's one of your favorite Connect moments since you joined?" garnering responses such as "connecting to Connect and help billions of people" noting again the idea of forging a team to help the world with place-based information. This idea of "connecting" with people all over the world, "helping" people all over the world, and making Google Maps "better and better" is repeated throughout the comments. They share that this is the "agora" or the "public space" for Local Guides.

Google Moderators suggest topics for discussion such as "easy tips for rating venues and writing structure reviews" as it relates to experience, ambience, cost, and tips. These Moderators also initiate campaigns for the types of content Google Maps wants to see on the map. For example, since 2020, we have tracked campaigns to support Women-Led Businesses and Asian-owned businesses where Local Guides are told to add these places to the map along with a badge that indicates these labels (Bhandari and Noone forthcoming; Reece 2021). Interestingly, Google Moderators on the platform are not available for direct messages asking for help from Local Guides "due the volume of private messages Google Moderators receive, I do not read or respond to private messages. Please post publicly so others may benefit from your discussion. If you require urgent assistance, please tag a Google Moderator. Thank you!" (TraciC 2016, 2022). Instead, Local Guides are asked to join contributor conversations and help other Local Guides with questions or share how they make helpful material. Indeed, those Local Guides who respond to these questions and help maintain an "active maps community" in this way are celebrated. This is another possible incentive for Local Guides.

Local Guides contribute Google Maps data both in ways that are visible and ways that are less easy to detect: they add business information such as hours, addresses, and phone numbers. They moderate other reviews and importantly report fake businesses. They also add place listings to the map that are otherwise missing. They refine, clean up, and build – maintaining Google Maps information and user experiences of that information, but also their place-based and location-based affordance of the map. We echo critical geographers Alexander Tarr and Luis F. Alvarez León (2019)'s description of Local Guides work called "digital piecework appealing to a sense of community" (92). Local Guides organizes their distributed network of local expertise through the logic of "insiders" and "community." This work of "watching the map" is driven by the platform's imperative to make maps "better," to protect the map from wrong information, maintain a general positive engagement within the platform, and to support "their local" communities off the map. Local Guide contributions, however motivated, are fed into Google's recentralized map providing more and more place-based information for Google to draw from. This is content moderation *and* database management where ordinary people are treated as Google "insiders" tasked with adding information but also upholding Google Maps' mission, authority, and values. But how does (so-called) misinformation filter into the map in the first place?

OUTSIDERS AS SOURCES OF MISINFORMATION

Alongside the construction of altruistic local experts who "help their communities" (Slabin 2017) and "share information on Google Maps to help others" (Kaplan and Vallis 2018), we noted both implicit and explicit positioning of misinformation on Google Maps as the product of outsiders (Leader 2022). In reviewing Google's literature about Local Guides found on the Google Maps official website as well as their official blog, The Keyword, where members of Google's management team share information about Google products, we tracked four types of classifications of misinformation that Local Guides are to look out for: The misinformation that is the result of unskilled contributors, the content vandal, the easily bribed, and the bad actor galvanized by political events. Threaded through these conceits is the assumption that "the outsider" is central to Google's characterization of misinformation on the map rather than produced by Google directly in their mapping project.

According to the Local Guides Platform, this "bad" information on Google Maps can sometimes be blurry photos or duplicate reviews. Blurry photos or uploading something twice does not meet the standards the platform establishes in its Community Guidelines. While not framed as malicious content but nevertheless unhelpful, blurry, or improperly rotated photos (i.e., on their side) are flagged by a Local Guides and removed by the moderators. The implicit reading of this form of content moderation is that no Local Guide would knowingly upload this content. Instead, this is outsider information since it does not reflect the skillsets of the Local Guides who are poised through careful onboarding processes to upload "clear and accurate" information.

Within these guidelines the community of Local Guides is expected to possess both digital literacy skills and technological savvy to take and upload a high-quality photo with optimal lighting, framing, and size. Local Guides must be able to make their profiles seem authentic and enticing, which demands proficiency not just with the platform but with auxiliary tools like cameras, and a clear and succinct writing style. Moreover, Local Guides are expected to be proficient in a language that makes place desirable through photographic, video, and text-based content. Therefore, the outsider who uploads this so-called unhelpful information is not proximate to the same cultural capital and platform know-how that the Local Guides insiders possess. While there are efforts to help people through discussion forums on Connect, this type of moderation and training becomes the responsibility of the "community" and a means to bring well-intentioned outsiders into the fold of the Local Guides Platform's community of insiders.

Beyond misinformation as "mistake," we also found that the platform frames misinformation on the map as the result of deliberate attacks on Google coming from outside Google Maps itself. Google marks this work as the product of "content vandals" who have infiltrated the crowdsourcing service with the intention to misinform (Pritchett 2021). According to Dan Pritchett, principal software engineer at Google Maps, these content vandals post "fake reviews" as well as "deliberately faked edits, copied or stolen photos, off-topic answers, defamatory language, personal attacks, and unnecessary or incorrect edits all in violation of our policy." Google frames this work as vandalism from the outside that marks the maps' pristine information environment (see Figure 6.6 for Local Guides' process on "How to Report a Review").

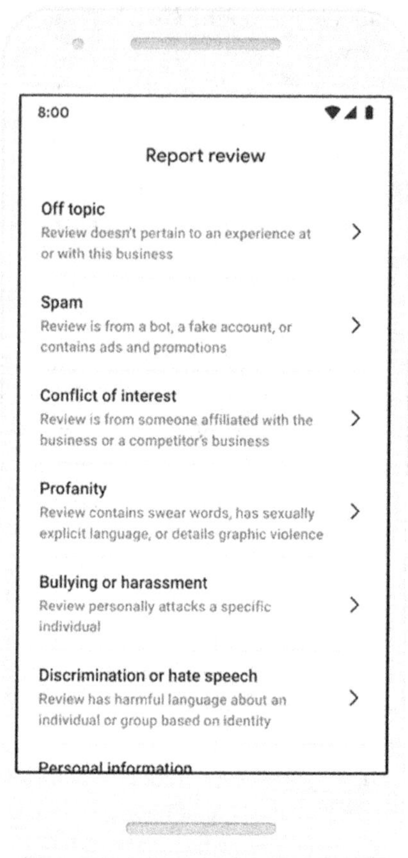

Google Maps users and businesses can easily report reviews that they feel violate one of our policies.

FIGURE 6.6 Screenshot of Google's guidance on how to report a review.
*The list of reasons why a report might be necessary including "off topic," "spam," "conflict of interest," "bullying," and "discrimination."

To further distance their map from this threat of misinformation, Google makes a clear distinction between the Local Guides adopting the language of public service like "help their community" and the content vandals who are, according to Google, "ultimately motivated by money." For Google Maps, these are bad actors who are "gaming the system" because they "try to trick people with scams like fake reviews to attract customers" or "fake listings to generate business leads" (Pritchett 2021). But here the Local Guides' role to "support local business" by putting the business on the map and attracting customers becomes difficult to distinguish from some of the work of content vandals. How do Local Guides moderators distinguish between the two types of mapping content in Google's polarizing framing? Moreover, what are the subtexts of a fake review that mark it as fake, not from the well-intentioned Local Guide who wants to help a local business. Here the limits of the role of community as a system designed to read space through commercial transactions becomes apparent. The Local Guides Platform's framing of content moderation is based on/perpetuates the framing of misinformation as an exterior threat that is infiltrating an otherwise healthy information system of Google (Kuo and Marwick 2021) and fails to account for how Google Maps' commercial priorities ultimately shape the types of participation that take place on the map, for better or worse (Plantin 2018). Therefore, we see how the knowledge commons is an action arena that is not available for all but only accessible to a few.

To situate misinformation, Google Maps frames misinformation as event based in which content vandals are "motivated by social and political events" or leave their mark online. So while content vandals and gamers of the system are often random and difficult to track, Google also marks misinformation as outside the system. Google states that "These patterns can be anything from a group of people leaving reviews on the same cluster of Business Profiles to a business or place receiving an unusually high number of 1- or 5-star reviews over a short period of time" (Leader 2022). Google's response to these issues is preemptive. These are restricted reviews for certain places where Google reports to have seen "higher rates of policy-violating content." Such places include schools in the US where teachers, students, and staff are targeted or polling stations in America during elections where misinformation about processes often take place (Leader 2022). Google's response is to preempt and modify their algorithm "to block racist reviews when we observed anti-Chinese xenophobia associated with COVID-19." And "To avoid the spread of election-related misinformation, we limited the ability for people to edit the phone numbers, addresses and other information for places like voting sites" (Pritchett 2021).

In this framing, Google Maps calls attention to its predictive capabilities. It is ready for the "content vandals" and "bad actors" and can anticipate where they will act. In consequence, anything that does filter through into the map appears like a glitch in the system (Nakamura 2013). Moreover, the glitch could be the fault of the Local Guides who did not act fast enough in reporting the misinformation. The glitch in this case works to isolate misinformation as one-off events and an anomaly through the antagonism of outsiders. However, as Kuo and Marwick (2021) note, "when the content of disinformation is ignored and treated as a mysterious and externally created pollutant, we cannot understand why it resonates, where it comes from, or how it spreads" (Kuo and Marwick drawing from del Vicario et al. 2016; Vosoughi et al. 2018). Moreover, since Google is the one in charge of the terms and conditions, it is clear that the interests these serve are Google's.

DISCUSSION: LOCAL GUIDES AS A PROJECT OF MISINFORMATION

Google's construction of insider and outsider knowledge frames how and when "misinformation" plays out on Google Maps. The Local Guides Platform plays a central role in constructing insider knowledge and Google Maps, by extension, as a supportive place to foster online communities. Google leverages Local Guides as benevolent citizens within the Google Maps ecosystem. Reflecting on the processes of constructing insiders, the question then becomes: Who has access to this citizenship? As evidenced in our finding, digital literacy, access to hardware, and social capital are all implicit requirements of gaining entry to Local Guides citizenship. While not within the scope of this case study, these currencies of Local Guides citizenship have been shown to be historical barriers to online access (Gibson and Martin 2019) as well as differential impacts of disinformation (Noble 2018) and extended definitions of misinformation. Such limitations already foreclose the question of "misinformation" when thinking about spatial information. Such normative constructions of misinformation fundamentally limit not only how tech companies look to fight misinformation but also who they are ultimately working for and whose ideals they uphold.

Through this case study, we illustrate the limitations of a misinformation conceit and that dominant framings of misinformation are in place to protect platforms and not people. Framing misinformation on the Local Guides platform through insider vs. outsider or bad vs. good actor binaries helps absolve Google Maps of responsibility for misinformation on the map (Kuo and Marwick 2021). This was evident in 2016 with the erasure of the names "Gaza" and "The West Bank" from Google Maps' representation of Palestine (Agha 2020; Graham and Dittus 2022) echoing the erasure of neighborhoods like The Fruit Belt, Rincon Hill, and South Beach from the Google Maps interface. Even if their names are "put back" on the map, the knowledge that their existence is precarious is shown through The Fruit Belt's ongoing campaign to preserve their name (Dewey 2019) or the ongoing petition campaign with over a million signatories for better mapping of Palestine (Graham and Dittus 2022). Google's processes of "adding information to the map" and "crowdsourcing information" illustrates the larger structural dynamics of power perpetuated by Google's mapping apparatus, which the Local Guides Platform is a distraction from.

Moreover, the Local Guides platform is designed to normalize commoditized relations to space (see Graham and Dittus 2022; McQuire 2019) while foreclosing what trustworthy and useful content is, through contrast to fake or fraudulent information. Kuo and Marwick (2021) note that characterizing disinformation in this way assumes an otherwise "shared, healthy information ecosystem, which ignores historical and ongoing ideological and political inequalities that center white viewpoints" (pp. 2). They call for the reframing of "disinformation" from a problem of information pollution to a form of knowledge that is often purposefully propagated and circulated to uphold existing structures of power (Bivens and Haimson 2020).

Local Guides platform is not simply representing space but actively shaping what counts as accurate spatial knowledge, who imparts this information, and what types of information get to inform everyday location-based decision-making processes. Reviews are made part of the map, not only in that they are visible when using Google Maps platform, but also in ways that are disassociated from the Local Guide reviewer and aggregated with other reviews. One user's rating becomes calculated into an "average rating" of place. For example, opening Google Maps and searching for "coffee," one will find an icon of a coffee cup to signal the location of a cafe along with the cafe's rating out of five immediately visible. Local Guides reviews feed how a place is seen on the map, based on the potential of how one consumes a place. We see how this opens a possible new line of map dynamics where businesses rely on the inputted comments of Local Guides to be seen through the map. Thus, localized and contextually dependent place-based knowledge becomes stripped from its local context, aggregated, and subsumed into Google's map.

Taking together the construction of the benevolent insider Local Guide watch-guard alongside the framing of misinformation as an always external threat from outside the system, it becomes apparent that the Local Guides platform constructs *enclosures* of authenticity that run counter to the projection of a "global community" that is "for everyone" (Reece 2021). In the words of Plantin (2018), "it can decide what needs to be included or excluded from the cartographic representation, who determines society's spatial representation, and what form and shape public participation will take". These enclosures presume a relationship to space based on the assumption that all space can be known through review and these reviews are helpful for everyone without considering how review opens the door to online harassment which is often missed in normative models of misinformation (Del Vicario et. al. 2016). Returning where we started, misinformation on the map is more than simply the fraudulent businesses preying upon innocent Google Maps users. Furthermore, misinformation is not a glitch but a feature of Google Maps' project to create and protect its total map of the world that prioritizes commercial and transactional orderings of space. Local Guides provides an enclosed participatory infrastructure for Google Maps that presents as open, community-driven and for the people. Yet, through this case study of the platform we see how the Local Guides platform is founded on a fantasy of a benevolent insider, insulated from the politics of mapping, and keeping out the menacing outsider ready to infiltrate the map with their hacks, fakes, and hot takes.

REFERENCES

Agha, Zena. 2020. "Maps, Technology, and Decolonial Spatial Practices in Palestine." Al-Shabaka, The Palestinian Policy Network 14: 1–9. https://al-shabaka.org/briefs/maps-technology-and-decolonial-spatial-practices-in-palestine/.

AriMar [Google Editor]. 2019. "Questions about a Local Guides Reward Email? Read This First." *Local Guides Connect.* Retrieved from: www.localguidesconnect.com/t5/Help-Desk/Question-about-a-Local-Guides-reward-email-Read-this-first/bc-p/3180901#M41698.

Benjamin, Ruha. 2019. *Race After Technology.* Cambridge, UK: Polity,

Bhandari, Aparajita, & Rebecca Noone. (2023). Support Local: The Exclusionary Configurations of 'Local Love' on Google Maps' Local Guides Platform. Communication, *Culture and Critique*, 1 (3), pp. 198–207. https://doi.org/10.1093/ccc/tcad018

Bivens, Rena, and Oliver L. Haimson. 2016. "Baking Gender into Social Media Design: How Platforms Shape Categories for Users and Advertisers." *Social Media + Society* 2 (4): 1–12.

Bourdieu, Pierre. 1986. "The Forms of Capital." In *Handbook of Theory and Research for the Sociology of Education*, ed. J. C. Richardson, 241–258. Greenwood New York. https://doi.org/10.1177/2056305116672486

Bucher, Taina. 2018. *If . . . Then: Algorithmic Power and Politics.* New York: Oxford University Press.

Carmi, Elinor. 2019. "The Hidden Listeners: Regulating the Line from Telephone Operators to Content Moderators." *International Journal of Communication* (online): 440–458. https://ijoc.org/index.php/ijoc/article/view/8588

Copeland, Rob, and Katherine Bindley. 2019. "Millions of Business Listings on Google Maps Are Fake: And Google Profits; Google Maps Is Filled with False Business Addresses Created by Firms Pretending to Be Nearby." *The Wall Street Journal. Eastern Edition.* New York: Dow Jones.

Del Vicario, Michela, Alessandro Bessi, and Walter Quattrociocchi et al. (2016) "The Spreading of Misinformation Online." *Proceedings of the National Academy of Sciences – PNAS* 113 (3): 554–559. https://doi.org/10.1073/pnas.1517441113.

Dewey, Caitlyn. 2019. 'Fruit Belt Fights for Its Name Over Fears Big Tech Is Erasing It.' *The Buffalo News.* March 17. Buffalo: Lee Enterprises.

Duffy, Brooke Erin. 2017. *(Not) Getting Paid to Do What You Love: Gender, Social Media, and Aspirational Work.* New Haven: Yale University Press.

Duguay, Stefanie, Jean Burgess, and Nicolas Suzor. 2020. "Queer Women's Experiences of Patchwork Platform Governance on Tinder, Instagram, and Vine." *Convergence* 26 (2): 237–252. https://doi.org/10.1177/1354856518781530

Fairclough, Norman. 2013 "Critical Discourse Analysis." *The Routledge Handbook of Discourse Analysis.* New York: Routledge. 9–20.

Francis, S. 2019. "Become a Google Local Guide in 2020." *Digital Maas* (December 16). www.digitalmaas.com/blog/become-google-local-guide/.

Frischmann, Brett M., Michael J. Madison, and Katherine J. Strandburg. 2014. *Governing Knowledge Commons.* New York: Oxford University Press.

Gerrard, Ysabel, and Helen Thornham. 2020. "Content Moderation: Social Media's Sexist Assemblages." *New Media and Society* 22 (7): 1266–1286. https://doi.org/10.1177/1461444820912540.

Gibson, Amelia N., and John D. Martin. 2019. "Re-situating Information Poverty: Information Marginalization and Parents of Individuals with Disabilities." *Journal of the Association for Information Science and Technology* 70 (5): 476–487. https://doi.org/10.1002/asi.24128.

Gillespie, Tarleton. 2015. "Platforms Intervene." *Social Media + Society* 1 (1): 1–2.

2017. "Governance of + by Platforms." In *The SAGE Handbook of Social Media*, ed. Jean Burgess, Thomas Poell, and Alice Marwick. Thousand Oaks, CA: Sage, 254–278.

2018. *Custodians of the Internet: Platforms, Content Moderation, and the Hidden Decisions That Shape Social Media.* New Haven: Yale University Press.

Google, a. 2022. "Our Approach to Search." *Google Search*. Last updated 2022. www.google.com/search/howsearchworks/our-approach/.

Google, b. 2022. "Community Policy." *Local Guides Help*. Last updated 2022. https://support.google.com/local-guides/answer/7358351?hl = en.

Google, c. 2022. "Points, Levels, Badges." *Local Guides Help*. Last updated 2022. https://support.google.com/local-guides/answer/6225851?hl = enandref_topic = 6225845.

Google, d. 2022. "Get Recognized." *Local Guides Help*. Last updated 2022. https://support.google.com/local-guides/answer/6281793.

Google, e. 2022. "Organize a Meet-Up." *Local Guides Help*. Last Updated 2022. https://support.google.com/local-guides/answer/6225579?hl = en-.GBandref_topic = 6214612.

Gorwa, Robert. 2019. "What Is Platform Governance?" *Information, Communication and Society* 22 (6): 854–871.

Graham, Mark, and Martin Dittus. 2022. *Geographies of Digital Exclusion: Data and Inequality*. London: Pluto.

Kaplan, Max., and Alex Vallis. 2018. "Seven Kinds of Local Guides You Might Spot on Google Maps." *Google: The Keyword*. (January 10). Retrieved from: https://blog.google/products/maps/seven-kinds-local-guides-you-might-spot-google-maps/.

Kiene, Charles, Jialun Jiang, and Benjamin Mako Hill. 2019. "Technological Frames and User Innovation: Exploring Technological Change in Community Moderation Teams." *Proceedings of the ACM on Human–Computer Interaction* 3 CSCW 1–23. https://doi.org/10.1145/3359146.

Kuo, Rachel, and Alice Marwick. 2021. "Critical Disinformation Studies: History, Power, and Politics." *Harvard Kennedy School Misinformation Review* 2 (4). https://misinforeview.hks.harvard.edu/article/critical-disinformation-studies-history-power-and-politics.

Leader, Ian. 2021. "How Reviews on Google Maps Work." *Google: The Keyword*. (February 2) Retrieved from: https://blog.google/products/maps/how-google-maps-reviews-work/.

Lee, Dami. 2019. "Google Maps Has a Fake Business Listing problem." *The Verge*. (June 20). Retrieved from: www.theverge.com/2019/6/20/18693144/google-maps-fake-business-listings-investigation-report.

Luca, Michael, and Georgios Zervas. 2016. "Fake It Till You Make It: Reputation, Competition, and Yelp Review Fraud." *Management Science* 62 (12): 3412–3427. https://doi.org/10.1287/mnsc.2015.2304.

Luque-Ayala, Andrés, and Flávia Neves Maia. 2019. "Digital Territories: Google Maps as a Political Technique in the Re-Making of Urban Informality." *Environment and Planning. D, Society and Space* 37 (3): 449–467.

Massey, Doreen B. 1994. *Space, Place and Gender*. Cambridge: Polity.

McCormack, R. 2015. "Google Maps Reviewers Can Now Earn 1 TB of Free Drive Storage." *The Verge*. (November 16). Retrieved from: www.theverge.com/2015/11/15/9741624/google-maps-local-guide-earn-drive-storage.

McMillan Cottom, Tressie. 2020. "Where Platform Capitalism and Racial Capitalism Meet: The Sociology of Race and Racism in the Digital Society." *Sociology of Race and Ethnicity* 6 (4): 441–449.

McQuire, Scott. 2019. "One Map to Rule Them All? Google Maps as Digital Technical Object." *Communication and the Public* 4 (2): 150–165.

Myers West, Sarah. 2018. "Censored, Suspended, Shadowbanned: User Interpretations of Content Moderation on Social Media Platforms." *New Media and Society* 20 (11): 4366–4383.

Nakamura, Lisa. 2013. "Glitch Racism: Networks as Actors within Vernacular Internet Theory." *Culture Digitally*. https://culturedigitally.org/2013/12/glitch-racism-networks-as-actors-within-vernacular-internet-theory/.

Nicas, Jack. 2018. "As Google Maps Renames Neighborhoods, Residents Fume." *The New York Times*. (August 2). www.nytimes.com/2018/08/02/technology/google-maps-neighborhood-names.html.

Noble, Safiya. 2018. *Algorithms of Oppression: How Search Engines Reinforce Racism*. New York: New York University Press.

Noone, Rebecca. 2024. *Location Awareness in the Age of Google Maps*. London: Routledge.

Plantin, Jean-Christophe. 2018. "Google Maps as Cartographic Infrastructure: From Participatory Mapmaking to Database Maintenance." *International Journal of Communication* (online): 489–506.

Pritchett, Dan. 2021. "A Look at How We Tackle Fake and Fraudulent Contributed Content." *Google: The Keyword*, February 18. Retrieved June 14 from: https://blog.google/products/maps/google-maps-101-how-we-tackle-fake-and-fraudulent-contributed-content/.

Reece, Kevin. 2021. "Three New Ways Anyone Can Update Google Maps." *Google: The Keyword*. March 11. Retrieved June 14, 2022 from: https://blog.google/products/maps/three-new-ways-anyone-can-update-google-maps/.

Roberts, Sarah T. 2019. *Behind the Screen: Content Moderation in the Shadows of Social Media*. New Haven: Yale University Press.

Russel, Ethan. 2019. "How We Fight Fake Business Profiles on Google Maps." *The Keyword*, June 29, 2019. https://blog.google/products/maps/how-we-fight-fake-business-profiles-google-maps.

Schoenebeck, Sarita, Oliver L. Haimson, and Lisa Nakamura. 2021. "Drawing from Justice Theories to Support Targets of Online Harassment." *New Media and Society* 23 (5): 1278–1300. https://doi.org/10.1177/1461444820913122.

Singh, Rianka. 2018. "Platform Feminism: Protest and the Politics of Spatial Organization." Ada: A Journal of Gender, New Media, and Technology, No. 14. https://doi.org/10.5399/uo/ada.2018.14.6

Slabin, Laura. 2017. "More Levels, and More Ways to Contribute for Local Guides." *Google: The Keyword*. https://blog.google/products/earth/more-levels-and-more-way-contribute-local-guides-l/.

Squirrell, Tim. 2019. "Platform Dialectics: The Relationships between Volunteer Moderators and End Users on Reddit." *New Media and Society* 2 (9): 1910–1927. https://doi.org/10.1177/1461444819834317

Tarr, Alexander, and Luis F. Alvarez León. 2019. "Will Review for Points: The Unpaid Affective Labour of Placemaking for Google's 'Local Guides." *Feminist Review* 123 (1): 89–105. https://doi.org/10.1177/0141778919879763

Tiidenberg, Katrin, Natalie Ann Hendry, and Crystal Abidin. 2021. *tumblr*, Cambridge: Polity.

TraciC. 2016. "Connect Is Growing and So Is our Connect Moderators Group." Local Guides connect message board posted by TraciC "Googler." (November 17). www.localguidesconnect.com/t5/General-Discussion/Connect-is-growing-and-so-is-our-Connect-moderators-group/m-p/66870#M11278.

2022. "Celebrating Connect's 5th Birthday." *Local Guides Connect*. (June 30). www.localguidesconnect.com/t5/Announcements/Celebrating-Connect-s-5th-Birthday/bcp/2974257/emcs_t/S2h8ZW1haWx8Ym9hcmRfc3Vic2NyaXBoaW9ufEw4WVBXSDk3V1U1ToVJfDI5NzQyNTd8U1VCU0NSSVBUSU9OU3xoSw#M87350.

Vosoughi, Soroush, Deb Roy, and Sinan Aral. 2018. "The Spread of True and False News Online." *Science (American Association for the Advancement of Science)* 359 6380: 1146–1151. https://doi.org/10.1126/science.aap9559.

Washington, Anne L., and Rachel S. Kuo. 2020. "Whose Side Are Ethics Codes On? Power, Responsibility and the Social Good." In *arXiv.org*. Ithaca: Cornell University Library. arXiv.org,. htps://doi.org/10.1145/3351095.3372844.

Wells, Allen. 2019. "Discover New places with Help from Top Local Guides." *Google: The Keyword*. November 15. www.blog.google/products/maps/discover-new-places-with-help-from-local-guides/.

Wilken, Rowan. 2019. *Cultural Economies of Locative Media*. New York: Oxford University Press.

7

The Human Infrastructure of Misinformation

A Case Study of Brazil's Heteromated Labor

David Nemer and William Marks

INTRODUCTION

As the information landscape continues to evolve, the widespread dissemination of disinformation is proliferating at a rate and scale never seen before in a democratic society. In recent years, major technology companies have taken much of the public blame for this reality, given their algorithms facilitate the sharing of – and sometimes even promote – falsehoods. This, however, misses a key reality: social media, search engines, and messaging services are not fully automated technologies. Rather, they are heteromated: they are reliant on participatory humans to serve their economic goals. Focusing on users, and on the sharing rather than the origination of disinformation, we connect theories of heteromation with those surrounding the Human Infrastructure of Misinformation (HIM) with the express purpose of contributing to a more holistic understanding of how and why misinformation is so prevalent online. We also attempt to approach aspects of the HIM within the Governing Knowledge Commons (GKC) framework in order to contribute to the understanding of how shared knowledge, information, and data are governed in a collective setting.

In the past, to run an effective disinformation campaign, one often had to sneak around institutional gatekeepers. In the preinternet era in the western world, the media was largely centralized and bounded by ethical standards. So in order to promote a false story, one may need to persuade a news outlet of the truth of falsehood and then hope the organization deemed the story worthy of a column or segment. Utilizing the closed-off infrastructure of the news media was one way to reach the eyes and ears of millions (Shimmer 2021).

As more individuals began to use the internet, these traditional gatekeepers remained stewards of their domains. However, the information superhighway was steadily getting busier, and the road had few tolls. This internet epoch was the "Rights Era," a classification coined by Jonathan Zittrain (2019) to highlight the

FIGURE 7.1 Visual themes from the human infrastructure of misinformation: a case study of Brazil's heteromated labor.

feeling at the time that information deserves to be free, and individuals had the rights to disseminate and consume it.

In 2004, a Google search for the term "Jew" returned Jewwatch.com as its first result. The horribly antisemitic website denied the Holocaust and stated that it was "keeping a close watch on Jewish communities and organizations worldwide." A Google spokesman responded that while the company was disturbed by the page rankings, it would not make any changes to preserve "the objectivity of our ranking function" (Flynn 2004). The Anti-Defamation League, a Jewish nongovernmental organization, defended Google, saying that the search rankings were an unintentional, unfortunate reality (Noble 2018). To the public, and even to those with an interest in the issue, Google was simply a tool that helped users see what was popular online. Its rankings did not reflect any judgment, and the company claimed it could not control automatically generated search results. Google maintained the pipes serving users their information, but the pipes were dumb. Google's peers, including Facebook and Twitter, served a similar custodial role.

Over time, however, the pipes became smarter and more precise. Social media companies and other tech companies built out their capabilities. Now, if a post violated a company's rule or standard, the organization could do any of the following: delete the post or ban the poster; label the post as controversial, contextualize the post, perhaps by pinning a "fact-check" to it; or apply friction to it by refusing to promote the post or restricting users' ability to share it – as managed by platforms such as Twitter, Facebook, and Instagram (Mena 2020). As the capacity for intervention increased, so too did calls to intervene. Behind the black boxes of algorithms promoting or adding friction to posts, technical design decisions made to affect behavior, and institutions standing up to make decisions about content online, it can be easy to lose track of the heteromation involved: of the humans spreading disinformation and, on the other side, moderating or choosing not to moderate it.

This can be aptly shown in the case of the spread of misinformation on WhatsApp during Brazil's 2018 general elections.[1] Critics contend that social media like Facebook, Instagram, and Twitter are plagued by filter bubbles and echo chambers, many of which are promulgated through personalized algorithmic feeds. An echo chamber is what might happen when people are overexposed to news that they like or agree with, potentially distorting their perception of reality because they see too much of one side, not enough of the other, and they start to think perhaps that

[1] We define disinformation as false information with the intention to mislead. Disinformation is deliberately created and spread as truth to influence public opinion, obscure the truth, and get a reaction that assists its creator. Disinformation is often confused with fake news. However, fake news is an umbrella term that covers a range of falsehoods or lies, including misinformation and disinformation. Misinformation is not always intended to mislead; for example, it may be false or inaccurate information that was created by mistake or inadvertently spread. Misinformation can also be true information that misinforms when taken out of context (Nemer 2021).

reality is like this (Fletcher 2020). A filter bubble is "a state of intellectual or ideological isolation that may result from algorithms feeding us information we agree with, based on our past behaviour and search history" (Fletcher 2020). It was coined by Eli Pariser (2011). Facebook creates filter bubbles by using personal information and online behavior to curate the information that shows up on users' feeds.

One of the dangers of filter bubbles is that platforms become a hotbed for mis- and disinformation, since its distribution may not break out of the filter bubble at its point of origin (DiFranzo and Gloria-Garcia 2017). Several studies have shown that the spread of mis- and disinformation is more similar to epidemics than real news stories are and that such stories usually stay within the same communities (Jin et al. 2013); that is, mis- and disinformation tends to not reach or convince outsiders. Mis- and disinformation and selective news filtering gained attention from scholars and the general public after they contributed to social media platforms' ideological polarization that favored Donald Trump in the 2016 US presidential election and "Vote Leave" (known as Brexit) in Britain's European Union membership referendum (Spohr 2017; Vaidhyanathan 2018). But since WhatsApp runs on a peer-to-peer architecture, no algorithm curates content according to the characteristics or demographics of the users. Instead, a human infrastructure was assembled to create a pro-Bolsonaro environment on WhatsApp and spread misinformation to bolster his candidacy. In this paper, we articulate the labor executed by the Human Infrastructure of Misinformation as heteromated labor.

The field of disinformation studies has yet to address questions related to the relationship between disinformation and labor, identity, and morality. Thus, as claimed by Jonathan Ong and Jason Cabañes (2019), approaching disinformation as a culture of production expands the field into understanding the social conditions that entice people to this work and the creative industry practices that normalize fake news as a side gig. By framing the Human Infrastructure of Misinformation as heteromated labor, this chapter joins a growing body of literature that critically examines disinformation as a culture of production, which means "conceptualizing disinformation as both product and process emerging from organizational structures, labor relations, and entrepreneurial subjectivities" (Ong and Cabañes 2019, 5772). We also hope to highlight how disinformation follows a process of both collaboration and an intentional product of hierarchized and distributed labor, so we can better inform disinformation interventions and regulations that address concerns with political campaigns.

HETEROMATION

The division of labor between computer systems and human beings has changed along both human and technical proportions. As claimed by Hamid Ekbia and Bonnie Nardi (2014), there has been a shift from technologies of automation, those which are entirely managed by machines, to those of "heteromation," which push critical tasks to end-users as indispensable mediators. Unlike technologies of automation, those of heteromation benefit from or depend on often unrecognized or uncompensated human labor to complete tasks. As Ekbia and Nardi (2014) originally noted, they also do not all fit a singular mold; heteromation, like automation, is a process that can describe a wide range of business models and behaviors.

Users may or may not know they are serving as a cog in the heteromated machine. In some cases it is clear: Amazon Mechanical Turk, first launched in 2006, is a "human-as-a-service" crowdsourcing marketplace in which employees pay "MTurkers" to simulate automation through the completion of small, repetitive tasks. Sometimes it is hazier when users are working on behalf of a company: when they review a video game on YouTube or engage in community governance online, such as by serving as a nonemployee moderator on a subreddit, or a topic-based forum on the popular website Reddit, they are taking part in a heteromated process. As are those who, even without such a semi-formal role, simply "upvote" or "downvote" a post, thereby feeding the site's algorithm so that it may show other "redditors" well-liked or controversial posts or comments. Meta, which owns Facebook, Instagram, and WhatsApp, and Alphabet, which owns Google and YouTube, each generates billions of dollars of revenue by relying on users, whether they be social media influencers, everyday people, or businesses, to create data, posts worth keeping browsers' attention, and to ultimately view the advertisements that the companies make money from. Doug Laney (2012) wrote in the *Wall Street Journal* that "Facebook's nearly one billion users have become the largest unpaid workforce in history." He gathered that each user provided about $81 of worth to the site.

In *Heteromation, and Other Stories of Computing and Capitalism*, Ekbia and Nardi (2017) outline five varieties of heteromated labor: communicative, cognitive, creative, emotional, and organizing. The authors first define communicative labor as the act of engaging online in return for self-validation and self-promotion. This is the Facebook model, where user signals such as "likes" and "comments" – "social data," as the authors refer to it – are used to build an effective advertising model. Next, Ekbia and Nardi describe cognitive labor – the Amazon Mechanical Turk model – as when complex systems deliberately and knowingly use human intelligence and work to complete tasks and function properly. They then explain creative labor, when users produce creative outputs – like submitting a logo to a design contest or modifying a video game – in return for recognition, community, or a *chance* for monetary reward.

After that, the authors discuss emotional labor, which, in the context of heteromation occurs when individuals help machines "care" for people. As the authors note, human labor is often required to allow machines to work efficiently and effectively. Consider, for instance, the example of therapeutic robots. They may help individuals, including by helping to comfort and stimulate the elderly. Still, caregivers often must calibrate devices to the patients' needs. Lastly, the authors tell of organizing labor, which draws on "the human capacity to organize an activity." Examples include customer surveys.

Heteromation is a product of current conditions including: (1) historically high levels of profit in which no realm of human activity is too private, repetitive, or objectifying to escape capital's grip; (2) the "race to the bottom" of increasing economic disparity resulting in the precariat; (3) a cultural need for high levels of stimulation fed by decades of quality popular entertainment and the generalized anxiety of the neoliberal subject which renders such stimulation particularly desirable; and, (4) feelings of uselessness or futility experienced by growing segments of the population such as the elderly, the chronically ill, the impoverished, and healthy retired people faced with twenty to thirty years without the social rewards of employment (Ekbia and Nardi 2017). Heteromation, as described, is fundamentally a product of social dilemmas that aren't evident in the world since the community of people who manage or govern information are kept in a precarious state and are hidden from the public eye.

While some technologies of heteromation – or the companies that run them – ask or require individuals to conduct specific and defined forms of labor, others take advantage of what Zittrain (2008) calls "generativity," or "a system's capacity to produce unanticipated change through unfiltered contribution from broad and varied audiences." Generative technologies are often innovative themselves; still, they typically rely on the creativity of others to sustain economic prosperity and drive growth. As Zittrain (2008) writes, while generativity's output is innovation, its input is participation. Can you imagine if when you "googled" *New York City* and the only results returned were what an individual Google employee had to say about the city? Or if you logged onto Tiktok, only to see the content it paid directly for, as though it were a firm like Netflix? No, instead, Google and Tiktok help facilitate the conditions for which users produce their own content, for their own reasons, for free. And therein lies the value of platforms.

BACKGROUND: WHATSAPP IN BRAZIL

When approaching a knowledge commons, in this case pro-Bolsonaro WhatsApp groups, it is important to analyze the environment that contains the resources that are to be shared and managed. According to Madison et al. (2009), "we must take a step back before describing the relevant characteristics of the shared resources to ask how we should define the environmental backdrop against which a commons is constructed," thus, in this section we describe the background of where the knowledge common was nested.

WhatsApp has been popular in Brazil since it entered the market in 2009. In 2018, WhatsApp had about 120 million active users in Brazil, out of a total population of 210 million. About 96 percent of Brazilians with access to a smartphone used

WhatsApp as one of their main methods of communication (Nemer 2019). WhatsApp's popularity was driven by its low cost compared to SMS texting, which in Brazil could cost around fifty-five times what it would in North America. Another reason for the app's popularity was that after Facebook bought WhatsApp for US$19 billion, the company partnered with the telecoms to offer a zero-rating plan that allowed subscribers to use WhatsApp basically for free. Zero-rating (also known as sponsored data) "refers to the practice in which mobile networks offer free data to customers who use specific services (e.g., streaming videos) or smartphone applications (e.g., [Facebook Chat], WhatsApp). Thus, customers who access this zero-rated/sponsored content do not pay for the mobile traffic generated by such use" (Omari 2020, 7).

WhatsApp also makes it easy to create group chats and share content, including videos. The app became a hotbed for political campaigns that utilized disinformation not only because of its affordances and widespread reach in the country but also because of its end-to-end encryption, which ensured that nobody besides the sender and the receiver could read the content of messages. As such, it is nearly impossible for WhatsApp analysts to identify disinformation campaigns (Nemer 2022).

ATTRIBUTES: THE CREATION AND MAINTENANCE OF THE COMMONS

Madison et al. (2009) ask "what technologies and skills are needed to create, obtain, and use the resources?" and "what are the characteristics of the resources?" In this section, we describe some attributes of the knowledge commons (pro-Bolsonaro WhatsApp groups) and give hints of how it was created and maintained.

During Nemer's fieldwork in Vitória, Brazil, in December 2017, Neuza, a 27-year-old woman, showed him a short video that she got in a WhatsApp group called "Bolsonaro our president."[2] In the video, Jair Bolsonaro tells Maria do Rosário, a member of Congress with the Workers' Party (PT), "I would never rape you because you don't deserve it." The video came with the label: "This is how we

[2] This study draws on data from Nemer's ethnographic work. Fieldwork took place in online and offline spaces and in different phases. For the in-person (offline) fieldwork, there were two main phases: phase 1 from June to July 2012, and phase 2 from April to October 2013. Nemer also conducted several in-person follow-up observations and interviews: August 2014; May, August, October, and November 2015; March, April, and December 2016; November 2017 through January 2018; July and December 2018; June through July, and December 2019. For the WhatsApp data, Nemer joined four self-declared pro-Bolsonaro public WhatsApp groups via invitation links that were publicly listed in the descriptions of conservative YouTube videos. Nemer began monitoring WhatsApp groups in March 2018, which had an average of 160 members. At the peak of the election cycle, each group was posting an average of one thousand messages per day. Nemer visited these groups every day for one to two hours. For more information about Nemer's ethnography see *Technology of the Oppressed: Inequity and the Digital Mundane in Favelas of Brazil* (2022, MIT Press).

treat communists." Neuza vehemently condemned the video. Then when asked how she got into such a group, she simply told Nemer: "Someone added me to this group chat. [...] There are three people I know on here; that's why I stayed. I thought it was a group chat of friends, but all these other people do is talk about Bolsonaro. Nonstop! They send all kinds of videos and photos of him. It's annoying and I don't know how to leave [the group]."

Fatima (49 years old) and Regina (39 years old) also mentioned that the WhatsApp groups they were part of had become spaces to talk about the upcoming 2018 elections – with a focus on Bolsonaro. But instead of being added to a chat, they had the experience of new people being added to their existing groups. The new arrivals brought up Bolsonaro. "These two guys who I don't know entered our church group chat and always talk about politics," Fatima explained. "This is a church group, not politics," she added with a laugh. "They go on and on about how socialism is evil ... but wasn't Jesus Christ a socialist?"

Listening to Neuza, Fatima, and Regina describe their experiences on WhatsApp, Nemer could relate to what they said : The same thing was happening with his own WhatsApp groups. In his extended family's group chat, Nemer noticed that three cousins were constantly sharing homemade misinformation and pro-Bolsonaro memes and videos. Since he knew they were not the ones creating them, Nemer asked them who made the content. Their answers were always the same: "I don't know. I got it from another group chat," Giovane, 46 years old, said. Most of the political and misinformation content that the cousins were sharing matched what Regina and Neuza were seeing. Since WhatsApp runs on a peer-to-peer architecture, there was no algorithm curating content according to their characteristics or demographics, which is how filter bubbles work on Facebook. Spreading mis- and disinformation on WhatsApp required deliberate human labor to create and distribute this content.

GOVERNANCE ACTION: A SOCIAL AND POLITICAL MESSAGING MACHINE

If Brexit and Donald Trump's election were made possible by Facebook's algorithms, then the rise of Brazil's far-right firebrand Jair Bolsonaro was partly due to human machinations on WhatsApp. On the meta-owned messaging app, a Human Infrastructure of Misinformation (HIM) was assembled to make the environment on the app feel staunchly pro-Bolsonaro through messaging and misinformation campaigns.[3] The HIM turned pro-Bolsonaro WhatsApp groups as the action arena where governance was taking place. They became the social spaces "where

3 Although such human infrastructure worked on disinformation campaigns to favor Jair Bolsonaro, we cannot affirm that they were affiliated with Bolsonaro himself or his political party, Partido Social Liberal (PSL).

participants with diverse preferences interact, exchange goods and services, solve problems, dominate one another, or fight (among the many things that individuals do in action arenas)" (Ostrom 2009).

Infrastructures, as anthropologist Brian Larkin (2013) defines them, are "built networks that facilitate the flow of goods, people, or ideas and allow for their exchange over space" (328). While there has been increasing work on social engagements with infrastructures in technological systems, particularly in the Global South (see Dye et al. 2018; Jack, Chen, and Jackson 2017; Nguyen 2016; Sambasivan and Smyth 2010), here we build on an understanding of infrastructure that goes beyond technological artifacts and focuses on humans as central to such networks. To explore the human side of infrastructure, we look at how humans organize to perform labor and accomplish tasks. We also follow Nithya Sambasivan and Thomas Smyth's (2010) concern for "social practices, flows of information and materials, and the creative processes that are engaged in building and maintaining these substrates" (1).

Bolsonaro became known more for his controversial speeches rather than for having a robust and well-formulated policy platform. He celebrated dictatorship, glorified torture, promised to reverse policies that protect the Amazon Region, and threatened Brazil's women, Black, and LGBTQIA people. However, none of his threats had an impact on Bolsonaro's popularity, since his fragmented approach to governance attempted to please his electorate that was composed of a loose coalition brought together by the candidate's appeal to "bullets, bibles, and bulls" – which is a conservative political agenda that aims to favor evangelical and Pentecostal politicians, pro-guns policies, and pro-agriculture and anti-environment acts (Schipani and Leahy 2018). Bolsonaro ran a campaign based on the idea that his presidency was the only hope to end violence and corruption in Brazil. His supporters called him *Mito* (Legend or Myth) and expected him to reinstate law and order in the country. Given the hyperpolarization of the elections, Bolsonaro built his rise on people's distrust in politics and the wariness of politicians and political establishments in general. Bolsonaro presented himself as the anti-establishment candidate, even though he had been a member of the Brazilian Congress for twenty-seven years.

Although mis- and disinformation had spread in Brazil across all forms of social media, WhatsApp's impact was the most notable. Due to the app's popularity, about 44 percent of the voting public in Brazil used WhatsApp to discover political information, according to the polling institute Datafolha (G1 2018). The app's simple design allowed users to easily share text audio notes, images, and videos – which facilitated the spread of misinformation. A study of 100,000 WhatsApp images widely shared in Brazil during the election found that more than half contained misleading or blatantly false information (Tardáguila, Benevenuto, and Ortellado 2018). Another study conducted by fact-checking agencies involved in Comprova found that 86 percent of false or misleading content shared on WhatsApp benefited Bolsonaro by attacking his opponent, Fernando Haddad, and his party, PT (I. Macedo 2018).

TABLE 7.1 *Clusters of members across pro-Bolsonaro WhatsApp groups and their duties*

Cluster of actors	Duties
Influencers	Create and/or bring new misinformation to pro-Bolsonaro WhatsApp groups.
Bolso-Army	Reinforce the content of the Influencers' misinformation and spread them across WhatsApp and other platforms, such as Twitter and Facebook. They also managed pro-Bolsonaro WhatsApp groups.
Average Brazilians	Consume misinformation and bring them to their personal groups.

THE ACTORS OF THE HUMAN INFRASTRUCTURE OF MISINFORMATION

Given the prevalence of WhatsApp use and the intriguing way that misinformation was spread during the 2018 presidential election, Nemer decided to research who was creating the false content and sharing it with people like Neuza, Fatima, and Regina. Nemer joined four self-declared pro-Bolsonaro WhatsApp groups via invitation links that were publicly listed in the description of conservative YouTube videos.[4] Nemer began monitoring WhatsApp groups in March 2018 that had an average of 160 members. At the peak of the election cycle, each group was posting an average of 1,000 messages per day. In August, after conducting the first thematic analysis of the data, Nemer identified three clusters of actors across the groups: the Average Brazilians, the Bolso-Army, and the Influencers, as detailed in Table 7.1. He found that misinformation was spread in these groups through a pyramid structure, similar to the classic two-step flow of communication model (Katz and Lazarsfeld, 1966)[5] in which each cluster of actors occupied a level. Influencers were at the top and Average Brazilians were at the bottom.

The vast majority of the actors fit within the description of the Average Brazilian. According to sociologist Esther Solano, Bolsonaro's typical voter was male, White, and middle class and held a college degree (Moysés 2018). However, on these WhatsApp groups, as Nemer analyzed the actors' conversations, he noticed that they were coming from different social classes and were both men and women. They justified their vote for Bolsonaro by sharing their life experiences and difficulties. Before entering the groups, many of them mentioned that they didn't have a strong opinion about the candidate. However, they saw WhatsApp groups as safe spaces where they could learn more about the *Mito*, verify rumors and news, and obtain digital content to share on other social media accounts and groups. Many of

[4] On YouTube, Nemer searched for "Bolsonaro WhatsApp grupo" and looked for videos that had public invitation links. He joined the first four groups that showed up in his search in which the invitation links worked.

[5] The two-step flow of communication model hypothesizes that ideas flow from mass media to opinion leaders, and from them to a wider population.

them voted for a different right-wing candidate in the first round of voting and switched to Bolsonaro for the final runoff. One of these people was Carlos, who said in the groups that he "wasn't going to vote in the runoff, but [did] after learning that our country was under an imminent socialist attack." He claimed that he decided to vote for Bolsonaro on the basis of this information.

Bolsonaro's groups functioned as echo chambers governed by the Bolso-Army and Influencers. Every time a member posted disinformation – such as poll results or memes about Bolsonaro – members rallied by cheering with the Brazilian flag – a sign of Bolsonaro's new emphasis on Brazilian nationalism – or by posting a specific emoji. The backhand index finger pointing right or left (👉) was Bolsonaro's trademark handgun symbol, referring to his promise to relax gun controls and allow police officers to shoot suspects with impunity. The Bolso-Army was Bolsonaro's loyal fan base and the machinery that stood ready to attack anyone who insulted Bolsonaro on WhatsApp or other social media platforms. They began following him long before his campaign started because they were in fact part of the governance team of these WhatsApp groups and kept a vigilant eye out to promptly ban infiltrators or people who dared to challenge anything related to the candidate.

In these groups, debate or discussion about Bolsonaro's policies was difficult. People got kicked out because they asked questions related to Bolsonaro's refusal to participate in televised debates, his family's mysterious assets, and even his record as a congressman. Every time average users attempted to verify information or ask a question, they were flooded with passionate Bolso-Army messages to shut down any doubt about Bolsonaro's legacy. Their arguments were mostly founded on misinformation.

The Bolso-Army formed the glue that held the Human Infrastructure of Misinformation (HIM) together to actively disseminate the misinformation produced by influencers across pro-Bolsonaro WhatsApp groups and other social media platforms. Given their stance, which displayed extreme confidence and left no space for questions, average users felt secure with the information they were given. They recirculated it, helping to spread misinformation even further. Influencers had a decisive role in creating mis- and disinformation. There were only four or five influencers per group, and they were not the most outspoken or active participants. They worked backstage to create and share fake news in these groups and to coordinate online and offline protests. They used image-and video-editing software to create convincing and emotionally engaging digital content. They knew how to work content into memes and short texts that went viral.

The Influencers took advantage of the Bolso-Army's loyalty to quickly spread their fake news. They often used affect (satire, irony, and humor) to create their content, creating memes about "Bolsonaro the Oppressor" to ironically show Bolsonaro's human side. They also worked fast to create fake news to delegitimize anybody who criticized Bolsonaro before group members read the news in other venues. For example, Marine Le Pen – the iconic far-right politician from France – stated that

"Bolsonaro says extreme things, unpleasant things which are insurmountable in France." Within thirty minutes of this story coming out in a popular Brazilian publication, the Influencers pooled resources together and posted a meme saying that Le Pen was a Communist (Nemer 2021). Their strategy was to label everyone who might undermine Bolsonaro a Communist and to discredit mainstream news.

Traditional right-leaning news outlets such as *Veja* and *Estado de São Paulo* were actually labeled as socialist venues by members of pro-Bolsonaro groups. The mis- and disinformation produced in WhatsApp insidiously altered perception, but the absurdity of some stories was even more astonishing. A group of influencers created a flyer alerting their members that Haddad, the candidate from the Workers' Party (PT), would sign an executive order allowing men to have sex with twelve-year-old children. When David Duke endorsed Bolsonaro for thinking like the Ku Klux Klan (KKK), the influencers were quick to produce content depicting the KKK as a product of the left-wing party, to distance Bolsonaro's image from the KKK. During the first round of votes, they circulated fake videos that showed malfunctioning electronic voting machines to reinforce the idea that the elections were rigged.

Influencers also found public videos on YouTube and Facebook that challenged Bolsonaro and posted their links to the WhatsApp groups so the "Bolso-swarm" could descend on them to express their dislike and show support for their legend. Although these three types of members – the Average Brazilians, the Bolso-Army, and the Influencers – had different roles in the Bolsonarist WhatsApp ecosystem, they had a lot in common.

Many Influencers and the Bolso-Army were compensated between US$100 and US$250 per week to distribute pro-Bolsonaro content and manage WhatsApp groups, as claimed in 2019 by former members of the Bolso-Army. By revealing this, they implicitly criticized influential groups of businessmen who they said financed the network, and they suggested that virtual militias (known as the Virtual Activist Movement) were paid to infiltrate WhatsApp groups and spread misinformation. They didn't directly implicate Bolsonaro's campaign team, though they have said that at least one person who was an adviser in Bolsonaro's government was among those paid to feed fake news to his supporters.

What happened during the 2018 presidential election debunked the idea that WhatsApp is a level playing field. WhatsApp's peer-to-peer encrypted architecture may give users a sense of security and privacy since there is no algorithm intervening in their messages. It may also give them a sense of spontaneity since the app affords anyone the ability to produce and share content. However, as described earlier, Bolsonaro's campaign relied on mis- and disinformation that was systematically created and spread by the HIM that orchestrated a guided campaign. It is hard to verify the exact impact that digital populism had on the 2018 presidential elections. However, given Carlos's account and many like it, in which people felt motivated to get out and vote for Bolsonaro, the HIM behind the disinformation campaign on WhatsApp undeniably helped Bolsonaro become Brazil's next president.

TABLE 7.2 *Heteromation of the human infrastructure of misinformation*

System	Heteromated functionality	Beneficiaries	Participant compensation	Affective rewards for participants
Misinformation on WhatsApp	Human Infrastructure of Misinformation Influencers: Content-creation; cognitive and creative labor.Bolso-Army: Repetitive microtasks; communicative and organizing labor. Average Brazilians: emotional and communicative labor.	Politicians	Influencers: medium-high Bolso-Army: medium-high	Average Brazilians: high

DISCUSSION

The Human Infrastructure of Misinformation (HIM) comprises a sociotechnical arrangement where humans organize to perform labor with the aid of technological systems and accomplish tasks that pertain to the creation, distribution, and consumption of misinformation. The HIM sustained an information system, in this case, misinformation on WhatsApp, that functioned through the actions of heterogeneous actors – in other words, the members of such infrastructure were users that provided the heteromated labor necessary to make misinformation on WhatsApp work.

Hamid Ekbia and Bonnie Nardi (2014) examine heteromated systems according to their functionality and reward structure. They categorize systems in terms of who benefits from the heteromated labor relation, whether participant compensation (monetary) is offered, and whether the system produces affective rewards. Beneficiaries are social actors that reap the major benefits of heteromated labor. Participants may benefit from affective rewards. Following the authors' categorization, the heteromation system that HIM labored in can be organized as detailed in Table 7.2.

The heteromated labor performed by Influencers and the Bolso-Army was compensated in amounts of between US$ 100 and US$ 250 per week (between BRL 380 and BRL 900 approximately). Such an amount can be considered "medium-high" since the Brazilian minimum monthly wage in 2018 was set in at BRL 954 and the average monthly household income per capita was BRL 1.337 (IBGE 2019). Influencers were at the top of the HIM: They were responsible for producing and bringing new misinformation to pro-Bolsonaro groups on WhatsApp. Their cognitive labor comprised of understanding the current and complex political environment in order to promote the presidential candidate Jair Bolsonaro and/or to respond to the criticism that Bolsonaro was receiving during his campaign. Such cognitive labor led to the Influencers' creative labor which meant creating outputs – like pro-Bolsonaro memes and short videos. Differently from the common understanding of what an Online Influencer attempts to accomplish, HIM's Influencers were not seeking recognition or fame, instead, they wanted to remain unrecognizable. We categorize them as Influencers since they had access to a large audience in WhatsApp groups and were able to persuade others to act based on their content. They only participated once or twice a day, bringing new misinformation to the groups because they knew that they could rely on the Bolso-Army to reinforce the message embedded in the content they created.

As for the Bolso-Army, they did not engage in heteromated cognitive labor, as defined by Ekbia and Nardi (2017). Instead, they were in charge of repetitive microtasks that followed a predefined script and/or reacted to a previous trigger. The Bolso-Army performed communicative labor as they were scripted to reinforce the Influencers' content's message, as soon as new content was posted to those pro-Bolsonaro groups on WhatsApp, in order to make sure that the Average Brazilians actually internalized their message, as well as to spread the same content to other WhatsApp groups and platforms such as Twitter and groups on Facebook. Differently than claimed by Ekbia and Nardi, that described communicative labor as the act of engaging online in return for self-validation and self-promotion, members of the Bolso-Army were not necessarily after self-validation and/or self-promotion; their main goal was to validate and promote the message contained in the Influencer's posts.

Another heteromated labor performed by the Bolso-Army was to organize. Bolso-Army members were also part of the admin team that managed those WhatsApp groups – they stood ready to remove anyone from those groups that insisted on questioning Bolsonaro's legacy and campaign or whom they perceived as infiltrators – users who joined those groups and flooded them with messages criticizing Bolsonaro. They also worked hard to maintain the tone in those groups favorable to Bolsonaro. They organized online harassment by asking Average Brazilians to descend on posts on other social media platforms to express their dislike and show support for Bolsonaro.

The Average Brazilians, at the base of the HIM, performed communicative labor as they also helped spread the misinformation posted on those pro-Bolsonaro groups to their other personal groups that were not necessarily politically oriented, such as Family, Neighborhood, and Friend groups. They often used the content posted by the Influencers as evidence for their arguments in their online debates. Average Brazilians also performed emotional labor, but not the emotional labor defined by Ekbia and Nardi (1997) which occurs when individuals help machines "care" for others, for example: when caregivers help users adopt, and calibrate, technology such as therapeutic robots. Average Brazilians performed emotional labor as defined by Hochschild (1983), which refers to regulating or managing feelings to fulfill the emotional requirement of one's work role. They were expected to regulate their emotions during interactions with the Bolso-Army and other Average Brazilians. Fearing being kicked out of those WhatsApp groups, they often complied with the Bolso-Army's authority and avoided asking questions that could be interpreted as them challenging Bolsonaro's legacy. Instead, they often responded to the Influencers or Bolso-Army posts by cheering with the Brazilian flag, validating messages, or with Bolsonaro's trademark handgun emoji (☞). That validating or positive response was perceived by Average Brazilians as a way to have their loyalty recognized by the Bolso-Army and rewarded with their permanence in the groups. Such emotional labor was done so as to produce a certain feeling in the Bolso-Army or Influencers that their labors were being appreciated and that they were succeeding in promoting Bolsonaro. In each of these instances, humans participated in the heteromated process of spreading disinformation.

Studies on heteromated systems have focused on creating awareness around labor exploitation, compensation, invisibility, and advocating for labor rights (see Bailey et al. 2018; Ekbia and Nardi 2019; Irany 2019; Sambasivan et al. 2021). However, in the case of the Human Infrastructure of Misinformation, given its ethical and quasilegal practices, such awareness may not be wanted nor beneficial to its members. In many cases, we might call for unpaid heteromated labor to be compensated, or for the greater acknowledgment of the work of low-paid creators. However, when it comes to the heteromated process of the HIM, the actors involved are creating what is likely a societal net negative, calling into question the types of behaviors that should be incentivized.

As it turns out, money was a strong motivator for Influencers and members of the Bolso-Army to remain loyal and contribute to Bolsonaro's digital propaganda machine. While some were surely ideologues or political allies looking to push Bolsonaro's campaign and presidency, others were likely swayed by the economic opportunity of being paid up to $250 a week

Importantly, many of the labors performed by the HIM can be – and indeed often are – done by algorithms on other platforms. On Whatsapp, the Influencers, the Bolso-Army, and Average Brazilians moderate chats and spread disinformation. Those roles are, on social media like Twitter or Instagram, filled by algorithms, including recommendation engines and algorithmic content moderators. In these instances, in that a gap is created by a lack of automation, it is filled by heteromated labor. This labor may be adaptive and resilient too; when the ecosystem changes, the labor can adjust to still serve its goals (Marks 2021).

Approaching HIM as heteromated labor and as a culture of production, which means "conceptualizing disinformation as both product and process emerging from organizational structures, labor relations, and entrepreneurial subjectivities" (Ong and Cabañes 2019, 5772), exposes the broader systems of practices that normalize and incentivize its everyday work arrangements. Our study attempts to highlight how disinformation is a process of collaboration and an intentional product of hierarchized and distributed labor. It joins Jonathan Ong and Jason Cabañes' call for proposing new "questions of ethics and accountability that consider broader institutional relations at play" (2019, 5772). Our study also sheds light on the responsibility of disinformation actors working in different capacities – such as Influences, the Bolso-Army, or low-level misinformation consumers and spreaders like the Average Brazilians. Each group's motivations and intentions may vary, but taken together, they constitute the key players in the pro-Bolsonaro WhatsApp misinformation ecosystem. Understanding HIM as a product of everyday work arrangements "pushes us to think of how we can expand our notion of disinformation interventions from content regulation and platform banning of bad actors and their harmful content to include process regulation of the ways political campaigns and consultancies are conducted" (Ong and Cabañes 2019, 5772).

In framing HIM as a culture of production, our goal is to understand the actors involved in the work arrangements that promote misinformation and to engage in a detailed study on challenging moral issues involving "complicity and collusion" in our media ecosystem (Silverstone 2007). By subverting expectations of people about who exactly paid disinformers are, we critique the workings of the system at large and the different levels of responsibility at play. We hope this approach contributes to broader debates about the political value of representing perpetrators' narratives, which in media scholar Fernando Canet's (2019) view can deepen "understand[ing of] both the personal and situational dimensions that could lead a human being to commit such acts [and] help us to determine how to prevent future cases" (15).

CONCLUSION

In this chapter, we analyze the labor within the Human Infrastructure of Misinformation as heteromated labor through the case study of WhatsApp groups in Brazil and highlight ties with the Governing Knowledge Commons (GKC) framework. On WhatsApp, Brazilian President Jair Bolsonaro benefited from misinformation that was used to garner support and ensure that people felt that he was a popular candidate. While the key to messaging campaigns on social media is often to appeal to an algorithm that will in turn broadcast a message, on WhatsApp there is no algorithmic feed. As such, the pro-Bolsonaro campaign relied on the HIM – specifically, Influencers, the Bolso-Army, and Average Brazilians.

The work the HIM performed represented heteromated labor: the influencers provided cognitive and creative labor through their content creation; the Bolso-Army provided communicative and organizing labor by acting as administrators of large group chats and performing repetitive microtasks like forwarding and promoting messages; and the Average Brazilians provided emotional and communicative labor by spreading misinformation and regulating their emotional responses while complying with norms established by the Influencers and the Bolso-Army.

Building on the work of Ong and Cabañes, the chapter also looks at how HIMs emerge from "organizational structures, labor relations, and entrepreneurial subjectivities." The complex misinformation ecosystem – with Influencers at the top of the pyramid, the Bolso-Army in the middle, and Average Brazilians at the bottom – was formed by an alliance of actors with varying motivations, whether they be for money, fame, ideology, or because they believed the information that they were sharing was true. Developing a further understanding of the HIM and the heteromated labor it employs at both the personal and organizational levels will allow future researchers and policymakers to better understand, and address, misinformation. We hope that future research will continue to address these important topics.

REFERENCES

Bailey, D. E., E. H. Diniz, B. A. Nardi, P. M. Leonardi, and D. Sholler. 2018. "A Critical Approach to Human Helping in Information Systems: Heteromation in the Brazilian Correspondent Banking System." *Information and Organization* 28 (3): 111–128.

Canet, F. 2019. "Documenting Atrocities around the World: Why Engage with the Perpetrators?" *International Journal of Cultural Studies* 22 (6): 804–822. https://doi.org/10.1177/1367877919840042.

DiFranzo, D., and K. Gloria-Garcia. 2017. "Filter Bubbles and Fake News." *XRDS: Crossroads, The ACM Magazine for Students* 23 (3): 32–35.

Dye, M., D. Nemer, J. Mangiameli, A. S. Bruckman, and N. Kumar. 2018. "El Paquete Semanal: The Week's Internet in Havana." In *Proceedings of the 2018 CHI Conference on Human Factors in Computing Systems.* (April). (1–12).

Ekbia, H. R., and B. Nardi. 2014. "Heteromation and Its (Dis) Contents: The Invisible Division of Labor between Humans and Machines." *First Monday* (online) 19 (6).

2017. *Heteromation, and Other Stories of Computing and Capitalism.* Cambridge, MA: MIT.

2019. "Keynes's Grandchildren and Marx's Gig Workers: Why Human Labour Still Matters." *International Labour Review* 158 (4): 653–676.

Fletcher, R. 2020. "The Truth behind Filter Bubbles: Bursting Some Myths." *RISJ Review.* https://reutersinstitute.politics.ox.ac.uk/risj-review/truth-behind-filter-bubbles-bursting-some-myths.

Flynn, L. 2004. "Google Says It Doesn't Plan to Change Search Results." *The New York Times.* April 13. www.nytimes.com/2004/04/13/business/google-says-it-doesn-t-plan-to-change-search-results.html.

G1. 2018. "Datafolha: Quantos Eleitores de Cada Candidato Usam Redes Sociais, Leem e Compartilham Notícias Sobre Política." *Globo.com.* October 3. https://g1.globo.com/politica/eleicoes/2018/eleicao-em-numeros/noticia/2018/10/03/datafolha-quantos-eleitores-de-cada-candidato-usam-redes-sociais-leem-e-compartilham-noticias-sobre-politica.ghtml.

Hochschild, A. R. 1983. The Managed Heart: Commercialization of Human Feeling. Oakland, CA: University of California Press.

IBGE. 2019. "PNAD Contínua 2018: 10% da população concentram 43,1% da massa de rendimentos do país." *Agência IBGE Notícias.* October 16. https://agenciadenoticias.ibge.gov.br/agencia-sala-de-imprensa/2013-agencia-de-noticias/releases/25700-pnad-continua-2018-10-da-populacao-concentram-43-1-da-massa-de-rendimentos-do-pais.

Irani, L. 2019. "Justice for Data Janitors." In *Think in Public*. New York: Columbia University Press. 23–40.

Jack, M., J. Chen, and S. J. Jackson. 2017. "Infrastructure as Creative Action: Online Buying, Selling, and Delivery in Phnom Penh." In *Proceedings of the 2017 CHI Conference on Human Factors in Computing Systems*. (May) (6511–6522).

Jin, F., E. Dougherty, P. Saraf, Y. Cao, and N. Ramakrishnan. 2013. "Epidemiological Modeling of News and Rumors on Twitter." In *Proceedings of the 7th Workshop on Social Network Mining and Analysis*. (August). (1–9).

Katz, E., and P. F. Lazarsfeld. 1964. *Personal Influence: The Part Played by People in the Flow of Mass Communications. A Report of the Bureau of Applied Social Research, Columbia University*. Glencoe, IL: Free Press of Glencoe.

Laney, D. 2012. "To Facebook You're Worth $80.95." *The Wall Street Journal*. (May 3). www.wsj.com/articles/BL-CIOB-298.

Macedo, I. 2018. "Das 123 Fake News Encontradas Por Agências de Checagem, 104 Beneficiaram Bolsonaro." *Congresso Em Foco, UOL*. (October 26). https://congressoemfoco.uol.com.br/eleicoes/das-123-fake-news-encontradas-por-agencias-de-checagem-104-beneficiaram-bolsonaro/.

Madison, Michael J., Brett M. Frischmann, and Katherine J. Strandburg. 2009. "Constructing Commons in the Cultural Environment." *Cornell Law Review* 95: 657–710.

Marks, W. 2021. "De-platforming Is a Fix, But Only a Short-Term One." *Just Security*. www.justsecurity.org/74342/de-platforming-is-a-fix-but-only-a-short-term-one/.

Mena, Paul. 2020. "Cleaning up Social Media: The Effect of Warning Labels on Likelihood of Sharing False News on Facebook." *Policy and Internet* 12 (2): 165–183.

Moysés, A. 2018. "Eleitor Típico de Bolsonaro é Homem Branco, de Classe Média e Superior Completo." *CartaCapital*. (September 19). www.cartacapital.com.br/politica/eleitor-tipico-de-bolsonaro-e-homem-branco-de-classe-media-e-superior-completo/.

Nemer, D. 2019. "WhatsApp Is Radicalizing the Right in Bolsonaro's Brazil." *HuffPost*. (August 16). www.huffpost.com/entry/brazil-jair-bolsonaro-whatsapp_n_5d542b0de4b05fa9df088ccc.

2021. "The Human Infrastructure of Fake News in Brazil." *Social Science Research Council: Items*. https://items.ssrc.org/extremism-online/the-human-infrastructure-of-fake-news-in-brazil/.

2022. *Technology of the Oppressed: Inequity and the Digital Mundane in Favelas of Brazil*. Cambridge, MA: MIT.

Nguyen, L. U. 2016. "Infrastructural Action in Vietnam: Inverting the Techno-politics of Hacking in the Global South." *New Media and Society* 18 (4): 637–652.

Noble, S. U. 2018. *Algorithms of Oppression: How Search Engines Reinforce Racism*. New York: New York University Press.

Omari, J. 2020. "Is Facebook the Internet? Ethnographic Perspectives on Open Internet Governance in Brazil." *Law and Social Inquiry* 45 (4): 1093–1112.

Ong, J., and J. Cabañes. 2019. "When Disinformation Studies Meets Production Studies: Social Identities and Moral Justifications in the Political Trolling Industry." *International Journal of Communication* 13: 5771–5790. Retrieved from: https://ijoc.org/index.php/ijoc/article/view/11417/2879.

Ostrom, Elinor. 2009. *Understanding Institutional Diversity*. Princeton, NJ: Princeton University Press.

Pariser, E. 2011. *The Filter Bubble: How the New Personalized Web Is Changing What We Read and How We Think*. London: Penguin.

Sambasivan, N., and T. Smyth. 2010. "The Human Infrastructure of ICTD." In *Proceedings of the 4th ACM/IEEE International Conference on Information and Communication Technologies and Development.* (1–9).

Sambasivan, N., S. Kapania, H. Highfill, D. Akrong, P. Paritosh, and L. M. Aroyo. 2021. "Everyone Wants to Do the Model Work, Not the Data Work": Data Cascades in High-Stakes AI." In *Proceedings of the 2021 CHI Conference on Human Factors in Computing Systems.* (May). (1–15).

Schipani, A., and L. Leahy. 2018. "Brazil's Bolsonaro Aims to Rule with 'Bull, Bullet and Bible' Bloc." *Financial Times.* www.ft.com/content/611f150e-cb51-11e8-b276-b9069bde0956.

Shimer, David. 2021. *Rigged: America, Russia, and One Hundred Years of Covert Electoral Interference.* New York: Vintage.

Silverstone, R. 2007. *Media and Morality: On the Rise of the Mediapolis.* Cambridge, UK: Polity.

Spohr, D. 2017. "Fake News and Ideological Polarization: Filter Bubbles and Selective Exposure on Social Media." *Business Information Review* 34 (3): 150–160.

Tardáguila, C., F. Benevenuto, and P. Ortellado. 2018. "Fake News Is Poisoning Brazilian Politics. WhatsApp Can Stop It." *New York Times.* (August 17). www.nytimes.com/2018/10/17/opinion/brazil-election-fake-news-whatsapp.html.Taylor,TinaLynn.2012.

Vaidhyanathan, S. 2018. *Antisocial Media: How Facebook Disconnects Us and Undermines Democracy.* Oxford University Press.

Zittrain, J. 2008. *The Future of the Internet: And How to Stop It.* New Haven, CT: Yale University Press.

2019. "Three Eras of Digital Governance." *SSRN.* (October 2). https://dx.doi.org/10.2139/ssrn.3458435.

8

Hidden Virality and the Everyday Burden of Correcting WhatsApp Mis- and Disinformation

Britt Paris and Irene Pasquetto

INTRODUCTION

Epistemic burden (Pierre et al. 2021), a concept drawn from social epistemology, black feminist thought, and other branches of social and cultural theory, is a way to describe how dominant culture actors discredit and blame the knowledge and knowledge practices of minoritized groups to further disadvantage these already burdened groups in a way that compounds their oppression. Burden is a unique theme when it comes to governance of many types. Generally, policy only attends to burden when it is too large or visible to ignore, and otherwise ignores and perpetuates it through a number of mechanisms. Elsewhere, we discussed epistemic burden in the context of collecting evidence of police brutality in the US (Paris et al. 2022). Lawmakers' expectation that more and more data is required to act on police brutality has the effect of indefinitely stalling interventions that could effectively fight it. Requiring "more data" can hardly be the solution to the problem as statistics on police brutality are coproduced by systems of power and oppression to specific ends (Paris et al. 2022). Survivors of police brutality and their families organize to collect various forms of evidence aimed at challenging or completing official

Britt Paris is a critical informatics scholar studying how groups build, use, and understand information systems according to their values, and how these systems influence evidentiary standards and political action. Paris is an assistant professor at Rutgers University in the Department of Library and Information Science. Her current work focuses on critical perspectives around techno solutionism for political problems, and practices and discourse around alternative internet infrastructure. She did her postdoctoral research at the Data and Society Research Institute focusing on audiovisual disinformation, and remains affiliated there.
Irene Pasquetto is a scholar in the field of information and communication science. She is an assistant professor at the University of Michigan School of Information where she teaches Ethics of Information Technologies and Digital Curation. Her most recent research work focuses on issues of science mis- and disinformation, open science practices, and public understanding and use/misuse of science products and infrastructures. As a research affiliate at the Harvard Kennedy School's Shorenstein Center on Media, Politics, and Public Policy, she cofounded the Harvard Kennedy School Misinformation Review.

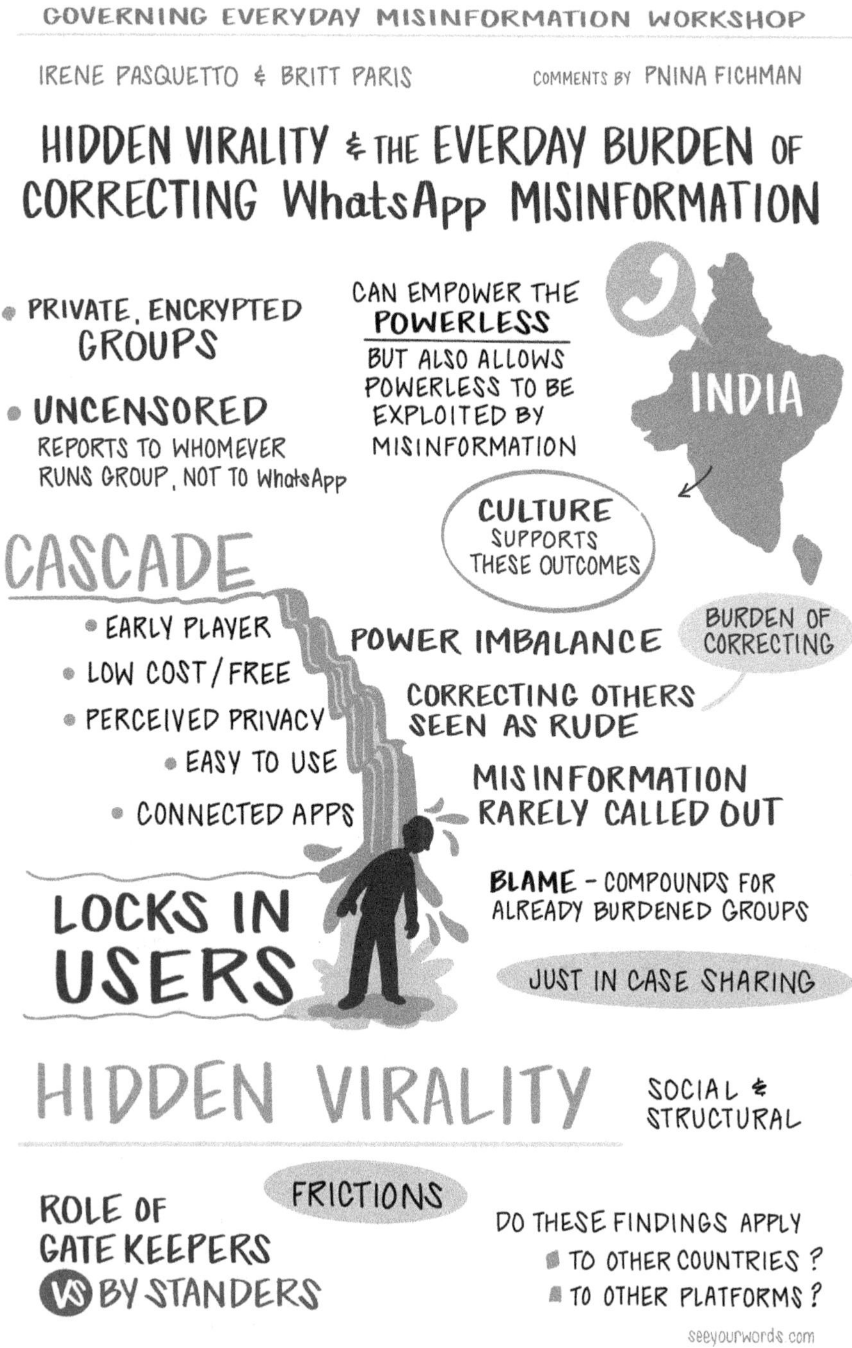

FIGURE 8.1 Visual themes from hidden virality and the everyday burden of correcting WhatsApp mis- and disinformation.

government-produced data. However, they find themselves overwhelmed by such practices and, most importantly, by the lack of recognition given to them by law enforcement agencies and the legal system. Local, community-led data collection and analyses efforts are disregarded, discredited, or not taken seriously, mainly because they start from different epistemological assumptions and, most importantly, point to the necessity of considering radical solutions that make those in power uncomfortable (i.e., the abolition of carceral structure). In this sense, the victims of police brutality are shouldered with the epistemic burden of making the phenomenon itself visible to people who will never fix it (Paris et al. 2022).

In terms of misinformation governance, burden is also largely unaddressed. Work to make online spaces better, with less harassment, hate speech, and false and misleading information, takes resources – epistemic resources like time, education, and the ability to present arguments, but also social resources like higher social profiles, which are commonly predicated on wealth, privilege, and notoriety (Fricker 2007; Paris et al. 2022; Pierre et al. 2021). Often, those with the fewest resources are the most targeted and, paradoxically, the ones who disproportionally bear the burden of making online spaces better (Citron 2016; Collins-Dexter 2020; Kuo and Marwick 2021; Noble 2013). This reality is rarely taken into consideration in conversations around misinformation and attendant concerns of using media literacy, information literacy, and fact-checking to mitigate it. Such interventions – like in the case of requesting more top-down data to prove police brutality – can hardly solve problems of false and misleading information given that power imbalances actively discourage users from sharing verified information that goes against local norms, and from correcting those who hold power. Most importantly, by failing to frame issues of misinformation into the broader cultural and economic dynamics that lead to them (tech colonialism *in primis*), literacy and fact-checking approaches leave the public with the false and unfair impression that users are at fault for causing, and at the same time responsible for countering, the problem of misinformation, while they are only one part of the story.

In this chapter, we apply the concept of epistemic burden to describe how Meta, a powerful US-based company, shapes the technical affordances that lead to cumbersome everyday practices of correction for mis- and disinformation on WhatsApp for users in India. In Latin America, Africa, Europe, the Middle East, and Asia, private WhatsApp chats are an irreplaceable tool for the organization and maintenance of social life (Rossini et al. 2020; Wardle 2020). Multiple factors seem to have led to WhatsApp's success worldwide. It is affordable and user-friendly, and it provides users with great flexibility in terms of which format to choose for sharing content (text, audio, video, etc.), full control over the selection of who sees such content (there is no algorithmic content curation), and a perceived feeling of privacy (mostly given by the presence of an encryption protocol). In addition, WhatsApp users tend to know each other personally, suggesting a prevalence of close contacts – a factor that further contributes to a perceived sense of privacy (Pasquetto et al. 2022). Taken together, these affordances have made WhatsApp the most popular messaging app and preferred means of everyday communication and information sharing in many countries including India, where we conducted our research.

While information on WhatsApp can theoretically be shared freely without prejudice, empowering minoritized groups to share their experiences and amplify their voices (Lim 2020), many use WhatsApp in ways that negate and sometimes directly inhibit enacting or expressing these freedoms. WhatsApp is a notorious vehicle for mis- and disinformation, and uncivil and dangerous speech. The use of WhatsApp has fostered pressing concerns around elections, ethnic violence, and other damaging consequences. False and misleading information spread through WhatsApp has been linked to tipping Brazilian, Indian, and Nigerian elections to authoritarian candidates (Benghani 2019; Cheeseman et al. 2020; Garimella and Eckles 2020; Machado et al. 2019), the genocide and forced migration of Rohingya Muslims in Myanmar (International Crisis Group 2017; Siddiquee 2020), and deaths due to disinformation about a global kidnapping ring (Banaji et al. 2019). During the COVID-19 outbreak, hoaxes, anonymous rumors, and conspiracy theories spread widely on mobile instant messaging platforms (Naeem and Bhatti 2020). In Lombardy – the Italian region most affected early in the pandemic – traffic on WhatsApp and Facebook Messenger quickly doubled after its onset, and with it the viral sharing of false information and conspiracy theories (Facebook Inc. 2020). In India, where we conducted our research, mis- and disinformation on WhatsApp not only caused health concerns by projecting alternative medicines as potential cures for COVID-19, but also led to panic situations through unverified claims about internet shutoffs and shortages of essential commodities during the lockdown period (Khan 2020; Pal 2020).

WhatsApp is particularly concerning when it comes to mis- and disinformation and uncivil and dangerous speech as we see the activation of what is known as the phenomenon of *hidden virality*, a construct originally theorized by Paris and Donovan (2019) that refers to unvetted, insular discourse on encrypted, private platforms that takes on a character of truth and remains mostly unnoticed until causing real-world harm. In this chapter, we discuss how and why hidden virality is activated on WhatsApp. We investigate how WhatsApp's users perceive the problem of mis- and disinformation on the platform and deal with it daily, and we also discuss how WhatsApp's sociotechnical affordances (either actual or perceived) can encourage users to share dangerous speech.

The chapter draws from digital ethnographic work that included semi-structured interviews and chat texting with forty self-reported WhatsApp users and group administrators who are residents of India. All who reported their location were either in or near (within fifty miles of) large cities. We conducted thirty interviews between January 2019 and March 2020. Fifteen formal interviews were conducted between January and March 2019, and fifteen more between January and March 2020. Chat conversations with ten more individuals were conducted between January and

March 2020. After the official interviews ended, we maintained chat conversations with some of the users for about six months. Employing an anticolonial, feminist praxis, we actively worked to create transparent relationships with our participants over time, building reciprocal trust and respect. Foundational in establishing these relationships was the acknowledgment of our positionality as western researchers with no lived experience of Indian historical and sociocultural dynamics. Chats and video calls were conducted in English. Whenever participants shared content in other languages (Hindi, Bengali, Tamil, Malayalam), it was translated into English by research assistants from the region, who also helped to contextualize shared information culturally and socially.

We found that for Indian WhatsApp users, the burden on the powerless of correcting everyday misinformation is exacerbated due to high power distance (Fichman and Rathi 2021) between administrators and users in small private encrypted groups, especially when users are not part of the dominant hegemonic subculture but the administrator is. In these cases, individuals, specifically women, who are targeted by misinformation or harassment do not correct it and struggle like bystanders in other settings (Fichman and Sanfilippo 2015; Herring et al. 2002; Maltby et al. 2016; Shachaf and Hara 2010). We found that promoting information accuracy is challenging in our research context with Indian WhatsApp users, but this phenomenon and difficulties surrounding it are neither unique to encrypted messaging, as they even occur on open social media platforms (Citron 2016), nor to the Indian context (Collins-Dexter 2020).

WHATSAPP'S INITIAL DEPLOYMENT AND USAGE IN INDIA

WhatsApp's 2010 introduction to India came just as 3G connectivity grew in the country (Kumar et al. 2012), causing a meteoric rise in use. Rather than having to buy a certain number of SMS messages from Indian telecommunications carriers, users could download the mobile app and pay a flat, very low fee for unlimited messaging that resembled SMS. In 2013, Facebook was underutilized by large segments of the global market (Prasad 2018). Trying to remedy this, that same year Facebook launched Free Basics, a mobile app allowing users to navigate a handful of platform-selected apps and services without using their data allowance. During its first year, Free Basics faced strong government backlash, while WhatsApp's popularity surged. WhatsApp gained 40 million users from April to September 2014, and by the end of that period, 10 percent of WhatsApp users worldwide were Indian (Diwanji 2019).

WhatsApp initially ran on a subscription model of $1/year; sometimes the first year's subscription was free, but users paid later. Facebook purchased WhatsApp for $19 billion in February 2014, and from 2016 WhatsApp was offered completely free of charge (WhatsApp 2016). That same year, the Telecommunication Regulatory Authority of India banned Free Basics in the country, stating that it, and similar zero-rating programs that allow the use of a narrow set of apps and services without cost to data plans, violated net neutrality (Telecom Regulatory Authority of India 2016). Facebook's WhatsApp does not offer the same infrastructure as Free Basics, but it is used for voice calling and texting and is used to organize everyday social life, allowing users to pay bills, book hotels, make restaurant reservations, send greeting cards, schedule doctor's appointments, post or answer real estate ads, and share information informally, the last of which has become the topic of much research and debate.

Adding to WhatsApp's appeal in India, the app uses Signal Protocol, developed by Open Whisper Systems (OWS), an open-source nonprofit organization now part of the Signal Foundation. OWS received funding from the Open Technology Fund, started in 2012 by US-backed Radio Free Asia; it has ties to the US Agency of Global Media (USAGM), formerly the Broadcasting Board of Governors (BBG), which encourages support of the United States' version of democracy abroad through its communication channels (BBG 2014, 2017; Roose 2018). Indeed, pro-democracy and human rights activists across the South Pacific rely on these encrypted technologies for mobilization (Lim 2020). WhatsApp's Signal Protocol boasts end-to-end encryption (E2EE), where both sender and receiver have keys to encrypt and decrypt messages that they share. No one else – not even the service provider (WhatsApp) or government – can decrypt message contents (WhatsApp 2020).

A "CASCADE OF AFFORDANCES" LOCKED USERS IN

In addition to its early adoption in the region and the use of E2EE protocols, a series of other factors have contributed to the successful adoption of WhatsApp in India and worldwide. While enabling limited broadcast communication, WhatsApp provides users with great control and flexibility over information sharing and management overall. Even groups on WhatsApp constitute highly regulated spaces as admins can remove or ban members at any time at their discretion (Valeriani and Vaccari 2018). The admins tend to set a group's tone and, using feedback from other members, decide what is and is not allowed to be discussed within the group. Therefore, when used for political campaigning, WhatsApp groups support microsegmentation and microtargeting of audiences (Evangelista and Bruno 2019). Contrary to social media platforms, WhatsApp is not subjected to algorithmic curation or platform content moderation. If users want to share a particular message, they can explicitly state and select the audience for that message. WhatsApp also provides delivery and read receipts for each message, which act as a further indicator of whether the intended audience has received/read the message. In addition to all this, users can also quickly and easily share content in different modalities (text, audio, video, images). Images are often shared in the form of "forwarded messages," which report no or little information about the source of that message or where it originated. This is unlike social media platforms such as Facebook or Twitter, where sharing or retweeting also carries information about the original post and its source, even if these can be easily manipulated (Acker and Chaiet 2020). Except for political campaigners, WhatsApp users typically know most of their contacts personally, as WhatsApp's users need a person's phone number to add them as their contact, suggesting a prevalence of close and personal connections. One respondent shared:

> WhatsApp is a good tool for communication because it is a free app that every friend of mine has in their hand at any given time. I receive messages only from people I am in touch with and not everyone who wants to contact me or send anonymous messages. Again, it is so easy to voice call, video call, and even voicemail or send and receive photos/pictures whenever … Compared to other apps, WhatsApp is easy, quick, reliable, and can be used by anyone at any time to communicate with people you know. (Study participant #10, interview round II, February 2020)

Ease of use seems a key factor that drove participants to the platform in the first place. Participants also reported having started to use WhatsApp for its unrestricted availability among many demographics and its inexpensive service. "Everyone already being there" plays a role in keeping them hooked on it. As participants noted – at the time of collecting our interviews – all their close family members and friends are on WhatsApp. Participants employed it mainly to communicate with family, friends, and colleagues in private chats and on calls or in small groups. Our conversations focused on the use of WhatsApp for one-on-one communication and small-group communication.

> I'm in four or five groups total on WhatsApp. One is my family group, the immediate family, another one is for high school friends (we formed the group two years ago, 46 years after graduation), one is for my wife's family, one for university friends (the 1974 graduating class). (Study participant #1, interview round II, February 2020)

> I use WhatsApp to communicate with my family members and friends. We have one local group for the family, mainly used for sharing information. I ask about their difficulties when I'm not at home. I travel quite a lot. … I'm in three groups, all about family or close friends … One is family, one for my community, relatives, other extended family, cousins. (Study participant #2, interview round II, February 2020)

Participants were generally enthusiastic about WhatsApp and reported feeling gratitude: It enables them to talk with their loved ones at no cost, which is perceived to be a great advantage, especially by older generations.

> I like WhatsApp very much. I like that I do not have to pay anything to talk with people, only my Internet bill. My wife can call my daughter in Europe every day; that would have been very expensive years ago. Life was terrible those days. … I have seen my grandson growing up on WhatsApp. It is a very useful tool. I'm glad it exists; I use it at least two or three hours per day. (Study participant #1, interview round II, February 2020)

Our respondents in India use WhatsApp for numerous pro-social activities, including communicating with distant family and friends. For those users, WhatsApp's accessibility, ability to interface with different people in different ways, and low cost make it the right tool for many tasks. Interview data also suggest that participants' preferred means of communication on WhatsApp is voice messages. Participants seemed to perceive audio as the fastest and easiest way of sharing information on the platform, as they can record and listen to voice messages while doing other activities. Indeed, WhatsApp allows users to lock the recording button until the recording is over, freeing their hands for other activities. Participants explained that they record and receive multiple chains of voice messages per day, which can be auto-played in sequence, without interruption. They seem to perceive audio messages to be an effortless way of consuming information. It might be that voice-based communication gives a more personal experience than text- or image-based content that does not directly involve another individual.

Individually, each one of these affordances has made WhatsApp attractive to users in India (as well as worldwide) but might not have been sufficient to ensure their long-term commitment to the platform. However, in combination, these affordances have made it costly for users to leave the platform for a new information-sharing space. In a sense, WhatsApp's affordances work as a "cascade" that keeps users from exiting its usage flow and routine information practices (Figure 8.2)

WhatsApp's free-to-use services in India initially drew people to the platform. Its technical features of connected apps and merging information with users' phone contacts enabled them to connect widely and often with their close friends and family. Once the app had been adopted by the majority of mobile users and their close contacts, it became the place "where everyone already is" – a factor that made it costly for users to leave the platform for another service. On top of this, users' "perceived privacy" – resulting from their awareness of E2EE protocols on the platform – encouraged continued usage of WhatsApp as it promotes feelings of privacy, closeness, and ease of expression. Like a powerful cascade of water incessantly falling over a person beneath a waterfall, WhatsApp's affordances "trap" users and keep them from swimming away.

"JUST-IN-CASE" INFORMATION SHARING ON WHATSAPP

While ensuring its long-term adoption, on the one hand, WhatsApp's affordances also set the platform up as an environment in which users feel particularly comfortable sharing information, including that which they might consider controversial (such as misinformation, disinformation, and uncivil and dangerous speech) and that they might be afraid to share in other spaces. It has been shown how, generally speaking, great flexibility in terms of audience selection and the possibility of retaining control over self-presentation is an important motivator for sharing news online (Ihm and Kim 2018). Valeriani and Vaccari (2018) noted that "network selection and message control allowed by messaging apps facilitate the circulation of controversial information within closed circles"; as a result, users who politically

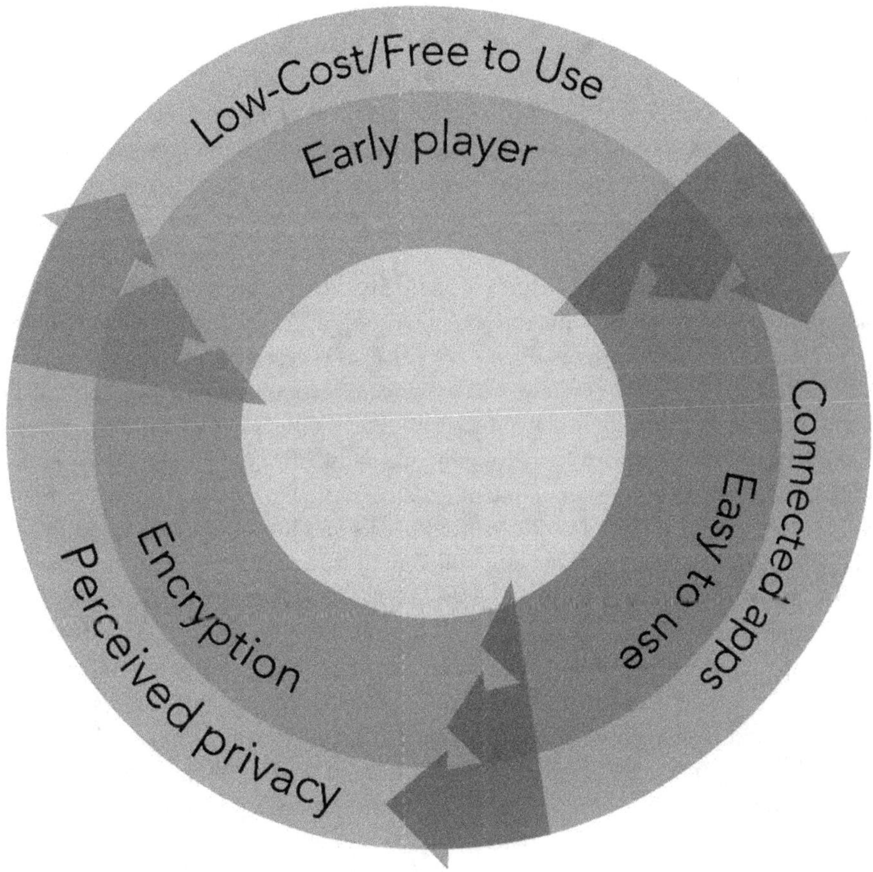

FIGURE 8.2 Cascading WhatsApp affordances.

censor themselves on social media are more likely to engage in political discussions on WhatsApp. Also, communication on messaging apps increases as a response to distrust in mainstream media and reluctance to express oneself in online forums for fear of repercussions (Kuru 2019).

Our investigation suggests that WhatsApp's E2EE feature, and the resulting feeling of "perceived privacy," also played a crucial role in encouraging the sharing of controversial content. Nearly all our participants expressed feeling freer to have "more honest" conversations on WhatsApp's various communication structures than on other platforms, most notably Facebook. The secure communications promised by WhatsApp's E2EE play a key role in this perception. All participants were aware of the E2EE protocol and mentioned it as the main factor for liking and using the platform. They reported being particularly appreciative of the idea that WhatsApp's conversations cannot be accessed and monitored by authorities; phone calls cannot be recorded and outsiders cannot join chat groups. However, E2EE is not nearly as impenetrable as it is marketed to be. There are several hacks and third-party apps that can be used to access, download, and analyze WhatsApp data and metadata, even when conversations have been deleted, or to record phone calls. In addition to this, groups set as "public" can in fact be joined by outsiders, including researchers (when an invite link exists on the web). None of the individuals we talked to reported being aware of any difference between "closed groups" and "open groups." All participants reported exclusively being members of what they perceived to be closed, private groups that can only be joined with the permission of the group administrator. Given these limited capabilities of E2EE to protect WhatsApp users' privacy, we refer to the privacy offered by E2EE as "perceived privacy."

The fact that WhatsApp allows for audience selection and message control, in addition to users' perception that WhatsApp's E2EE allows free and honest conversations, manifests in various uses. Respondents most commonly reported using the app "to share jokes and fun content" that are often dark in tone or contain adult content, and often related to current events and breaking news. When asked "How is content shared on WhatsApp different from content shared on Facebook?" they consistently reported that content on WhatsApp is funnier and more engaging than content shared on Facebook. The second most common answer was that WhatsApp content presents more useful or helpful information compared to Facebook content, typically related to current events and urgent situations.

While most participants claimed awareness that content circulating on WhatsApp might be inaccurate or misleading ("risky"), this did not stop them from sharing it. Our conversations with participants revealed that when exposed to an unverified rumor, they seemed to value its potential utility more than its potential inaccuracy. In other words, when they receive a piece of content on WhatsApp and believe that it is potentially useful or interesting to someone they care about, they share it despite knowing that it could be inaccurate. Accuracy is not the primary concern, especially in times of crisis. Instead, users engage in what we call *just-in-case sharing* in which they place more importance on the potential benefits for friends and family should the information be true than the social repercussions they might face if the rumor is false or inaccurate (e.g., losing credibility). Signs of this behavior emerged during multiple interactions with our study participants. During an informal conversation via chat between one of the authors and a user, the user stated that "I re-share what they sent me even if it might be not accurate, just in case it is true."

ADMIN-LED MODERATION OF DANGEROUS SPEECH IN SMALL, CLOSED GROUPS

A significant amount of offensive, dangerous, and explicit content proliferates within WhatsApp's groups. Adult content is particularly common in men-only chats. Nationalistic or religious content inciting hatred toward others is also shared quite often, and, as our participants noted, it "goes both ways," with Hindus sharing offensive content about Muslims and vice versa. Women and members of religious minorities actively reach out to group administrators to ask for the removal of accounts that share inappropriate or offensive content. A user reported being ostracized by group members for refusing to share nationalistic and religious content that he perceived as offensive toward the Muslim minority:

> To join each group, you either know the administrator or you need a recommendation from someone. You have to conform to the norm of that community. There is that expectation if you try to criticize. Once, they emotionally blackmailed me. I was ostracized from that community. "You don't love your country," they told me. These groups of colleagues and friends are closely guarded communities.

Everyday communications on WhatsApp happen in small groups that are perceived as "closed," meaning that they are managed by someone users personally know. Participants voiced concerns about unmoderated content spread in such groups, independently of age and sex. Closed WhatsApp groups' information-sharing practices operate on what we refer to as a *membership model*; group administrators add members to the group, controlling who is in and who is out. They set the group's tone and, sometimes using feedback from other members, decide what is and is not allowed to be posted. Group members concerned with the spread of dangerous speech observed that the issue with offensive and inappropriate content is not only the content itself but *where* it is shared. From this perspective, if controversial content is shared where it is allowed, WhatsApp users tolerate it.

THE BURDEN OF CORRECTING EVERYDAY MIS- AND DISINFORMATION

Participants also noted that while they are willing to fact-check WhatsApp forwards for their own sake, they rarely engage in correcting other users by sharing evidence. Most users expressed the desire to correct other users on a daily basis. However, when asked to provide a concrete example of such behavior, only two respondents were able to do so. This specific finding might help with the interpretation of results from survey research based on self-reported correction behavior that found that corrections are quite common on WhatsApp, suggesting that such pro-social behavior might be over-reported (Rossini et al. 2020). When asked to elaborate further on the challenges of correcting others, participants indicated that while they think that correcting others is important, doing it might be considered impolite or rude because of cultural factors, especially if the sharer is senior or "outranks" them in terms of social status, clearly displaying the concept of power distance and how it reduces to bystanders those targeted by or wishing to act in solidarity with those who are targeted by misinformation or harassment (Fichman and Rathi 2021; Fichman and Sanfilippo 2015; Herring et al. 2002; Maltby et al. 2016). A participant clearly spelled out what seemed to be a feeling shared by most of the individuals we talked to: that the onus should not be on users to address information problems with the app but on WhatsApp itself.

> I think it is WhatsApp's responsibility to clean up these messages, not mine. Otherwise, their [WhatsApp's] credibility will go down; people are getting tired of WhatsApp. There is no mechanism for fact-checking; it is not right.

Due to the platform's E2EE, total share limits are the only method of moderating speech. This is a source of frustration for users we spoke with, including a few young participants who would like to be able to report false, offensive, and dangerous content on WhatsApp.

> We are a democratic country, but our central government … has brought fights between castes and religions. … Offensive speeches and whatnot; people are dying u know in Delhi! At JNU University, people share hate speech through WhatsApp statuses … India is getting messed up 😣 I fear what's gonna happen … I like WhatsApp because it makes my life easier. But one thing I would like to change is if the report button was there and WhatsApp could have a check when someone reports a profile (due to offensive content).

Because the platform lacks fact-checking mechanisms and content moderation strategies, participants perceive that the burden of cleaning up the everyday sharing of misinformation on the platform is on them. However, correcting others is a practice that goes against their cultural upbringing and makes them very uncomfortable.

We have already noted that the burden of keeping WhatsApp information safe in groups as well as in one-to-one conversations is on users. Minorities or individuals sympathetic toward minorities, such as women, students, and religious minorities find themselves in charge of conducting this delicate work. But in order to have content removed or accounts blocked, these individuals have to convince the administrators of the gravity of the situation. Administrators deliberately decide what should or should not be moderated. WhatsApp groups, then, create situations in which dissent is silenced in favor of the perceived group interests because members are unable to report offensive, false, or misleading content anonymously and face the threat of being questioned, harassed, or ostracized by other members, both online and offline. In these circumstances, we suggest that Meta's WhatsApp has shaped information practices within local, and now global, contexts without much knowledge of these contexts or consideration of the risks faced by users, such as the dangerous consequences of false and dangerous speech mentioned in the introduction (Benghani 2019; Cheeseman et al. 2020; Garimella and Eckles 2020; Machado et al. 2019).

Unpacking the sociotechnical affordances of WhatsApp in India that have led to hidden virality ("perceived privacy" above all) suggests how infrastructural design and deployment have allowed the company and its owners to engage in these goals in ways

TABLE 8.1 *Definitions of key concepts from the chapter*

Hidden virality	Refers to how unvetted, insular discourse on digital media can take on a character of truth and remain unnoticed until causing real-world harm.
Epistemic burden	Describes how dominant culture actors discredit and blame the knowledge and knowledge practices of minoritized groups to further disadvantage these already burdened groups in a way that compounds their oppression.
Cascading affordances	Like a powerful cascade of water incessantly falling over a person beneath a waterfall, cascading affordances "trap" users and keep them from swimming away (i.e., moving to a new service).
Just-in-case sharing	Certain users place more importance on the potential benefits for friends and family should a piece of information be true than the repercussions they and others might face if the rumor is false or inaccurate (e.g., losing credibility). As a result, users are aware of the potential falsity of the information that they encounter online, but they share it anyway.

that have exonerated them of accountability for entering global markets and wreaking havoc in the name of profit, entering markets as infrastructural actors in countries such as Myanmar and India, and allowing disinformation content to circulate unchecked until it resulted in genocide against Rohingya Muslims in Myanmar (Mozur 2018) and lynching in rural India (Liao 2018). Attempts to mitigate the negative effects of hidden virality in India have done little to curb its spread, as our respondents reported seeing no changes in the amount of disinformation that crossed their feeds after the platform limited sharing, with one user sharing one message to just five users or groups. Meta's interventions to counter misinformation include labeling viral forwards and chain messages, limiting forwards, and the design of on-platform Tip-Lines and ChatBots. Little evidence exists on whether any of these interventions work, or to what extent. Due to WhatsApp's encrypted nature, it is difficult for researchers to investigate the efficacy of any on-platform fact-checking efforts. Meanwhile, platforms and researchers alike have tried to shift the blame to users' misuse and lack of education (Chakrabarti, Stengel, and Solanki 2018). However, users are only part of the story; this paper lays bare that they are engaging with the platform exactly as it was designed (Table 8.1). While the sample for the qualitative study may have been biased toward better-educated individuals living in or in the proximity of large cities, the fact that our respondents feel they cannot directly influence how the platform works reveals much about structural power at work.

POSSIBLE SOLUTIONS: CHANGING NORMS AROUND DESIGN

Given all these difficulties in responding to misinformation on WhatsApp, how can hidden virality be captured and addressed? Now that we have given a bit of background and described the everyday burden for information curation that is offloaded to users on WhatsApp and similar platforms, thinking with the Governing Knowledge Commons framework and governance strategies focusing on changing norms around design offer some promising avenues. Here we outline solutions to change norms around design, and the benefits and shortcomings of each of these action areas, who would be responsible for enacting solutions, and who would be affected.

Meta shows that it "values free speech" by making moderation technically impossible, all the while cultivating economic power from WhatsApp's widespread use. This is an entrenched norm in the tech industry and with Meta and WhatsApp in particular. One simple but likely superficial way is for users and lawmakers across geopolitical borders to demand a change in platform norms, for example by ordering more transparency from these platforms about what content is actually shared and through which mechanisms, and allowing users to easily opt in and out of sharing certain types of data with certain parties, if they so wish. The institution of legislative policy, namely the General Data Protection Regulation (GDPR) and the California Consumer Privacy Act (CCPA) in a few key geopolitical areas, has effectively ushered in opt-in design features across a number of websites and platforms. Solutions for WhatsApp might follow similar protocols.

But this would not necessarily address the problem of the epistemic burden foisted upon those who are minoritized within private WhatsApp groups. More appropriately framing WhatsApp's onslaught of offensive and dangerous content as a platform problem would require concrete design- and norm-changing measures, such as redesigning the platform to reduce the spread of such content, likely decrease engagement, and reduce the ability to leverage the economic power of the user base. WhatsApp has limited sharing messages to keep information from spreading too quickly. While this is a positive step for the platform, the respondents in our study say they see just as much disinformation as before these share limits were enacted.

While it wouldn't necessarily limit the amount of disinformation that people see, requiring users themselves to label manipulated or satirical information might be useful to some degree in terms of limiting the work that bystanders do to debunk certain types of problematic information, as we have seen with legislation and moderation around manipulated audiovisual content (Paris 2021). But this intervention would likely not decrease the amount of problematic content that is offensive and troublesome to certain disenfranchised bystanders, nor would it reduce the tendency of the most persistent offenders to see this as censorship, as already happens with social media posts labeled as false (Ognyanova et al. 2020). Moreover, without oversight and content moderation from the platform, this intervention would be impossible to enforce.

Taking note from respondents of this study, adding an anonymous reporting feature might be a useful intervention for WhatsApp to institute. But it would need accountability and oversight measures to keep it from falling into the reporting traps that are so prevalent on other platforms – namely that nothing happens unless the person reporting, or offending, has some level of notoriety or social capital. Some examples of this are harassment reporting around caste in India, where higher caste members at tech companies ignore the harassment claims brought by those who are of lower caste (Soundarajan et al. 2019), or when women who are not public figures report manipulated nude and sexualized images of themselves being spread across platforms (Paris 2021). In these and other cases, it is precisely this resource of social capital that respondents lack, as do others on Twitter and other platforms who encounter difficulties in having their reports reviewed.

Then, also, there are strong normative expectations around existing WhatsApp affordances coming from within the user base to keep the platform as is. Redesigning the platform and instituting anonymous reporting might detract from the perceived benefits of the platform for parties who need or like it because people can share whatever they want with whomever they want, with no formalized oversight. Tech companies and their adherents often argue that changing the platform would decrease WhatsApp's pro-social possibilities; as Lim (2020) notes, there are pro-democracy and human rights groups that use the platform only because of these affordances.

But the deluge of antisocial content on WhatsApp resulting in real-world consequences need not be justified as problems necessarily generated and sustained by users' information and social practices. Instead, drawing on science and technology studies (STS) and Costanza-Chock's (2020) design justice branch of critical informatics, we must problematize this framing and envision new sociotechnical solutions that better suit user needs. This reimagination of sociotechnical systems must take the politics of knowledge into account and demand difficult discussions of design goals to focus on the goals of users, including them in the design process as experts on their informational needs, not just as data sources to exploit. This work demands that technologists grapple with the complex cultural and political contexts these technologies will be used in and develop tools that are useful, not exploitative. There are many possible avenues ahead, but all require rethinking our relationships with one another and the global community, how we communicate and cooperate within groups, and what roles technology should play in our lives.

CONCLUSIONS

As encrypted messaging apps remain one of the last uncensored spaces on the internet, it becomes vital to disentangle the dynamics of hidden virality, or when, how, and why dangerous content goes viral in these closed spaces while remaining unnoticed to outsiders. This chapter has provided an overview of the dynamics that made false and dangerous content widespread on WhatsApp in India between 2020 and 2021 while discussing users' takes on such issues. While we focus on WhatsApp in India, such methods of ethnographic research can be employed with users of other encrypted, private messaging platforms like Telegram or Signal in other countries across the globe. Certain issues around the political economy of the platform would be different in other countries and app contexts, each having their own political-economic concerns and issues around social capital (Abubakar, Hafiz, and Dasuki 2018; Chauchard and Garimella 2022; Soares et al. 2021). Further studies in this vein would provide bases for rich comparisons and a better understanding of the interlinked phenomena we see at play in this study.

The wide-scale, long-term adoption of WhatsApp in India results from the deployment of not one but a "cascade" of sociotechnical and interdependent affordances (Figure 8.1). This cascading effect has clear positive consequences for Meta, the corporate entity behind this technology. WhatsApp's affordances were not designed in a vacuum or by means of neutral intentions: They result from aggregating data across applications on user phones to leverage contacts and services and increase engagement. As Cecere, Le Guel, and Lefrere (2020), Glick and Ruetschlin (2019), and Tang (2016) note, the promise of corporate benefits drove the inclusion and successive maintenance of E2EE as a key affordance of WhatsApp in the first place; the goal of offering aggregated services at a low cost is to generate networked data that is extremely valuable to WhatsApp and its owners, making the sociotechnical affordances a tool that furthers profit-based goals. However, the same cascading effect might not be equally beneficial to users.

Users feel they must stay on the platform even though they may not like everything about it, all the while interacting with others and with information, and engaging in other activities, generating data stored on WhatsApp's servers that holds promise for revenue. WhatsApp's revenue model centers on extracting data from WhatsApp and using it to grow and market other Meta products. For example, WhatsApp has access to the phone owner's contact list. Contacts are used to suggest "new friends" on Facebook, grow Facebook's user base, and increase its market value. E2EE offers corporate benefits as it bypasses both external and internal platform oversight, which results in diminished accountability for powerful stakeholders (platform owners, shareholders, and government entities) and shifts responsibility to users, as they are the only arbiters of what content is acceptable.

The possibility of sharing information in what is perceived to be a private environment, combined with the possibility of selecting specific audiences and retaining control over messages, encourages the sharing of what is perceived to be "risky but useful" content. Such content rarely comes from official sources or mainstream media, and it is shared through a text message or an audio file and closely resembles "rumors," which have been defined as public-facing statements imbued with private hypotheses about the workings of the world (Rosnow 1991), and products of sense-making that people generate to cope with uncertainty and concomitant anxiety (Rosnow 1988). These definitions both suggest that rumors offer a "collective problem-solving opportunity to individuals who participate" (Kwon, Cha, and

Jung 2017). A key contribution of this study of how rumors spread on messaging apps is our proposal that WhatsApp users are often aware of the potential falsity of the information that they encounter on WhatsApp, but they share it anyway. This is what we call the "just-in-case" sharing practice: WhatsApp users place more importance on the potential benefits for friends and family should the information be true than the repercussions they and others might face if the rumor is false or inaccurate (e.g., losing credibility). This practice is made possible by a combination of factors and dominant cultural norms, which include the urge to care for close ones by sharing potentially useful content, the lack of awareness of the potential risks involved in amplifying false or misleading content, and the widespread preference among our participants for not correcting family members and close friends out of respect and politeness.

Our work speaks to the need to address many types of epistemic burden that manifest themselves as users engage in information and social practices through WhatsApp and other similar encrypted apps. In these examples, we see how Meta, a powerful US-based technology company, has shaped technical affordances that lead to more work in the everyday practices of correcting mis-and disinformation for those who are already disadvantaged. Typically, tech companies, the popular press, and sometimes even academics blame the presence of disinformation and hate speech on WhatsApp on the knowledge and knowledge practices of the users. This practice fails to acknowledge the role played by western companies and, most importantly, how such blame (intentional or not) further disadvantages these already burdened groups in a way that compounds their oppression. Perceived privacy, when combined with the impossibility for users to report inappropriate content anonymously on WhatsApp, actively encourages the spread of offensive content on the platform. Offensive content is particularly prominent in small, closed groups, which are tightly controlled by administrators. Group members need to personally contact administrators every time they are exposed to offensive content and ask them to remove it. The burden of flagging offensive content falls on the group members who might feel hurt by it, but once again must challenge dominant cultural norms to request moderation.

REFERENCES

Abubakar, Naima Hafiz, and Salihu Ibrahim Dasuki. 2018. "Empowerment in Their Hands: Use of WhatsApp by Women in Nigeria." *Gender, Technology and Development* 22 (2): 164–183. https://doi.org/10.1080/09718524.2018.1509490.

Acker, Amelia, and Mitch Chaiet. 2020. "The Weaponization of Web Archives: Data Craft and COVID-19 Publics." *Harvard Kennedy School Misinformation Review* 1 (3): 1–11. https://doi.org/10.37016/mr-2020-41.

Banaji, Shakuntala, Ramnath Bhat, Anushi Agarwal, Nihal Passanha, and Mukti Sadhana Pravin. 2019. *WhatsApp Vigilantes: An Exploration of Citizen Reception and Circulation of WhatsApp Misinformation Linked to Mob Violence in India*. London: London School of Economics. http://eprints.lse.ac.uk/104316/1/Banaji_whatsapp_vigilantes_exploration_of_citizen_reception_published.pdf.

Benghani, Priyanjana. 2019. "India Had Its First 'WhatsApp Election.' We Have a Million Messages from It." *Columbia Journalism Review*. October 16. www.cjr.org/tow_center/india-whatsapp-analysis-election-security.php.

Broadcasting Board of Governors (BBG). 2014. *BBG Board Meeting, Part 4*. Broadcasting Board of Governors headquarters in Washington, DC. April 11. http://archive.org/details/BBGBoardMeeting04112014.

2017. "2016 Annual Report." https://2016.bbg.gov/.

Cecere, Grazia, Fabrice Le Guel, and Vincent Lefrere. "Economics of Free Mobile Applications: Personal Data and Third Parties." Preprint, Social Science Research Network, last revised: March 15 2020. https://doi.org/10.2139/ssrn.3136661.

Chakrabarti, Santanu, Lucile Stengel, and Sapna Solanki. 2018. "Fake News and the Ordinary Citizen in India." *Beyond 'Fake News'*. London: British Broadcasting Corporation. http://downloads.bbc.co.uk/mediacentre/duty-identity-credibility.pdf.

Chauchard, Simon, and Kiran Garimella. 2022. "What Circulates on Partisan WhatsApp in India? Insights from an Unusual Dataset." *Journal of Quantitative Description: Digital Media* 2 (March): 1–76. https://doi.org/10.51685/jqd.2022.006.

Cheeseman, Nic, Jonathan Fisher, Idayat Hassan, and Jamie Hitchen. 2020. "Social Media Disruption: Nigeria's WhatsApp Politics." *Journal of Democracy* 31 (3): 145–159. https://doi.org/10.1353/jod.2020.0037.

Citron, Danielle Keats. 2016. *Hate Crimes in Cyberspace*. Reprint ed. Place of publication not identified: Harvard University Press. www.hup.harvard.edu/books/9780674659902.

Collins-Dexter, Brandi. 2020. "Canaries in the Coal Mine: COVID-19 Misinformation and Black Communities." Technology and Social Change Research Project. The Shorenstein Center on Media, Politics and Public Policy. https://doi.org/10.37016/TASC-2020-01.

Costanza-Chock, Sasha. 2020. *Design Justice: Community-Led Practices to Build the Worlds We Need*. Cambridge, MA: MIT.

Diwanji, Sanika. 2019. "India – Monthly Active Users on WhatsApp 2017." Statista. Accessed 2019. www.statista.com/statistics/280914/monthly-active-whatsapp-users-in-india/.

Evangelista, Rafael, and Fernanda Bruno. 2019. "WhatsApp and Political Instability in Brazil: Targeted Messages and Political Radicalisation." *Internet Policy Review* 8 (4): 1–23. https://doi.org/10.14763/2019.4.1434.

Facebook, Inc. 2020. "Facebook Press Call Transcript." Final Transcript March 18, 2020. https://about.fb.com/wp-content/uploads/2020/03/March-18-2020-Press-Call-Transcript.pdf.

Fichman, Pnina, and Maanvi Rathi. 2021. "Cross-Cultural Analysis of Trolling Behaviors." *Proceedings of the Association for Information Science and Technology* 58 (1): 716–717. https://doi.org/10.1002/pra2.539.

Fichman, Pnina, and Madelyn Rose Sanfilippo. 2015. "The Bad Boys and Girls of Cyberspace: How Gender and Context Impact Perception of and Reaction to Trolling." *Social Science Computer Review* 33 (2): 163–180. https://doi.org/10.1177/0894439314533169.

Fricker, Miranda. 2007. *Epistemic Injustice: Power and the Ethics of Knowing*. Oxford: Oxford University Press. https://academic.oup.com/book/32817.

Garimella, Kiran, and Dean Eckles. 2020. "Images and Misinformation in Political Groups: Evidence from WhatsApp in India." *Harvard Kennedy School Misinformation Review* 1 (5): 1–12. https://doi.org/10.37016/mr-2020-030.

Glick, Mark, and Catherine Ruetschlin. 2019. "Big Tech Acquisitions and the Potential Competition Doctrine: The Case of Facebook." Working Paper Series. Institute for New Economic Thinking, October. www.ineteconomics.org/research/research-papers/big-tech-acquisitions-and-the-potential-competition-doctrine-the-case-of-facebook.

Herring, Susan, Kirk Job-Sluder, Rebecca Scheckler, and Sasha Barab. 2002. "Searching for Safety Online: Managing 'Trolling' in a Feminist Forum." *The Information Society* 18 (5): 371–384. https://doi.org/10.1080/01972240290108186.

Ihm, Jennifer, and Eun-mee Kim. 2018. "The Hidden Side of News Diffusion: Understanding Online News Sharing as an Interpersonal Behavior." *New Media and Society* 20 (11): 4346–4365. https://doi.org/10.1177/1461444818772847.

International Crisis Group. 2017. "Myanmar's Rohingya Crisis Enters a Dangerous New Phase." Report N° 292. Belgium: International Crisis Group. www.crisisgroup.org/asia/south-east-asia/myanmar/292-myanmars-rohingya-crisis-enters-dangerous-new-phase.

Joyojeet, Pal, 2020. "Temporal Patterns in COVID-19 Misinformation in India." *Joyojeet Pal* (blog). April 16. http://joyojeet.people.si.umich.edu/temporal-patterns-in-covid-19-misinformation-in-india/.

Khan, Nabeela. 2020. "Trends in Covid-19 Misinformation in India." *Health Analytics Asia* (blog). June 12. www.ha-asia.com/trends-in-covid-19-misinformation-in-india/.

Kumar, Harish, Pushpneel Verma, Rudrapratap Ojha, and G. Venkat Babu. 2012. "Study Analysis of Capacity Enhance and Coverage Increase with Repeaters in 3rd Generation for Rural/Semi Urban Mobile Communication System in India." In *2012 Second International Conference on Digital Information and Communication Technology and It's Applications (DICTAP)*, 368–372. https://doi.org/10.1109/DICTAP.2012.6215377.

Kuo, Rachel, and Alice Marwick. 2021. "Critical Disinformation Studies: History, Power, and Politics." *Harvard Kennedy School Misinformation Review* 2 (4): 1–12. https://doi.org/10.37016/mr-2020-76.

Kuru, Ozan. 2019. "Understanding Informational Processing in WhatsApp Groups: A Comparative Study of User Perceptions and Practices in Turkey, Singapore, and the USA." Conference Presentation at the American Political Science Association 2019, Washington DC., August 28. www.asc.upenn.edu/news-events/news/annenberg-presentations-apsa-2019.

Kwon, Sejeong, Meeyoung Cha, and Kyomin Jung. 2017. "Rumor Detection over Varying Time Windows." *PLoS ONE* 12 (1): e0168344. https://doi.org/10.1371/journal.pone.0168344.

Liao, Shannon. 2018. "WhatsApp Tests Limiting Message Forwarding after Violent Lynchings in India." *The Verge*, July 20. www.theverge.com/2018/7/20/17595478/whatsapp-message-forwarding-end-violent-lynching-india.

Lim, Gabrielle. 2020. "Securitize/Counter-Securitize." Data and Society Research Institute, March 25. https://datasociety.net/library/securitize-counter-securitize/.

Machado, Caio, Beatriz Kira, Vidya Narayanan, Bence Kollanyi, and Philip Howard. 2019. "A Study of Misinformation in WhatsApp Groups with a Focus on the Brazilian Presidential Elections." In *WWW '19: Companion Proceedings of The 2019 World Wide Web Conference*. New York: Association for Computing Machinery, 1013–1019. https://doi.org/10.1145/3308560.3316738.

Maltby, John, Liz Day, and Caren A. Frosch et al. 2016. "Implicit Theories of Online Trolling: Evidence That Attention-Seeking Conceptions Are Associated with Increased Psychological Resilience." *British Journal of Psychology* 107 (3): 448–466. https://doi.org/10.1111/bjop.12154.

Mozur, Paul. 2018. "A Genocide Incited on Facebook, With Posts From Myanmar's Military." *The New York Times* (sec. Technology), October 15. www.nytimes.com/2018/10/15/technology/myanmar-facebook-genocide.html.

Naeem, Salman Bin, and Rubina Bhatti. 2020. "The Covid-19 'Infodemic': A New Front for Information Professionals." *Health Information and Libraries Journal* 37 (3): 233–239. https://doi.org/10.1111/hir.12311.

Noble, Safiya. 2013. "Google Search: Hyper-Visibility as a Means of Rendering Black Women and Girls Invisible." *InVisible Culture: An Electronic Journal for Visual Culture*, no. 19 (October 29). http://ivc.lib.rochester.edu/google-search-hyper-visibility-as-a-means-of-rendering-black-women-and-girls-invisible/.

Ognyanova, Katherine, David Lazer, Ronald E. Robertson, and Christo Wilson. 2020. "Misinformation in Action: Fake News Exposure Is Linked to Lower Trust in Media, Higher Trust in Government When Your Side Is in Power." *Harvard Kennedy School Misinformation Review* 1(4): 1–19. https://doi.org/10.37016/mr-2020-024.

Paris, Britt. 2021. "Configuring Fakes: Digitized Bodies, the Politics of Evidence, and Agency." *Social Media + Society* (online) 7 (4). https://doi.org/10.1177/20563051211062919.

Paris, Britt and Joan Donovan. 2019. "Deepfakes and Cheap Fakes: The Manipulation of Audio and Visual Evidence." *Data and Society*. September 18. https://datasociety.net/output/deepfakes-and-cheap-fakes/.

Paris, Britt, Jennifer Pierre, Morgan Currie, and Irene Pasquetto. 2022. "Data Burdens: Epistemologies of Evidence in Police Reform and Abolition Movements." In *Data Justice and the Right to the City*, ed. Morgan Currie, Jeremy Knox, and Callum McGregor, 301–325. Edinburgh: University of Edinburgh Press.

Pasquetto, Irene V., Eaman Jahani, Shubham Atreja, and Matthew Baum. 2022. "Social Debunking of Misinformation on WhatsApp: The Case for Strong and In-Group Ties." *Proceedings of the ACM on Human-Computer Interaction* 6 (CSCW1), Article 117: 1–35. https://doi.org/10.1145/3512964.

Pierre, Jennifer, Roderic Crooks, Morgan E. Currie, Britt S. Paris, and Irene V. Pasquetto. 2021. "Getting Ourselves Together: Data-Centered Participatory Design Research and Epistemic Burden." In *CHI '21: Proceedings of the 2021 CHI Conference on Human Factors in Computing Systems*. Yokohama, Japan. Article 406: 1–11. https://doi.org/10.1145/3411764.3445103.

Prasad, Revati. 2018. "Ascendant India, Digital India: How Net Neutrality Advocates Defeated Facebook's Free Basics." *Media, Culture and Society* 40 (3): 415–431. https://doi.org/10.1177/0163443717736117.

Roose, Kevin. 2018. "U.S.-Funded Broadcaster Directed Ads to Americans." *The New York Times* (sec. Technology), July 19. www.nytimes.com/2018/07/19/technology/facebook-ads-propaganda.html.

Rosnow, Ralph L. 1988. "Rumor as Communication: A Contextualist Approach." *Journal of Communication* 38 (1): 12–28. https://doi.org/10.1111/j.1460-2466.1988.tb02033.x.

1991. "Inside Rumor: A Personal Journey." *American Psychologist* 46 (5): 484–496. https://doi.org/10.1037/0003-066X.46.5.484.

Rossini, Patrícia, Jennifer Stromer-Galley, Erica Anita Baptista, and Vanessa Veiga de Oliveira. 2020. "Dysfunctional Information Sharing on WhatsApp and Facebook: The Role of Political Talk, Cross-Cutting Exposure and Social Corrections." *New Media and Society* 23 (8): 2430-2451. https://doi.org/10.1177/1461444820928059.

Shachaf, Pnina, and Noriko Hara. 2010. "Beyond Vandalism: Wikipedia Trolls." *Journal of Information Science* 36 (3): 357–370. https://doi.org/10.1177/0165551510365390.

Siddiquee, Md Ali. 2020. "The Portrayal of the Rohingya Genocide and Refugee Crisis in the Age of Post-Truth Politics." *Asian Journal of Comparative Politics* 5 (2): 89–103. https://doi.org/10.1177/2057891119864454.

Soares, Felipe Bonow, Raquel Recuero, Taiane Volcan, Giane Fagundes, and Giéle Sodré. 2021. "Research Note: Bolsonaro's Firehose: How Covid-19 Disinformation on WhatsApp Was Used to Fight a Government Political Crisis in Brazil." *Harvard Kennedy School Misinformation Review* 2 (1): 1–12. https://doi.org/10.37016/mr-2020-54.

Soundarajan, Thenmozhi, Abishek Kumar, Priya Nair, and Josh Greely. 2019. "Facebook India Report." NYC: Equality Labs. www.equalitylabs.org/facebookindiareport.

Tang, Ailie K. Y. 2016. "Mobile App Monetization: App Business Models in the Digital Era." *International Journal of Innovation, Management and Technology* 7 (5): 224–227. https://doi.org/10.18178/ijimt.2016.7.5.677.

Telecom Regulatory Authority of India. 2016. "Prohibition of Discriminatory Tariffs for Data Services Regulations." February 8, 2. www.trai.gov.in/sites/default/files/Regulation_Data_Service.pdf.

Valeriani, Augusto, and Cristian Vaccari. 2018. "Political Talk on Mobile Instant Messaging Services: A Comparative Analysis of Germany, Italy, and the UK." *Information, Communication and Society* 21 (11): 1715–1731. https://doi.org/10.1080/1369118X.2017.1350730.

Wardle, Claire. 2020. "Monitoring and Reporting Inside Closed Groups and Messaging Apps." In *Verification Handbook for Disinformation and Media Manipulation*, ed. Craig Silverman, 94–97. DataJournalism.com. https://datajournalism.com/read/handbook/verification-3/investigating-platforms/7-monitoring-and-reporting-inside-closed-groups-and-messaging-apps.

WhatsApp. 2016. "Making WhatsApp Free and More Useful." *WhatsApp.Com* (blog). September. https://blog.whatsapp.com/making-whats-app-free-and-more-useful.

2020. "WhatsApp Privacy." *WhatsApp.Com* (blog). 2020. www.whatsapp.com/privacy.

9

"Do Your Own Research"

Everyday Misinformation and Conspiracy in Online Information Worlds

Gary Burnett and Shannon Williams

INTRODUCTION

Access to everyday information has evolved rapidly over the last twenty years and so have information-seeking behaviors. With the ever-increasing amount of data made available to individuals in this Information Age, there is a trending movement to filter information and tailor it to one's individual interest and algorithms help us achieve this goal.

The Pew Research Center reports the Internet of Things (IoT) is making it virtually impossible to disconnect from all forms of internet access making the ability to avoid undesirable information near impossible. Pew researchers Rainie and Anderson (Rainie and Anderson 2017) report that despite wide concern about cyberattacks, outages, and privacy violations, most experts believe the Internet of Things will continue to expand successfully over the next few years, tying machines to machines and linking people to valuable resources, services, and opportunities. Connection begets connection.

All this, with the ubiquity of social media and the constant stream of all different kinds of information flowing through one's daily feed, the concept of "everyday information" as a category distinct from more consequential (or "serious") information becomes blurred – all information is constantly accessible and constantly within view, so all information becomes, in a sense, "everyday information."

Conspiracy theories are predicated on the spread of misinformation and disinformation in such a way that it becomes an integral part of people's everyday information practices and how they view and understand all aspects of the world around them, affecting family and workplace relationships (Rothschild 2021).

The findings below were gathered through close monitoring and open-ended interpretivist analysis of numerous public Facebook pages over several months, augmented by extensive searching of websites, web archives, YouTube videos, and more, for related content. Our analysis of the online activities of conspiracy theorists and their related posts can help uncover the mis- and disinformation being

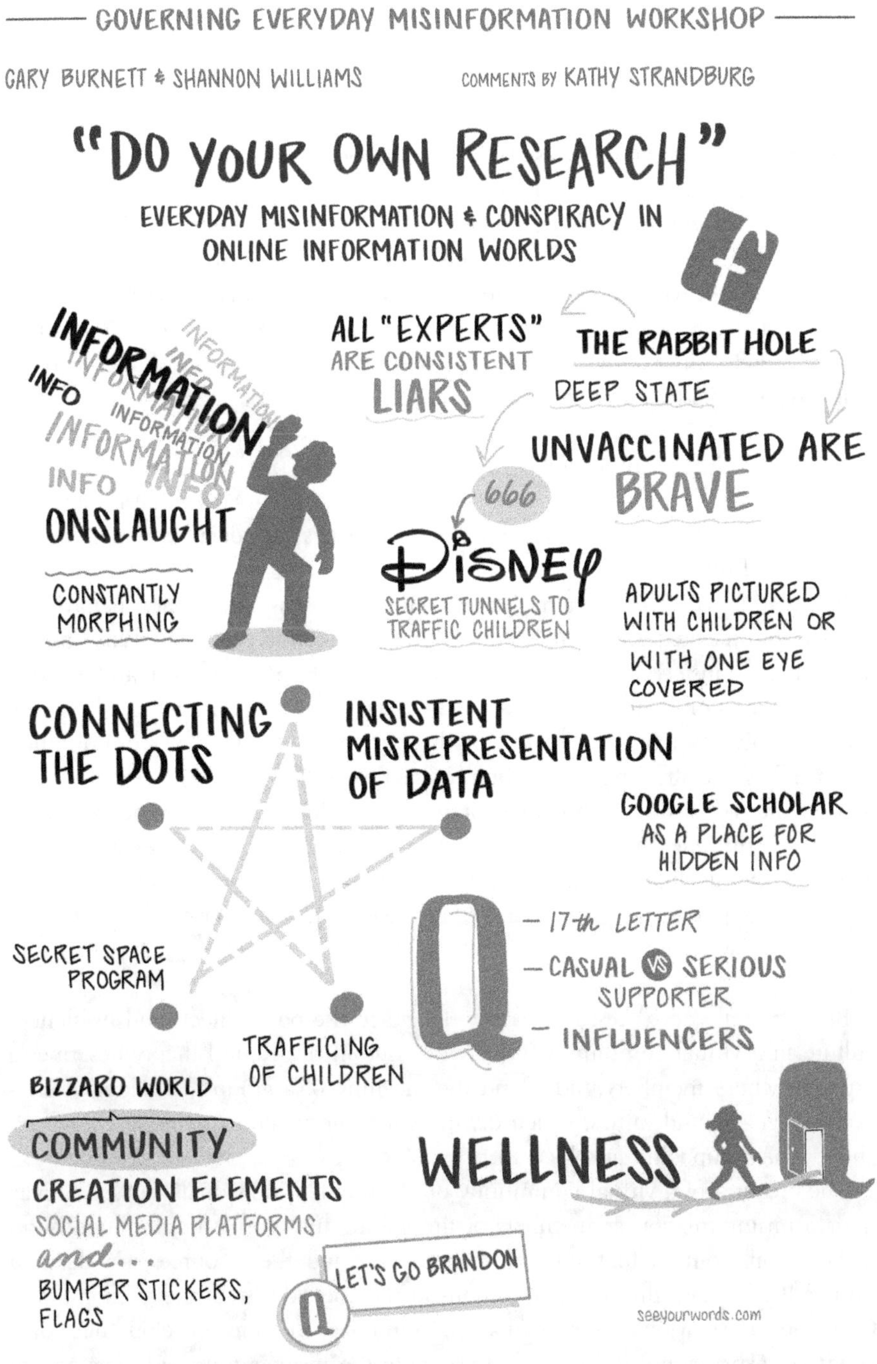

FIGURE 9.1 Visual themes from "do your own research": everyday misinformation and conspiracy in online information worlds.

promoted and, in some cases, and can help us gain insights about the connections between the creating, sharing and the receiving of misinformation as "fact."

Our analysis was influenced by theory of Information Worlds and by the Governing Knowledge Commons (GKC) framework. The GKC framework, which is discussed more extensively elsewhere in this book, investigates how the rules-in-use of a particular community are codetermined by the background environment, including rules and norms determined at higher contextual and societal levels, and that helped influence our thinking (Sanfilippo, Frischmann, and Strandburg 2021).

For some, the continuous onslaught of everyday information is managed by filtering what one wants to encounter as opposed to information adverse to one's beliefs, therefore inadvertently and sometimes purposefully creating filter bubbles that generate small worlds that generate information worlds.

Small worlds, a theory developed by Chatman (1991), erects barriers against unwelcomed information and restricts information sharing thus creating an insider/outsider environment. Jaeger and Burnett further process Chatman's theory by connecting the ways in which information creates the social worlds of people (Jaeger and Burnett 2010).

Information Worlds is a framework for conceptualizing information behavior; it offers a multifaceted argument involving the intersection of personal and public information transfer (Burnett, Jaeger, and Thompson 2008); Jaeger and Burnett 2010). Information Worlds is rooted in the social study of information, taking into account activities that are situated in the customary practices of communities, specifically those affecting information interactions.

The internet has been available for public use for over twenty-five years and virtual communities have existed as a way to exchange pleasantries and argue, engage in intellectual discourse, conduct commerce, exchange knowledge, share emotional support, make plans, brainstorm, gossip, feud, fall in love, find friends and lose them, play games, flirt, create a little high art and a lot of idle talk (Rheingold 1993).

Burnett et al. (2003) researched how language use both reflects and influences culture in a virtual community. Chatman's Life in the Round theory describes a situation where members understand the meaning of a group's particular expressions, language, and values, which defines what information is considered acceptable by the group (Zhu and Liao 2020).

The creation of a virtual community of like-minded individuals can easily play host to misinformation as members of the community primarily introduce information from sources that support their beliefs, and these sources are deemed trustworthy because they are harmonious to member ideals (Burnett, Besant, and Chatman 2001). Members of virtual communities, in time, develop their own language, expressions, and symbols to identify the insiders from outsiders and to solidify a bond or comradery amongst each other (Burnett, Besant, and Chatman 2001; De Vynck and Lerman 2021; Squirrell and Sonnad 2017).

The algorithms of social media sites monitor the interests of their users and develop custom filter bubbles, which are the result of the careful curation of social media feeds that enable users to be surrounded by like-minded people and information that is congruent with their existing beliefs (Cooke 2017). A Facebook user who likes a post concerning flat-earthers will see ads and promotions for similar interests and conspiracies. Filter bubbles are further developed by confirmation bias, which suggests that users play a role in the creation of their virtual communities by actively seeking and using information that already coincides with existing mental schema, as opposed to seeking information from a variety of potentially conflicting sources (Cooke 2017).

A possible outcome of collective virtual information activities, like those associated with the QAnon movement, is information avoidance. Cooke (2017) states it is possible for people to avoid distasteful or upsetting information while in their social media filter bubbles. Case and Given (2016) refer to this as selective exposure, or selective information seeking, which is defined as the tendency to seek information that is congruent with prior knowledge, beliefs, and opinions, and to avoid exposure to information that conflicts with those internal states.

One such group of like-minded individuals refer to themselves as QAnon. QAnon is an American political conspiracy theory and political movement that originated in the American far-right political sphere in 2017 (Martineau 2017). The core of the QAnon movement centers around false claims made by an anonymous individual or individuals known as "Q" and the central QAnon theory is that a cabal of Satanic, cannibalistic sexual abusers of children operating a global child sex trafficking ring conspired against former US President Donald Trump during his term in office (Bracewell 2021). QAnon has direct roots in Pizzagate, an internet conspiracy theory that appeared one year earlier and which we discuss below (Roose 2021). It also incorporates elements of many other conspiracy theories, and some experts go so far as to describe QAnon as a cult (Davies 2021; Mulkerrins 2021; Polantz 2021; Roose 2021; Stanton 2020; Uscinski 2018, 1–32)

Despite efforts to ban or slow the sharing of QAnon-related information by Facebook, Instagram, and other social media platforms, the movement continues to thrive primarily because it weaves a myriad of conspiracies, from antivaccine and anti5G conspiracies to antisemitic and antimigrant tropes (Haimowitz 2020). Unlike other conspiracy theories that are linear and hard focused, studying the QAnon movement in an effort to combat everyday misinformation is ideal because this particular conspiracy theory has the fluidity to morph, accommodating the needs of each individual suspicious mind. While the main focus on Q is to stop the cabal of elites battling Trump, the conspiracy theory has managed to strengthen, grow appendages, and even diverge to different countries such as Japan where the conspiracies have included local issues on whether complaints that the prime minister's media ratings are inflated, to even suggesting that a Japanese movie, *Demon Slayer: Kimetsu no Yaiba*, is warning the

Japanese people of the pedophilia and cannibalism of the international elite (Haimowitz 2020).

It has become conventional wisdom in public discourse that misinformation and conspiracy theories have become more widespread since the advent and growth of social media platforms (Suciu 2020). In response, information consumers are forming behaviors and practices in an attempt to cope with undesirable exposure to content that can potentially become knowledge. This filtering or control of information on the internet has aided in the spread of conspiracy theories through social media platforms (Stecula and Pickup 2021).

Some of the most popular social media groups (specifically Facebook and Twitter) have taken steps to ban/remove major QAnon activity. As social media tightens restrictions on certain hate, violent, and sexual language, individuals develop alternate language to thwart algorithms and continue to communicate their message to insiders within their circle and newly interested individuals. Outsiders must do their own research to decipher the encoding in order to develop an understanding.

The infant formula crisis in 2022 is believed to be a conspiracy by QAnon followers. Parents across the US struggled to find baby formula after a national product recall triggered empty store shelves, higher prices, and retailers limiting sales of scarce products (Baldwin 2022). An example of the "encoding" practices related to the shortage shows on a public social media page as "M@N^UF^@CTo^R€D CR^!$^!$" and interprets as "Manufactured Crisis." The formula shortage, in fact, stemmed from a product recall by an Abbott Nutrition facility in Michigan, where unsanitary conditions and contaminated products led the US Food and Drug Administration (FDA) to temporarily shut down the facility (Pathak et al. 2022). The encoding practices haver been formulated as an attempt to circumvent speech restrictions on Facebook and other social media sites.

As a result of speech restrictions on major social media platforms (Facebook, Instagram, Twitter, etc.) alternative sites that are more amenable to QAnon and related conspiracy practices such as Telegram, Donald Trump's Truth Social, and so on, have developed to support believers. Telegram's appeal for fringe communities can be explained by its ability to reconcile private messaging and broadcasting, offering users both protected messaging and public channels to attain publicity (Rogers 2020). Unlike other social media platforms that are public-facing and require aliases to maintain anonymity, Telegram is at its core an instant messaging app with an emphasis on privacy (Urman and Katz 2020).

RELATED LITERATURE

The ease of accessing news and other data in this Information Age allows those seeking information to retrieve an abundance of data within seconds. Multiple sources of technology used daily bringing information to our fingertips have become an everyday part of most people's lives. So much so, that those individuals seeking to avoid information will find it difficult to do so.

There has been a surge in group activity on social media linking like-minded individuals to others who share the same beliefs, creating a comfortable "small world" atmosphere where one does not need to confront or defend alternative views, ostensibly creating spaces where everyday information is shared and received unchecked as truth from trusted community sources (Chatman 1991).

A report from NYU's The Governance Lab on the role online communities play in peoples' lives states that Facebook groups are used by 1.8 billion people every month with these users citing a sense of belonging and a strong sense of community in a virtual group as an attraction even when members are physically far apart (Noveck et al. 2021).

Virtual communities can not only be a positive place to collaborate with others, but it can also become a protected place to foster misinformation and breed conspiracy theories and other rhetoric on an everyday basis. The definition of a reliable source changes in these online groups as members develop trust in highly active information providers in the virtual community.

But virtual communities are not new to the internet. Virtual communities have been in existence since the widespread use of the internet in homes. America Online (AOL) provided space for publicly accessible themed chat groups and private chat rooms for group communication in the mid to late 1990s and the concept of virtual communities has survived through the decades. These virtual communities essentially transform into information worlds. Information worlds reflect the dual role of the individual in society, and it reflects an individual's day-to-day information habits (Jaeger and Burnett 2010).

Information Worlds is a very versatile theory that can be used with a variety of methods, and focuses on describing information in social contexts, ranging from very small and local contexts (e.g., an academic department) to the larger contexts in which those are embedded (e.g., a university) (Jaeger and Burnett 2010).

Information Worlds focuses on the social aspects of information in settings of all sizes, from very localized contexts to broader social contexts, and the interactions between those multiple worlds. It draws upon Elfreda Chatman's concept of small worlds (Burnett, Besant, and Chatman 2001) and Jurgen Habermas' concept of the lifeworld (Habermas 1992) – the "collective information and social environment that weaves together the diverse information resources, voices, and perspectives" across a culture as a whole (Jaeger and Burnett 2010). The theory argues that individual information worlds are never isolated. Instead, they overlap, intersect, and interact in a variety of ways, all of which has an impact on how information is conceptualized and used within and across worlds.

Burnett and Jaeger's theory of Information Worlds is comprised of a set of five interconnected concepts: Social Norms, Social Types, Information Value,

Information Behavior, and Boundaries. The concepts of Social Norms, Social Types, and Information Behavior are derived directly from the work of Chatman (Burnett et al. 2001), while Information Value significantly revises her concept of Worldview. The concept of boundaries is the newest concept to the theory.

Social Norms refers to those agreed-upon observable behaviors that are common and accepted within a world. Within a virtual group of QAnon followers, the social norms would include anti-establishment ideologies rooted in a quasi-apocalyptic desire to destroy the existing, "corrupt" world order and usher in a promised golden age (GNET Team 2020).

Social Types refers to perceptions of the roles played by individuals within a world which may be explicit or implicit. QAnon followers perceive the enigmatic figure "Q" as the leader of the movement. Q is an anonymous poster claiming to be a high-level government insider who sporadically "drops" a series of cryptic posts, often written in the form of a set of questions on various aspects of US domestic and foreign affairs known as Q drops (Zihiri et al. 2022). The identity of Q has also been expanding; at times, online posts have implied that more than one person is involved with the account, and more specifically, that former US President Trump may be either Q or an even more powerful figure identified as Q+ (Zihiri et al. 2022). At the broadest level, QAnon adherents engage in social typing through their collective identity as "anons" and their typecasting all non-adherents as "sheep" or "normies."

Information Value highlights the notion that not only does each world have its own agreed-upon (if often implicit) scale for assessing the importance of different kinds of information but also that the kind of value attached to information (and the appropriate metrics for weighing that value) may differ from world to world (Burnett 2015).

As an ideology, the QAnon belief system allows for the development of symbolic resources that enable those who believe to place an information value of the highest level on "Q drops" viewed as "intelligence leaks" issued by "Q" online with the firm belief that these Q drops are based on empirical evidence and consequently, they rely on the creation of elaborate, often labyrinthine productions of said "evidence" in order to substantiate and decipher QAnon's coded claims (GNET Team 2020). See Figure 9.2.

For true believers in the movement, this reliance on the essential, coded "truths" of QAnon render it unfalsifiable and largely for this reason, QAnon adherents are not passive consumers of QAnon content; they are also online activists, content creators who generate memes, videos, texts, music, and films, much of which in turn goes on to have its own life, feeding and generating more "Q" claims (GNET Team 2020).

Information Behavior refers to the full range of normative activities and practices related to information within a world. A QAnon virtual community confronts information contrary to their beliefs by, first, delegitimizing voices of dissent both within and outside of the movement by utilizing primarily less credible information sources (Zihiri et al. 2022).

Boundaries are the places at which different worlds come into contact with each other in one way or another (Burnett 2015). Defining a boundary for the QAnon conspiracy movement is nigh impossible as QAnon has often been described as the umbrella term for a sprawling spiderweb of right-wing internet conspiracy theories with antisemitic and antiLGBTQ elements that falsely claim the world is run by a secret cabal of pedophiles who worship Satan and are plotting against President Trump (Hatewatch Staff 2020). QAnon adherents can include antimaskers, COVID-19 deniers and antivaxxers to Pizzagate and flat-earthers. The exact nuances comprising the conspiracy are challenging to pinpoint as the theory posits that a range of political, social, and cultural elites are waging war against freedom and decency while Trump leads a counterattack against these malicious forces (Zihiri et al. 2022).

CONSPIRACY THEORIES AND INFORMATION BEHAVIOR

The QAnon conspiracy theory has been linked to numerous violent acts since 2018, with QAnon supporters arrested for threatening politicians, breaking into the residence of the Canadian prime minister, an armed standoff near the Hoover dam, a kidnapping plot and two kidnappings, and at least one murder (Beckett 2020).

Hannah (2021b) argues that despite the fact that Americans are navigating data at an astounding rate, we have still not developed national initiatives to promote media or information literacies to debunk bad information derived from that data. As a result of this exposure to massive amounts of data, Americans have become increasingly accustomed to interpretations of that data provided via visualizations and dashboards.

The QAnon movement comprised a group of anonymous online posters who organized a crowdsourced conspiracy theory believing they were coordinating with President Trump to expose and punish a cadre of globalist, Satanic pedophiles. In the QAnon eschatology, a "Great Awakening" was imminent in which President Trump, with the help of a mysterious intelligence operative known as Q, would unseal indictments of these globalists thereby destroying the conspiracy.

Through a uniquely modern fusion of dark Web anonymity, social media communities, distrust of mainstream media, and hostility toward academic expertise, we are now witnessing a "Great Awakening" as Q refers to it, of online organizing around particular notions of power and paranoia (Hannah 2021a). This theory of power relies on disdain toward government/media/academia and a belief that entrenched elites manipulate national policy for their own perverse desires, as followers believe that powerful politicians and celebrities are involved in trafficking children, pedophilia, harvesting organs, and worshipping Satan (Hannah 2021a).

INFORMATION AUTHORITY AND GATEKEEPING

While we have seen claims on publicly accessible QAnon-related Facebook pages that the movement proper has *only* to do with the series of Q "drops" having to reveal the secrets of the elite "deep state" and the associated push to wrest power from those elites, other conspiracy theory beliefs, sometimes associated with QAnon, circle around what nonadherents would consider arcane or far-fetched ideas.

These include, for example, the notion that the elite are "really" shape-shifting lizard people or the association of flat-earth theory with complex "maps" of the firmament and the geographic regions of heaven, as well as frankly gruesome fantasies such as the core QAnon narrative about the "harvesting" of adrenochrome through the torture, rape, and eating of young children (much of the more recent – and even widespread – hysteria surrounding the ostensible "grooming" of children by gays, transexuals, or even Disney employees seems related to this particular narrative).

Still other beliefs are linked to more mundane issues of day-to-day life in the modern world. Many of these are couched in rhetoric or ideas related to questions of information authority or governance and questions of source reliability. Given that the growth of QAnon has largely coincided with the spread of COVID-19, it is perhaps not surprising that one of the most persistent real-world conspiracy themes has centered not only around the efficacy and safety of the vaccination, but also more generally around information sources and the credibility and/or motives of these sources. This section reviews a few of such situations, in which authoritative information sources have been overtly challenged or rejected by conspiracy adherents, or in which such adherents have drawn upon a different understanding of what constitutes informational authority.

For example, the Vaccine Adverse Event Reporting System (VAERS) has been a persistent focus of attention. VAERS, which has been in existence since 1990 and has been used to monitor a wide range of vaccinations, explicitly disavows any causal relationship between vaccines and the "adverse events" it records. Rather, as their website says, "VAERS is not designed to determine if a vaccine caused a health problem" (VAERS n.d.), but only records chronology – adverse events of all sorts that take place at some point in time *after* a vaccination.

Still, in the QAnon information world – and in the conspiracy-oriented politically conservative information world more generally – the data reported by VAERS has been persistently, and loudly, misused to argue that COVID-19 vaccines had "caused" alarming numbers of dangerous reactions and even death; as one Facebook post, quoting an online article by the Truth and Liberty Coalition (an ultra-conservative religious group, not apparently linked to the QAnon movement, but sharing some of its political leanings) unambiguously puts it, "more than 13,000 vaccine-related deaths were reported in the United States through August 13 [2021], along with an additional 13,000 life-threatening conditions *caused*. More than

54,000 hospitalizations were also *attributed* to the vaccine" (Truth and Liberty Staff 2021; emphasis added).

Such claims rely on at least a semblance of "authority" and evidence rooted in source data, albeit filtered through active and persistent misreading and misrepresentation. Given the conspiracy-related aspects of such misreading, rooted in "anti-vax" rhetoric and political claims that COVID-19 was created in a lab, it is not surprising that similar approaches have been used in relation to Dr. Anthony Fauci, long the most visible public figure in efforts to control the pandemic. A recent series of Facebook posts, portraying him as suffering from the "little man syndrome (like Hitler)," links to a variety of conspiracy-oriented websites in order to misread some of his own statements in order to vilify him and to suggest that antipandemic measures were always and simply intended to "take away our freedoms." For instance, a linked article in the far-right now-deleted site *Communities Digital News*, titled "The arrogance of Dr. Anthony Fauci and his fallible science" used a quote from him saying that "detractors 'are really criticizing science'" as evidence of his ego and "self-proclaimed infallibility," and linking him and the Center for Disease Control and Prevention (CDC) to eugenics. Ultimately, the article dismisses science altogether, claiming that it "like Fauci himself [is] devoid of that singular requirement of those with souls – a moral compass." (Note: as of this writing, *Community Digital News* no longer exists, but has been replaced by an equally far-right site named *American Wire: America's Political News Service*. The article from which we have quoted is no longer extant, and *Community Digital News* does not appear to have been cached by the Internet Archive's Wayback Machine or elsewhere; for *American Wire*, see https://americanwirenews.com.)

Again, this argument rests on a semblance of informational authority in its presentation, including quotations and citations to information sources, filtered through quite radical misreading. Dr. Fauci's statement that they are "really criticizing science" is, as the article itself ultimately demonstrates, entirely accurate, but it is not a statement trumpeting his own "infallibility," but rather a defense of scientific standards of evidence and process – the scientific method. The article – and the Facebook posts sharing it – rhetorically gestures in the direction of following standards of evidence by providing quotes and naming sources but uses those quotes and sources to present pure misinformation in support of a radically tendentious set of political claims.

While these two instances – misinformation ostensibly based on VAERS data and the misrepresentation of Dr. Fauci – mix a global-level crisis with day-to-day information activities via social media, a third is more "down to earth," in the sense that it has direct implications for the management of daily life, with or without a pandemic. That is, it offers a clear intersection between conspiracy theories, online dissemination of misinformation, questions of informational authority, and everyday life. As reported in a May 2022 article in *Vice* (Lamoureux 2022), a Canadian woman – a QAnon adherent, and an active participant in QAnon-related

Telegram groups – decreed that she was the true "Queen of Canada," ruling the country behind the scenes, in the process garnering more than 70,000 followers. In this guise, she has issued a series of directives to her followers, clearly containing quite striking misinformation:

> Didulo has issued several "royal decrees" on her Telegram page, some regarding utility bills. The critical ones are "Decree 24," claiming that electricity is now free in Canada; "Decree 15," which abolishes income tax; and "Decree 23," which makes water bills illegal. Another decree, number 79, reverts the price of rent, housing, and propane back to 1955 levels. Other decrees issued by Didulo are that critical race theory is illegal in Canada (this was her very first decree, in fact) and that the age of consent was changed to 24 – which sparked an outcry from her followers.

In response, a number of her followers have stopped paying utility bills, as well as other standard expenses, with the predictable result of having such services suspended, resulting in no electricity, water, and so on.

A number of things are important here. First, and most obvious is the meaningful impact that blatant misinformation had on some people's daily lives. Second, as in both the VAERS and Fauci examples, this instance has clear implications for how information worlds can understand, interpret, and respond to informational authority. In the terms of information worlds, the value – and perceived accuracy – of information here appears to rest entirely on the "social type" of the source. That is, the information offered by the "Queen" was deemed to be more plausible than the established norms and practices of normal day-to-day life in an industrialized world precisely because of her role as an active and visible QAnon representative; in this role, she can be taken to be a trustworthy champion of "We the People" against the "criminal" enforcers of the shadowy and evil powers of the deep state and business-as-usual. The story communicated to her followers through her edicts fits the overall narrative of conspiracy in a way that standard information practices – particularly through "authorized" and expert channels – simply cannot. In other words, adherence to the narrative of conspiracy determines information governance in this particular information world.

"CONNECTING THE DOTS": CONSPIRACY AND (MIS)INFORMATION

In addition to the kinds of persistent misreading and misrepresentation discussed above, conspiracy theories rely on a particular persistent action arena – or a form of information behavior – in which chains of often dubious or tangential connections between people, events, and other phenomena are followed in order to discover evidence of and justify conclusions about conspiracies. This model is often accompanied by calls to "do your own research," a phrase which functions as a commonplace particularly in antivax information worlds (Maruf 2021). Also common are calls to "connect the dots" or claims that a particular narrative "connects the dots," providing evidence to support some kind of claim (JT Editorial Board 2020). Indeed, even cursory searches of right-wing and conspiracy Facebook pages will turn up numerous instances of both phrases and, indeed, "connecting the dots" appears to be the primary process through which adherents "do [their] own research."

In one sense, this can be seen as an instance of what Dervin (1992) has called "sensemaking," an ongoing process of information gathering and interpretation in order to bridge cognitive gaps and "make sense" of some aspect of the world in which we live. In the context of the everyday, this process is a way of inferring (or inventing) coherent and meaningful patterns in the flow of information as it goes by.

In Dervin's framework, such sensemaking is always provisional, and subject to further modification as new information comes along. In this regard, sensemaking functions as a kind of hermeneutic process, in which information and interpretation are both conceptualized as processes rather than conclusions; in the information social media worlds of conspiracy, this often manifests itself as a process of ongoing information sharing that purports to eliminate the provisionality of Dervin's model, substituting an unquestioned certainty in the veracity of the information uncovered in the process of "research"; as one common meme puts it, researchers "have no beliefs," but "choose to know things," no matter how far-fetched those things may appear to "normies" who do not buy into the tenets of the conspiracy. Despite such certainty in what one "knows," conspiracy is still imbued with process, as further research can simultaneously uncover new knowledge and reinforce already existing conclusions; that is, new "dots" reinforce already understood "connections" and patterns in knowledge. Participants engage in ongoing dialog to "connect the dots" between different bits of information (Burnett et al. 2003). To put it simply, there are always more "dots" to connect, and conspiracy "research" is a practice or process without a solid ending point except to use new discoveries to reinforce existing patterns.

When it comes to QAnon, such a paradoxical process seems to be built into the information world from the very beginning. The movement began as an online response to an ongoing set of "drops," made by a shadowy figure known only as "Q" to the website 4chan (Rothschild 2021). These "drops," which appeared between October 2017 and October 2020, when they stopped for almost two years (Tian 2021), are typically built out of a cryptic set of rhetorical and often leading questions, seemingly arcane references, vague prognostications, directives, and so on, as seen in the following partial extraction from the very first "drop" of October 28, 2017 ("/Pol/ - Politically Incorrect» Thread #146981635" 2017):

> Mockingbird
> HRC detained, not arrested (yet).
> Where is Huma? Follow Huma.
> This has nothing to do w/Russia (yet).
> Why does Potus surround himself w/ generals?

> What is military intelligence?
> Why go around the 3 letter agencies?
> What Supreme Court case allows for the use of MI v Congressional assembled and approved agencies?
> Who has ultimate authority over our branches of military w/o approval conditions unless 90+ in wartime conditions?
> What is the military code?

As such, Q "drops" function in a couple of different ways. First, through their curious combination of extreme vagueness coupled with a few specifics ("Huma," for example, referring to Huma Abedin, one of Hillary Clinton's 2016 campaign managers) hint at information without actually containing any clear or verifiable information at all. In this case, the ordering of the statements – one a common QAnon reference to an ostensible Cold-War era CIA program to manipulate the media (see Rothschild 2021 for a useful account of the program) and the other two qualified by an ominous "(yet)" – and questions hint at a tantalizing but irreducibly nebulous set of people and situations that one can only imagine have to do with great crimes and potential punishments.

The other function of such a list is that – precisely because it is so nebulous, hinting at much but never quite saying anything – it requires its readers to become active participants in the construction of whatever signification or information may be present. That is, the "drops" require interested readers to engage in a process of "connecting the dots" to "do [their] own research" if they are to glean anything at all from them. In this sense, information behaviors associated with conspiracy, embedded as they are in the ubiquitous daily flow of information of all sorts (including mis- dis- and mal-information) and interaction that is the online world, simply become part of one's "everyday" life activity. This activity can even be thought of as a kind of daily "game," like Wordle ("Wordle – a Daily Word Game" 2022), except that instead of trying to guess a five-letter word in six tries, "players" decipher sets of hints and clues that seem to point to an unnamed but ominous evil. As Berkowitz (2020), a game designer, has suggested, conspiracy work (and QAnon, in particular) is a kind of "game that plays people," taking advantage of game-like pleasures of "connecting the dots" and emphasizing a "tendency to perceive a connection or meaningful pattern between unrelated or random things" (or "apophenia").

In the context of conspiracy-related information practices, such a quest to uncover connections through an ongoing process of "connecting the dots" far outpaces the ways pattern recognition and the fun of putting things together function in actual games (that is, a jigsaw puzzle always has a finite number of pieces and a clearly recognizable final stage). For conspiracy information worlds such as QAnon, the "game" functions as an action arena in which information work (or, more accurately, misinformation work) can take place as an open-ended project, in which there is no recognizable point when the pattern is actually complete. Indeed, as

Rothschild (2021) notes repeatedly, QAnon adherents have a remarkable ability to simply move past unfulfilled predictions, seeing them merely as further proof-of-concept, as parts of an endlessly emerging pattern or inevitable narrative. Interestingly, a new "drop" from Q appeared on June 24, 2022, after a gap of more than a year, referring explicitly to the framing of the QAnon information project as such an endless game: "Shall we play the game again?" (Thompson 2022).

In part because of the very open-endedness of the game of "connecting the dots," the information products produced by conspiracy theorists often take quite elaborate form, as in the "Q-Web" or "Deep State Map" in Figure 9.2 (Hannah 2021b). A wide range of conspiracy theory maps and visualizations can be found as part of a blog titled "Through the Looking Glass," created by "The Infomniac" (The Infomaniac 2017).

As Hannah (2021a) puts it "anons have been engaged in a surprising amount of data collection and visualization, translating a labyrinth of speculation into a coherent visual narrative. That is, such conspiracy work can not only be conceptualized as a "game," but also as a large-scale, group-based and collaborative project to uncover, create, and share information widely, no matter how tenuous that information may be. In this example, the "dots" take the form of a set of nodes spread across a densely crowded visualized field, interwoven with arrows and lines, "proving" the connections between such disparate phenomena as the Pentagon (at the center of the image, immediately juxtaposed with "Nazi Germany"), the Protocols of the Elders of Zion, Antarctica, the Library of Alexandria, and the Jesuits ("God's Assassins").

This map shows an interconnected web of "dots" indicating more-or-less "real world" information related to banking, religious hierarchies, presidents, recognizable geographic locations, and historical narratives (some of which are, of course, fabrications, fakes, and, at best, misleading), all pointing toward the various "plots" and misdeeds of the "Deep State."

Such "dot-connecting" is not limited to the recognizable quotidian world. A second map, which explicitly includes references to both "Q" and "Q Anon," further situates conspiracy within a much more expansive and esoteric universe, leaping from the surface of the earth both down into subterranean manifestations such as an "Inner Earth Civilizations" and up into interstellar space like "Secret Space Program[s], an "Offensive Space Fleet" beyond the solar system as well as promises of an "Ascension into 5th Dimensional Earth Age of the Golden Race" (see Figure 9.3).

This second map combines motifs from various spiritual and mystical traditions ("Samsara," "Bardo," "Tao Te Ching") with psychedelics ("psilocybin mushrooms") and references to other conspiracy theories ("WTC7 Controlled Demolition," "Area 51," "Chemtrails") together with a variety of references that, in another context, would likely be thought of as science fiction or science as filtered through science fiction ("hyperdimensional physics," "Torsion 'Warp' Drives") and allusions to information, both hidden ("Top Secret," nestled between references to "levels" of

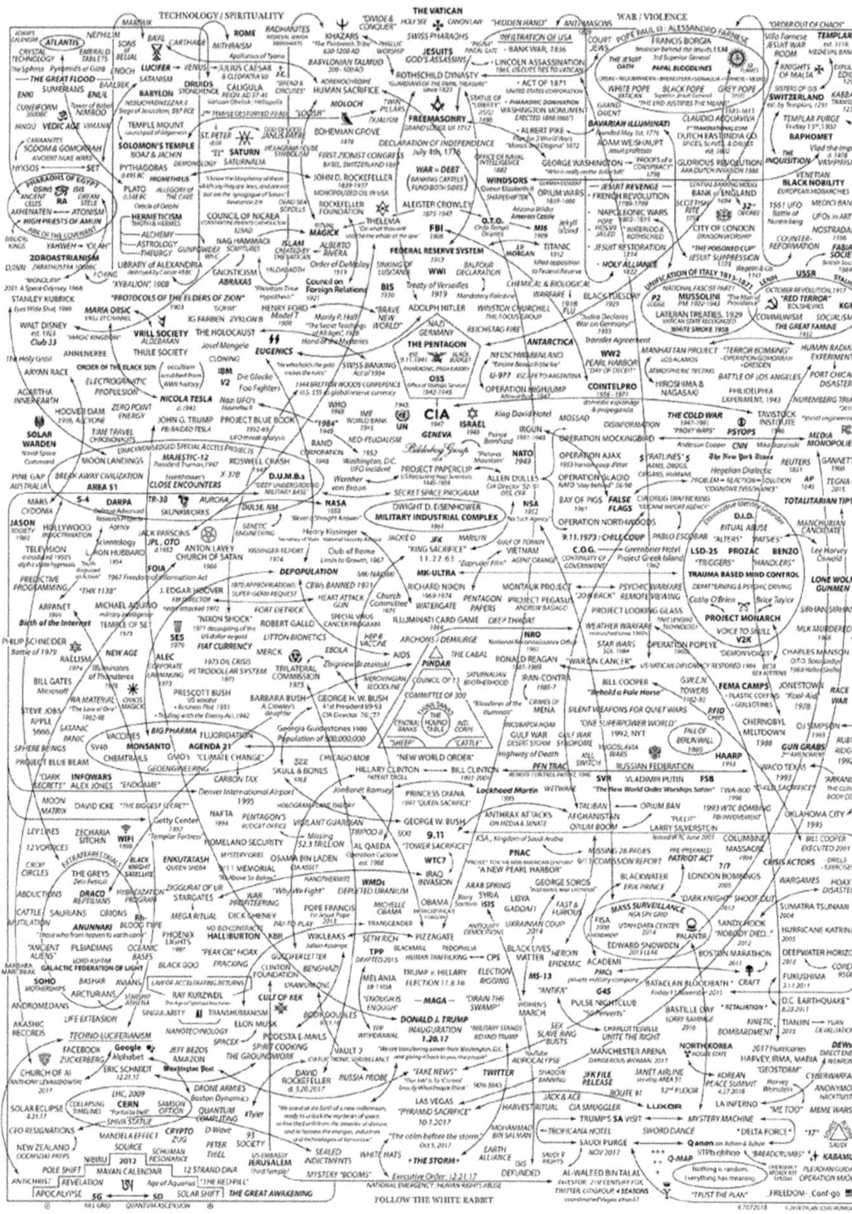

FIGURE 9.2 Key to the Q-Web or deep state map mapping project (Monroe 2018).

intelligence both above and below the "P.O.T.U.S.") and revealed ("Cosmic Disclosure," "[Secret Space Program Alliance] Whistleblowers"). Not only can some of the concerns of conspiracy theorists such as those associated with QAnon seem more closely aligned to science fiction novels than to the day-to-day world, but

FIGURE 9.3 Great awakening map exposing and connecting all hidden global information (TyGunnard 2020).

science fiction itself is woven into the fabric of the QAnon narrative, with *The Matrix* movie series in particular being a recurring motif with "red pill" often used to label the disclosure of previously "hidden" information. In at least one case, this connection was directly implicated in the brutal murder of two children by their

QAnon-following father; his decision grew “as he was lying in bed . . . ‘seeing all the pieces being decoded like *The Matrix*,’” and believing himself to be the character Neo (Gilbert 2022). And, indeed, much of the QAnon worldview, to the extent it can be coherently understood, bears a striking resemblance both to the world of *The Matrix*, and to the hallucinatory novels of Philip K. Dick, a linkage that has been noted both by analysists of the phenomenon (Pettipiece 2021) and QAnon adherents themselves in social media sites such as Facebook, often with observations that his novels seem to be “true” and accurate representations of the real world due to their depictions of a world made of “simulated reality” á la *The Matrix*.

Such representations of “connecting the dots” not only follow a spatial metaphor as in the maps above, but also involve time, as in the “Q-Clock” (see Figure 9.4), which has appeared in numerous permutations designed to demonstrate the

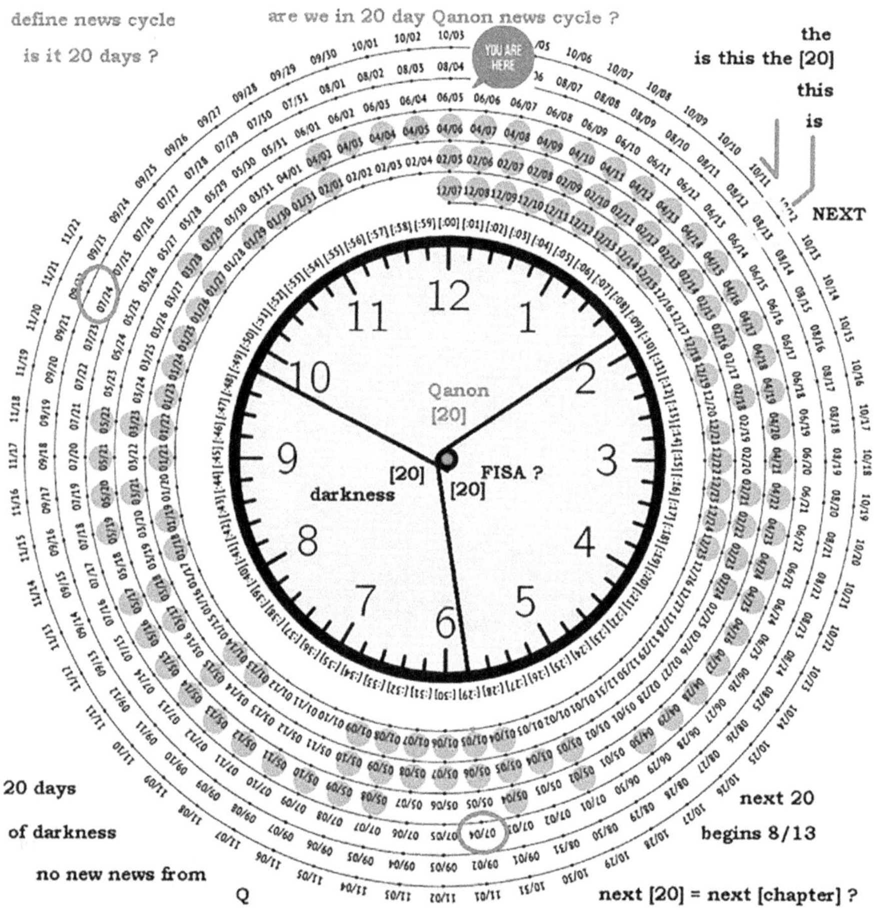

FIGURE 9.4 The Q-Clock (“how to read the Q-Clock” 2020).

common Q credo "future proves past" through a complex representation of cyclic temporality. (Other related practices involve a variety of numerological processes, most often Jewish "Gematria" for "analyzing" and "discovering" relationships between otherwise unrelated phenomena ("Trump and Qanon Value in Gematria Calculator" n.d.).

Such geographic and temporal metaphors function in a couple of different paradoxical ways. First, they draw upon very familiar information tools – maps and clocks – in order to metaphorically situate a complex set of real and ostensible relationships within a recognizable framework in a form that, if printed out, could even be hung on a kitchen wall. This has the effect of turning the seemingly chaotic web of conspiracies into a part of the "everyday" world; to put it simply, the implication is that if something can be represented as something as quotidian as a map or a clock, it must be true, no matter how far-fetched it may otherwise seem. Second, their sheer complexity, highlighted by numerous lines and arrows connecting nodes to each other, makes it possible to suggest (or, for those inclined to believe) even "prove" that disparate phenomena are, in fact, closely linked. Indeed, another common QAnon credo is that "There are no coincidences," and the use of such visualizations suggests that connections can be established between any two possible phenomena because they visually occupy the same physical or temporal space (Hannah 2021a; Papasavva et al. 2021).

In terms of the theory of information worlds, such mechanisms of conspiratorial representation function as both observable social norms and as markers of information value. As social norms, they serve as a preferred mode of communication, offering a recognizable (albeit jumbled) set of information-sharing practices that are meaningful to the participants – the "actors," in GKC terms – of this particular information world. As representations of information value, they provide visual representation of the particular interwoven aggregate of information that is of particular importance within the conspiratorial world, even if the connections between different bits of information does not translate to the outside world of "normies." To put it another way, they function as visual markers of the "knowledge commons" of this particular world and its actors, yoking them together through a shared set of commonly understood and interwoven references that are not accessible to those who are excluded from the commons. Such patterns are further discussed in the next section.

"SYMBOLISM WILL BE THEIR DOWNFALL": CONSPIRACY AND THE EVERYDAY

In a world in which everything is interconnected – in which researching the "dots" reveal vast networks of relationships in which those relationships are obfuscated by equally interconnected networks of "lies" and conspiracies – it follows that every image, every action, and even every gesture can be taken as meaningful; in other words, every act, no matter how thoroughly intertwined with normal everyday behavior is understood to be imbued with "symbolism." Implicitly, there are three ways of understanding the significance of such patterns of "meaning": either one is part of the conspiracy (as a member of the "Deep State," the Illuminati, or some other often vaguely defined conspiratorial group), using such imagery to simultaneously communicate and obfuscate complicity in the conspiracy; or one is a "normie," blindly participating in the world without realizing the depths of its depravity; or one is engaged in the research necessary to uncover and reveal such hidden meanings.

Such an information world of conspiracy reinforces a strong sense of insiders and outsiders – people, as information actors, are either actively or passively complicit in the conspiracy, or are working to unveil the "truth" through ongoing "research" as part of their everyday routine. This dynamic underpins the common QAnon rallying cry of "Where we go one, we go all" or "WWG1WGA" (Rahn and Patterson 2021). While this phrase rhetorically gestures toward radical inclusion ("we go all"), it also depends on radical exclusion, implicitly omitting those who do not share the conspiratorial knowledge that "we" possess.

Although "symbolism" or the use of code to communicate clandestinely is at least thousands of years old, the 2016 "Pizzagate" conspiracy was perhaps the source of the current conspiratorial focus. Growing out of an assertion on 4chan that the phrase "cheese pizza" referred to child pornography and a series of emails by John Podesta (along with Huma Abedin one of Hillary Clinton's 2016 campaign managers) this conspiracy asserted that a Washington, DC pizzeria was a front for a pedophilia ring with Hillary Clinton and other leading democrats at its center. Although this specific claim was tied to a very particular location and ostensible series of events, it grew into a much more elaborate theory focusing on a complex collection of "code words." At its most extreme, this theory cast doubt on such everyday words like "pizza," "hot dog," and "ice cream," any and all uses of which implied complicity in pedophilia. See the December 2016 *New York Times* article "Dissecting the #PizzaGate Conspiracy Theories" (Aisch, Huang, and Kang 2016) for a good overview of these theories (the article is, interestingly, accompanied by a graphic "map" similar to those discussed above, as well as a variety of simpler graphic images linked to "symbolism" and hidden meanings).

Just as the process of "connecting the dots," if taken far enough as an action arena, can find linkages between any two disparate things, a focus on "symbolism" can take even the most quotidian words or images and transform them into indicators or even "proof" of pervasive conspiracy; instances are far too numerous to discuss thoroughly here, so just a couple of quick examples must suffice.

Similar to the misinterpretations of VAERS data, discussed above, such readings of "symbolism" take intent to be a given – messages are often inferred to be proof of culpability with or without direct evidence that they are so. For example, one widely disseminated motif involves photographs of celebrities, politicians, or other popular culture figures with only one eye visible, taken to be prima facie evidence of Satanic, Illuminati, and/or pedophiliac involvement. This interpretation may be rooted in ancient images of the "Eye of Providence" or the "Eye of Horus" (Wilson 2020). See, for example, a YouTube video from a user named "Critical Thinker" (2020), which shows Madonna, Mr. Burns from *The Simpsons*, Snoop Dog, the television network Nickelodeon, covering an eye. This short video further explicitly invokes the Q pronouncement that "Symbolism will be their downfall" (Critical Thinker 2020). In many cases, sets of promotional photographs of entertainers are presented not as single pictures but as collages or groupings of numerous photographs, in each of which a child appears, where one of the entertainer's eyes is covered, or the entertainer is making some other kind of common gesture; such collages are often accompanied by explicit statements that they are direct evidence of the "evil," "monstrous," or "satanic" motivations of the individuals portrayed. See, for example, Manoj 2016, a typical example drawn from an atypical source: an online highschool publication editorial report about the Illuminati and conspiracy theories.

Other such "symbolic" images target corporations and corporate logos. For example, the famous Disney logo is interpreted as willfully incorporating the number "666" in the swirls of its letters (see Clipart Library n.d. for one example among many). As one collection of such corporate symbolism puts it, "Examine this of Disney carefully and observe the three six camouflaged in the writing of the Walt Disney corporate logo. …There is SO MUCH Symbolism in their movies, too much for here." See Pouissant (2019) for the Disney claim along with numerous other "readings" of such logos, explicitly tied to Q and condemnations of the ways "Luciferians" control the world and sacrifice "our children by the millions"; the site from which this was taken, rattibha.com, appears to be a repository, primarily in Arabic, that gathers and stores Twitter content; in this case, the original post is from a user whose Twitter account has been discontinued, presumably because of such Q-related posts).

This kind of conspiracy-laden commons clearly sets itself apart from the mainstream and from any kind of mainstream normative "governance," substituting instead its own set of conspiracy-related norms and values that make sense only within the framework of the alternative conspiracy outlook itself. As argued above, however, if one plays the "game" of connecting the dots, it is clear that the dots are everywhere and inextricably entwined in that particular way of viewing the world. If one follows down that particular path, such a viewpoint encompasses everything – not only "serious" information, but also all aspects of the everyday.

CONCLUSION

It is quite challenging to characterize participants in conspiracy-related information worlds in settings such as Facebook groups, as members of a clearly identifiable and coherent community. For one thing, while such groups may have tens of thousands of members, most are nothing more than consumers or "lurkers," and posts are often made by a very small handful of active members. It is, further, impossible to know how many others monitor the groups but do not subscribe. In addition, given how Facebook's algorithms work, the information experience on the platform – what one does or does not see, the order in which posts appear, and more – can be radically different from user to user; it can even be difficult to retrace one's own steps to find a post seen earlier, whether because of the algorithm or because the post has been removed or otherwise hidden.

Indeed, the most extreme example, QAnon, has as its figurehead an actor who, despite having some ostensible characteristics such as top-secret security clearance and despite being the object of speculation, remains fundamentally unknown, and even prone to disappearing for months at a time (Rothschild 2021). In other words, the online QAnon world lacks a definable or recognizable center; it is decentered, built around a largely absent figure seen only in a series of radically indirect and suggestive texts of hazy provenance.

Further, online conspiracy information worlds are not only geospatially decentered, but also temporally askew, with no clear linear relationship between the production and dissemination of information resources and their consumption. Such temporal complexity is metaphorized in the QAnon credo "future proves past" and explicitly embodied in the Q-Clock. Chronology becomes one more mechanism for "connecting the dots" between temporally separated events, as in the "proof" that someone falling ill a certain number of days – or weeks, or months – after being vaccinated is clear evidence of causality. Conversely, as Rotheschild (2021) repeatedly notes, the fact that specific events – such as the arrest of Hillary Clinton – fail to materialize as predicted does not result in any diminution of QAnon fervor. Within QAnon, the very conception of time appears to be somewhat "unstuck" (to use the famous Kurt Vonnegut quote).

In this sense, conspiracy worlds are functionally built around a particular narrative assumption: that no matter what might happen in the meantime, the story's end is always already known; as adherents put it "Nothing can stop what's coming." In the case of QAnon, the projected conclusion most often takes the form of the ascendency of Donald Trump and the downfall of the "deep state" and all of the "normies" who unwittingly support it, replete with military tribunals and public executions. Just as the world of QAnon revolves around the absent center of "Q," this implicit narrative relies on an ending that is forever deferred but also always painted as inevitable. The endpoint of conspiracy theory, thus, is simultaneously certain and endlessly deferred, a promise according to which adherents live their lives and which is impervious to disappointment because it is always still in play as the assured outcome. Disappointment and unexpected events simply reinforce the narrative promise of an already known ending.

Such a dynamic is built into the very structure of a system like Facebook, because the platform is, as Sanfilippo and Strandburg (2021, drawing upon the work of McGinnis 2011) note, intrinsically nested. That is, all Facebook groups, whether public or private, are embedded within the larger framework of Facebook itself, and subject to its "exogenous" policies and algorithms governing both access and information flow. This embededness influences much of what is possible in the "action arena" of groups in a couple of different ways. First, it is not uncommon for individuals to run afoul of Facebook's policies when posting, resulting in restrictions, removal of offending posts, thirty-day bans, or even account deletion. Second, both individuals and groups in the Facebook conspiracy worlds tend to define themselves in opposition to the very platform that makes it possible for them to share information, often referring to it as "Fakebook" and vowing repeatedly to abandon the platform while nevertheless returning after bans and sometimes creating alternate identities in clear violation of the system's "community standards." In the terms of the GKC framework, that is, conspiracy adherents situate themselves in open defiance of the very background environment that, through its affordances and information dissemination capabilities makes their quite oppositional information (or misinformation) work possible.

Such a structure – a distributed, decentered information world nested within, and in opposition to, a larger context – results in a set of action arenas and informational strategies marked by seemingly willful misreading and misdirection, the drawing of spurious interconnections between disparate "dots," and a predilection to see nefarious intent "hidden in plain sight" in the most innocent of images, actions, and texts. And, further, this information world defines itself as the only truly literate world, compared to the external world of the "normies" inhabited by unthinking "sheep" who simply can neither accept nor even understand the truth in front of them, and who will ultimately either disappear or be driven to desperation when they finally grasp the depth of their delusions. In such a situation it is, to put it simply, hard to see a way to undo the workings of conspiracy, since any such effort will always already and a priori be perceived as just a further step in the inevitable process leading to the promised ending. In other words, conspiracy theories – and QAnon in particular – may be particularly impervious to intervention and corrective measures, since any such attempt simply reaffirms the "truth" of the conspiracy. Such conspiracy theories have persisted through time, and likely will continue to do so, although the specific content and focus of the theories may morph and though the numbers of adherents may wax and wane.

ACKNOWLEDGMENTS

The authors wish to acknowledge Kate McDowell, Beth Patin, and Tyler Youngman for helpful conversations about aspects of this chapter.

BIBLIOGRAPHY

Aisch, Gregor, Jon Huang, and Cecilia Kang. 2016. "Dissecting the #PizzaGate Conspiracy Theories." *The New York Times*, December 10. www.nytimes.com/interactive/2016/12/10/business/media/pizzagate.html.

Baldwin, Shawn. 2022. "What Led to the U.S. Baby Formula Shortage and How Officials Are Trying to Prevent It from Happening Again." CNBC. June 15. www.cnbc.com/2022/06/15/what-led-to-the-baby-formula-shortage-how-lawmakers-are-trying-to-prevent-another.html.

Beckett, Lois. 2020. "QAnon: A Timeline of Violence Linked to the Conspiracy Theory." *Guardian*, October 16. www.theguardian.com/us-news/2020/oct/15/qanon-violence-crimes-timeline.

Berkowitz, Reed. 2020. "A Game Designer's Analysis of QAnon." *Medium*. November 21. https://medium.com/curiouserinstitute/a-game-designers-analysis-of-qanon-580972548be5.

Bracewell, Lorna. 2021. "Gender, Populism, and the QAnon Conspiracy Movement." *Frontiers in Sociology* 5 (January). https://doi.org/10.3389/fsoc.2020.615727.

Burnett, Gary. 2015. "Information Worlds and Interpretive Practices: Toward an Integration of Domains." *Journal of Information Science Theory and Practice* 3 (3): 6–16. https://doi.org/10.1633/jistap.2015.3.3.1.

Burnett, Gary, Michele Besant, and Elfreda A. Chatman. 2001. "Small Worlds: Normative Behavior in Virtual Communities and Feminist Bookselling." *Journal of the American Society for Information Science and Technology* 52 (7): 536–547. https://doi.org/10.1002/asi.1102.

Burnett, Gary, Michelle Kazmer, Michael Dickey, and Katherine Chudoba. 2003. "Inscription and Interpretation of Text: A Cultural Hermeneutic Examination of Virtual Community." *Information Research* (online) 9 (1). https://informationr.net/ir/9-1/paper162.html.

Burnett, Gary, Paul T. Jaeger, and Kim M. Thompson. 2008. "Normative Behavior and Information: The Social Aspects of Information Access." *Library and Information Science Research* 30 (1): 56–66. https://doi.org/10.1016/j.lisr.2007.07.003.

Case, Donald Owen, and Lisa M. Given. 2016. *Looking for Information: A Survey of Research on Information Seeking, Needs, and Behavior*. Bingley: Emerald.

Chatman, Elfreda A. 1991. "Life in a Small World: Applicability of Gratification Theory to Information-Seeking Behavior." *Journal of the American Society for Information Science* 42 (6): 438–449. https://doi.org/3.0.co;2-b">10.1002/(sici)1097-4571(199107)42:6<438::aid-asi6>3.0.co;2-b.

Clipart Library. n.d. "Walt Disney Logo 666 – Clip Art Library." Clipart Library. Accessed November 28, 2022. http://clipart-library.com/clipart/8izr7zbLT.htm.

Cooke, Nicole A. 2017. "Posttruth, Truthiness, and Alternative Facts: Information Behavior and Critical Information Consumption for a New Age." *Library Quarterly* 87 (3): 211–221. https://doi.org/10.1086/692298.

Critical Thinker. 2020. "Qanon – Symbolism Will Be Their Downfall All Seeing Eye." www.youtube.com. March 30.www.youtube.com/watch?v = aygrv2HNKHg.

Davies, Dave. 2021. "Without Their 'Messiah' QAnon Believers Confront a Post-Trump World." *NPR.org*, January 28. www.npr.org/2021/01/28/961507953/without-their-messiah-qanon-believers-confront-a-post-trump-world.

De Vynck, Gerrit, and Rachel Lerman. 2021. "Facebook and YouTube Spent a Year Fighting Covid Misinformation: It's Still Spreading." *Washington Post*, July 22. www.washingtonpost.com/technology/2021/07/22/facebook-youtube-vaccine-misinformation/.

Dervin, Brenda. 1992. "From the Mind's Eye of the User: The Sense-Making Qualitative-Quantitative Methodology." In *Qualitative Research in Information Management*, 61–84. Englewood: Libraries Unlimited.

Frischmann, Brett M., Michael J. Madison, and Katherine Jo Strandburg. 2014. *Governing Knowledge Commons*. Oxford: Oxford University Press.

Gilbert, David. 2022. "QAnon Surfer Who Killed His Kids Believed He Was Neo from 'the Matrix.'" www.vice.com. March 1. www.vice.com/en/article/n7n9gw/qanon-surfer-matthew-coleman.
GNET Team. 2020. "What Is QAnon?" Global Network on Extremism and Technology. October 15. https://gnet-research.org/2020/10/15/what-is-qanon/.
Habermas, Jurgen, and Peter Dews. 1992. *Autonomy and Solidarity: Interviews with Jurgen Habermas*. London: Verso.
Haimowitz, Ian. 2020. "No One Is Immune: The Spread of Q-Anon through Social Media and the Pandemic." www.csis.org. December 17. www.csis.org/blogs/technology-policy-blog/no-one-immune-spread-q-anon-through-social-media-and-pandemic.
Hannah, Matthew N. 2021a. "QAnon and the Information Dark Age." *First Monday* (online) 26 (2). https://doi.org/10.5210/fm.v26i2.10868.
2021b. "A Conspiracy of Data: QAnon, Social Media, and Information Visualization." *Social Media + Society* 7 (3): 205630512110360. https://doi.org/10.1177/20563051211036064.
Hatewatch Staffx Staff. 2020. "What You Need to Know about QAnon." Southern 2020. "What You Need to Know about QAnon." Southern Poverty Law Center. October 27. www.splcenter.org/hatewatch/2020/10/27/what-you-need-know-about-qanon.
"How to Read the Q Clock." 2020. Web.archive.org. March 19.
The Infomaniac. 2017. "QAnon: 'Learn to Read the Map': A Cartography of the Globally Organized Corruption Networks: A Treasure Trove of Maps, Diagrams, Org Charts, and Family Trees." *Through the Looking-Glass*. November 25. https://throughthelookingglassnews.wordpress.com/2017/11/24/q-anon-learn-to-read-the-map/.
Jaeger, Paul T., and Gary Burnett. 2010. *Information Worlds: Social Context, Technology, and Information Behavior in the Age of the Internet*. New York: Routledge.
JT Editorial Board. 2020. "QAnon: Connecting Nonexistent Dots." *Baltimore Jewish Times*. August 27. www.jewishtimes.com/qanon-connecting-nonexistent-dots/.
Lamoureux, Mark. 2022. "The QAnon Queen Told Followers They Didn't Need to Pay Bills. It Didn't End Well." www.vice.com. May 10. www.vice.com/en/article/z3na53/qanon-queen-bills-electricty-canada.
Manoj, Aparnna. 2016. "Magnified: The Illuminati." *The Hawk Eye*. October 18. www.hebronhawkeye.com/opinion/blogs/2016/10/18/magnified-the-illuminati/.
Martineau, Paris. 2017. "The Storm Is the New Pizzagate: Only Worse." *Intelligencer*. December 19. https://nymag.com/intelligencer/2017/12/qanon-4chan-the-storm-conspiracy-explained.html.
Maruf, Ramishah. 2021. "These Four Words Are Helping Spread Vaccine Misinformation | CNN Business." www.vice.com. September 19.
McGinnis, Michael D. 2011. "An Introduction to IAD and the Language of the Ostrom Workshop: A Simple Guide to a Complex Framework." *Policy Studies Journal* 39 (1): 169–183. www.cnn.com/2021/09/19/media/reliable-sources-covid-research/index.html.
Monroe, Dylan L.. 2018. "Key to the Q-Web." Dylan Louis Monroe. July 7. www.dylanlouismonroe.com/q-web.html.
Mulkerrins, Jane. 2021. "Life inside QAnon, the Cult That Stormed the Capitol." www.thetimes.co.uk, January 15, sec. magazine. www.thetimes.co.uk/article/life-inside-qanon-the-cult-that-stormed-the-capitol-v5q3hmbmb.
Noveck, Beth Simone, James Button, Dane Gambrell, Lex Paulson, Paolo Spada, and Lakshminarayanan Subramanian. 2021. "The Power of Virtual Communities." *The Gov Lab*. https://virtual-communities.thegovlab.org/files/DTR_report_en_EN.pdf.
Papasavva, Antonis, Jeremy Blackburn, Gianluca Stringhini, Savvas Zannettou, and Emiliano De Cristofaro. 2021. "'Is It a Qoincidence?': A First Step towards Understanding and Characterizing the QAnon Movement on Voat*." https://arxiv.org/pdf/2009.04885.pdf.

Pathak, Arohi, Osub Ahmed, Caroline Medina, Marc Jarsulic, Jill Rosenthal, and Emily DiMatteo. 2022. "The National Baby Formula Shortage and the Inequitable U.S. Food System." Center for American Progress. June 17. www.americanprogress.org/article/the-national-baby-formula-shortage-and-the-inequitable-u-s-food-system/#:~:text = equitable %20food%20system.

Pettipiece, Timothy. 2021. "History Repeats Itself: From the New Testament to QAnon." *The Conversation*. March 21, 2021. https://theconversation.com/history-repeats-itself-from-the-new-testament-to-qanon-156915.

Polantz, Katelyn. 2021. "US Takes Back Its Assertion That Capitol Rioters Wanted to 'Capture and Assassinate' Officials." CNN. January 15. www.cnn.com/2021/01/15/politics/capitol-capture-assassinate-elected-officials/index.html.

"/Pol/ - Politically Incorrect» Thread #146981635." 2017. Archive.4plebs.org. October 28. https://archive.4plebs.org/pol/thread/146981635/#q147023341.

Pouissant. 2019. "Rattibha." Rattibha.com. June 4. https://rattibha.com/thread/1135897991901044738?lang = en>_lang = en.

Rahn, Will, and Dan Patterson. 2021. "What Is the QAnon Conspiracy Theory?" www.cbsnews.com. March 29. www.cbsnews.com/news/what-is-the-qanon-conspiracy-theory/.

Rainie, Lee, and Janna Anderson. 2017. "The Internet of Things Connectivity Binge: What Are the Implications?" Pew Research Center: Internet, Science and Tech. June 6. www.pewresearch.org/internet/2017/06/06/the-internet-of-things-connectivity-binge-what-are-the-implications/.

Rheingold, Howard. 1993. *The Virtual Community: Homesteading on the Electronic Frontier*. Cambridge, MA: MIT.

Rogers, Richard. 2020. "Deplatforming: Following Extreme Internet Celebrities to Telegram and Alternative Social Media." *European Journal of Communication* 35 (3): 213–229. https://doi.org/10.1177/0267323120922066.

Roose, Kevin. 2021. "What Is QAnon, the Viral Pro-Trump Conspiracy Theory?" *The New York Times*, September 3, sec. Technology. www.nytimes.com/article/what-is-qanon.html.

Rothschild, Mike. 2021. *The Storm Is Upon Us: How QAnon Became a Movement, Cult, and Conspiracy Theory of Everything*. Brooklyn: Melville House.

Sanfilippo, Madelyn Rose, Brett M. Frischmann, and Katherine J. Strandburg, eds. 2021. Governing Privacy in Knowledge Commons. Cambridge, UK: Cambridge University Press. https://doi.org/10.1017/9781108749978.

Squirrell, Tim, and Nikhil Sonnad. 2017. "The Alt-Right Is Creating Its Own Dialect. Here's the Dictionary." *Quartz*. October 30. https://qz.com/1092037/the-alt-right-is-creating-its-own-dialect-heres-a-complete-guide.

Stanton, Gregory. 2020. "QAnon Is a Nazi Cult, Rebranded." Just Security. September 9. www.justsecurity.org/72339/qanon-is-a-nazi-cult-rebranded/.

Stecula, Dominik A., and Mark Pickup. 2021. "Social Media, Cognitive Reflection, and Conspiracy Beliefs." *Frontiers in Political Science* 3 (June). https://doi.org/10.3389/fpos.2021.647957.

Suciu, Peter. 2020. "Conspiracy Theories Have Gained Traction since 9/11 Thanks to Social Media." Forbes. September 11. www.forbes.com/sites/petersuciu/2020/09/11/conspiracy-theories-have-gained-traction-since-911-thanks-to-social-media/?sh = 7c3b79583ddb.

Thompson, Stuart A. 2022. "The Leader of the QAnon Conspiracy Theory Returns." *The New York Times*, June 25, sec. Technology. www.nytimes.com/2022/06/25/technology/qanon-leader-returns.html.

Tian, Edward. 2021. "The QAnon Timeline: Four Years, 5,000 Drops and Countless Failed Prophecies." Bellingcat. January 29. www.bellingcat.com/news/americas/2021/01/29/the-qanon-timeline.

"Trump and Qanon Value in Gematria Calculator." n.d. www.gematrix.org. Accessed September 19, 2022. www.gematrix.org/?word = trump%20and%20qanon.

Truth and Liberty Staff. 2021. "Liberty Counsel Aids in Fight against Vaccine Mandates." Truth and Liberty Coalition. August 25. https://truthandliberty.net/2021/08/liberty-counsel-aids-in-fight-against-vaccine-mandates/.

TyGunnard. 2020. "English: Map Exposing and Connecting All Hidden Global Information." Wikimedia Commons. December 30. https://commons.wikimedia.org/w/index.php?title = File:Great_Awakening_Map_.jpgandoldid = 623787676.

Urman, Aleksandra, and Stefan Katz. 2020. "What They Do in the Shadows: Examining the Far-Right Networks on Telegram." *Information, Communication and Society* 25 (7): 1–20. https://doi.org/10.1080/1369118x.2020.1803946.

Uscinski, Joseph E. 2018. *Conspiracy Theories and the People Who Believe Them.* Oxford Academic. https://doi.org/10.1093/oso/9780190844073.003.0001.

"VAERS." n.d. Vaers.hhs.gov. https://vaers.hhs.gov/about.html.

Vox Staff. 2020. "QAnon: The Conspiracy Theory Embraced by Trump, Several Politicians, and Some American Moms." Vox. October 14. www.vox.com/2020/10/9/21504910/qanon-conspiracy-theory-facebook-ban-trump.

Wilson, Matthew. 2020. "The Eye of Providence: The Symbol with a Secret Meaning?" www.bbc.com. November 13. www.bbc.com/culture/article/20201112-the-eye-of-providence-the-symbol-with-a-secret-meaning.

"Wordle – A Daily Word Game." 2022. www.nytimes.com. www.nytimes.com/games/wordle/index.html.

Zhu, Ming, and Xizhu Liao. 2020. "Chatman's Theory of Life in the Round Applied to the Information Seeking of Small Populations of Ethnic Minorities in China." *Informationr.net,* September. http://informationr.net/ir/25-3/paper868.html.

Zihiri, Saifeldeen, Gabriel Lima, Jiyoung Han, Meeyoung Cha, and Wonjae Lee. 2022. "QAnon Shifts into the Mainstream, Remains a Far-Right Ally." *Heliyon* 8 (2): e08764. https://doi.org/10.1016/j.heliyon.2022.e08764.

10

How to Manage Issues on Twitter

Perspectives from Twitter Users Concerned about Mis/Dis-Information

Joanna Sin Sei-Ching and Kyung-Sun Kim

INTRODUCTION

Social media has brought significant changes to people's lives. Some even suggest that social media rewires our brains (MSNBC 2022). While more evidence is needed on the latter, most would agree with the former, especially regarding changes in people's communication and information behavior. Social media platforms have become popular sources of information in everyday life, through which users stay informed and connected. A Pew Research Center study (hereafter, *Pew study*) found, for example, that almost half of US adults get news from social media at least sometimes (Walker and Matsa 2021). Unfortunately, problems on social media, such as mis/dis-information, have become increasingly pressing. Around 64 percent of US adults felt that social media has mostly negative effects on how things are going in the country (Auxier 2020). It is critical for scholars and policymakers to explore ways to tackle these issues. It is equally important to hear from the community of users themselves as they are more directly involved.

The current study focuses on Twitter, a popular microblogging platform. Among various social media platforms (e.g., Facebook, YouTube, Instagram), Twitter has the largest share of users (55 percent) who get news from the site regularly (Walker and Matsa 2021). Twitter is also a go-to platform for breaking news (Osborne and Dredze 2014) and plays a notable role in crisis communication and disaster management, such as during natural disasters and mass shootings (Acar and Muraki 2011). However, in part due to the short length, large volume, and easy and speedy distribution of postings, the information environment of Twitter can be volatile and challenging (Sankaranarayanan et al. 2009; Sin 2016). Like other social media, Twitter also faces extensive mis/dis-information issues including disinformation campaigns during elections, and vaccine misinformation (Bovet and Makse 2019;

Joanna Sin Sei-Ching is an Associate Professor at the Wee Kim Wee School of Communication and Information, Nanyang Technological University, Singapore. Kyung-Sun Kim is a Professor at the Information School, University of Wisconsin-Madison.

FIGURE 10.1 Visual themes from how to manage issues on Twitter: perspectives from Twitter users concerned about mis/dis-information.

Chamberlain 2010; Linvill and Warren 2020), which pose significant and long-lasting threats to individuals and societies. It is disconcerting that falsehoods often spread on social media faster and further than truth, and that they persist and resurge even after they have been debunked, as shown in other Twitter studies (Shin et al. 2018; Vosoughi, Roy, and Aral 2018). In a Pew study, 91 percent of US adults reported encountering at least some inaccurate or misleading information on Twitter, while 33 percent said they encountered such information "a lot" (Odabaş 2022b). It is thus of interest to examine how Twitter users (also called *Twitterers*) handle these mis/dis-information issues on the platform.

As a platform where a community of users creates and shares information, social media such as Twitter can be conceptualized as knowledge commons. Knowledge commons is a complex ecosystem where a resource is shared by a group of people that is subject to social dilemmas (Hess and Ostrom 2007). Focusing on the governance structure, the Governing Knowledge Commons Framework (GKC) (Frischmann, Madison, and Strandburg 2014; Madison, Frischmann, and Strandburg 2010) views the knowledge commons as "the institutionalized community governance of the sharing and, in some cases, creation, of information, science, knowledge, data, and other types of intellectual and cultural resources" (Frischmann, Madison, and Strandburg 2014, 3). So far, little is known about the governance of social media as knowledge commons, such as what users do and want for social media platform governance (Riedl, Whipple, and Wallace 2021).

Informed by the GKC (Frischmann, Madison, and Strandburg 2014; Madison, Frischmann, and Strandburg 2010), which modified and extended Ostrom's Institutional Analysis and Development framework (IAD), this study is designed to understand Twitter's user community and governance: how Twitter users manage day-to-day issues, and what actions and actors they see as vital to governing the platform – a knowledge commons.

The first set of research questions (RQs) focus on the *action arena*, exploring what *actions* have been taken by Twitter users and what actions and actors they deemed important:

RQ1a: How frequently do Twitter users *take various actions* when encountering problems on Twitter?
RQ1b: To what extent do Twitter users *think others should take various actions* for managing problems on Twitter?
RQ1c: To what extent do Twitter users think different *actors should take responsibility* for managing problems on Twitter?

GKC and IAD posit that the action arena can be affected by several underlying factors, including the attributes of the community. A heterogeneous community could mean that the knowledge commons is more challenging and costly to maintain, as community members are likely to have varying values and interests (Ostrom and Hess 2007). This is pertinent to Twitter, where the community is diverse (Auxier and Anderson 2021; Chaffey 2022). Moreover, a social media user may undertake multiple roles (e.g., sharers of existing resources, creator of new resources, cocreators of collective resources), which would add another layer of complexity to knowledge production and management (Frischmann, Madison, and Strandburg 2014). We cannot assume that community members will act the same way and share the same view regarding the governance of the commons. The second set of RQs thus examines whether individuals with different characteristics would act or think differently:

> **RQ2**: Are there demographic differences (gender, age, education, frequency of use) in: (a) how frequently Twitter users themselves take various actions; (b) the extent to which Twitter users want various actions to be taken by others; and (c) the extent to which Twitter users think different actors should take responsibility?

Scholars and policymakers will find the study findings useful when developing plans to tackle rising challenges on Twitter, such as mis/dis-information. Findings on demographic differences could also inform policies and system designs to be more attuned to individual differences. In addition, the findings could be of interest to researchers who are collecting empirical evidence to develop hypotheses and models on users' social media information behavior. Through the study findings, Twitter users can gain insights into their fellow community members regarding their actions and preferences in managing Twitter problems. Such insights could be instrumental for users in improving Twitter and other social media communities through collective actions.

BACKGROUND AND LITERATURE REVIEW

Social Media Platforms as Knowledge Commons

Social media, "online and mobile technologies or platforms people use to interact and share content" (Chandler and Munday 2016b), is often seen as heralding a new era in how information and content are produced, communicated, and consumed (Bruns 2007). Among the different types of social media, microblogs allow users to broadcast short text messages (Chandler and Munday 2016a). Twitter, one of the most popular microblog platforms, was estimated to have about 329 million users worldwide in 2022 (Statista 2022), posting over 690 million tweets per day on average (Internet Live Stats 2022). Many US users (89 percent) keep their accounts public (McClain et al. 2021), thus allowing people without Twitter accounts to read their public tweets. As a digital space and virtual community where a group of individuals come together to share content, social media platforms are shared informational resources (i.e., knowledge commons). Corresponding analytical frameworks, such as GKC and IAD, can offer insight into understanding people's decision-making and behaviors on these platforms (Ostrom and Hess 2007).

The Governing Knowledge Commons Framework (GKC)

Overall, GKC puts forth the following constructs and relationships. First, it identifies three groups of interacting variables – *resource characteristics*, *attributes of the community*, and *rules-in-use* – as underlying factors. These factors would affect the *action arena*, including *action situations* and *actors*. Specifically, the action arena captures how community members make decisions within a situation. These decisions would produce various *patterns of interaction*, which feed back to the three underlying factors and the action arena. *Evaluation criteria* can be leveraged to assess the interaction patterns. Regarding details of the GKC framework, readers may consult Chapter 1 of this volume.

The current study focuses on the action arena, which Ostrom and Hess (2007) identified as "often at the heart of the analysis" and "particularly useful in analyzing specific problems or dilemmas in the process of institutional change" (45). Before zooming into the action arena, it would be helpful to contextualize the study environment. Based on GKC and IAD, the following sections will present the description of the study context and relevant literature related to the three groups of underlying factors: *resource characteristics, attributes of the community*, and *rules-in-use*.

Resource Characteristics

On resource characteristics, the Twitter site (https://twitter.com) can be viewed as the *facilities*. It is where *artifacts* (e.g., the tweets) expressing various *ideas* are stored and made available to the public (Ostrom and Hess 2007). While it is possible that some materials (e.g., photos) posted on Twitter are copyrighted and thus have restrictions on reuse, the ideas communicated in digital form through Twitter are nonrivalrous (Frischmann, Madison, and Strandburg 2014). That is, a person reading a tweet and using its ideas does not deplete the resource pool nor prevent others from accessing the same tweet and ideas. Through Twitter, users can post and access different types of information and ideas, including news (Kim, Sin, and Yoo-Lee 2014; Walker and Matsa 2021), especially breaking news (Osborne and Dredze 2014), citizen journalism (Murthy 2011), social activism information (Sandoval-Almazan and Gil-Garcia 2014), opinions and popular trends (Kim and Sin 2016; Kim, Sin, and Yoo-Lee 2021), and entertainments (McClain et al. 2021). Twitter is also a source of expert and academic information (Mohammadi et al. 2018). Especially after 2017 when Twitter added threads as a feature, experts are known to write "tweetorials," a series of tweets to explain technical concepts to a wider public audience (Breu 2019). Overall, the informational and educational potentials of Twitter as a shared knowledge commons are considerable. Unfortunately, Twitter could not escape the "dark side" of social media (Baccarella et al. 2018). Mis/dis-information, hate speech, harassment, and doxing are among the problems plaguing the Twitter community (MacAllister 2016; McClain et al. 2021; Whittaker and Kowalski 2015).

Attributes of the Community

Regarding knowledge commons' attributes of the community, IAD identifies *information users*, *information providers*, and *information managers/policymakers* (Ostrom and Hess 2007). A defining characteristic of social media like Twitter is that they blur the line between information users and information providers (Bruns 2007). Such blurring of roles is highlighted by the GKC (Frischmann, Madison, and Strandburg 2014). Instead of being mere receivers, social media users now have the ability to broadcast user-generated content to large masses at low costs, allowing diverse and marginalized voices that have been traditionally sidelined by mass media to come through. The democratizing and transformative potential of social media has long been lauded. However, it is increasingly apparent that along with potential, significant dangers also exist (Picard 2015). The quality of information on social media is often found wanting (McClain et al. 2021), and there is evidence of coordinated disinformation campaigns (Beskow and Carley 2020; Hindman and Barash 2018). Exacerbating the problem is that credibility assessment of social media content is often challenging. For example, the original source of the content may be obscured, rendering it difficult to apply traditionally well-used heuristics such as checking source expertise and trustworthiness (Wathen and Burkell 2002). Users are often found to rely on superficial heuristics such as the date or the length of a message when evaluating social media content (Kim, Sin, and Yoo-Lee 2021). Furthermore, poor information evaluation awareness or skills are not always the culprit (Kim and Sin 2011). It appears that, sometimes, truthfulness simply is not a high priority in users' everyday social media sharing. Users' willingness to share a message on social media is not necessarily affected by the accuracy or trustworthiness of the message. Instead, a message is shared because it is perceived as novel, eye-catching, and a good topic of conversation (Chen et al. 2015a; Leeder 2019). The unfortunate outcome is that social media users often unknowingly become spreaders of mis/dis-information.

Not all Twitter "users" are humans. Some are bots, automated accounts programmed to perform specific tasks. Bots can provide useful functions, such as adding video captions (e.g., @HeadlinerClip). News bots (Lokot and Diakopoulos 2016) (e.g., @earthquakeBot, @FintechBot, @MagicRealismBot, and @parliamentedits) help share breaking news or topical information and commentary, while others send self-care (e.g., @tinycarebot) and positive messages (e.g., @TheNiceBot). Unfortunately, the values bots can bring to the Twitter community are often overshadowed by malicious bots, which play significant roles in spreading falsehood, spamming, and phishing (Alothali et al. 2018). Bots have been found to target influential users (e.g., Twitter users with many followers) by mentioning or replying to them with links to low-credibility information, hoping the information will be reshared (Shao et al. 2018). Removing those bots could improve the quality of social media information considerably. However, researchers have emphasized that bots alone cannot account for the spread of falsehood. The actions of human users still matter to the propagation of mis/dis-information (Shao et al. 2018; Vosoughi, Roy, and Aral 2018).

The management team of Twitter, Inc., the company that owns Twitter, is the primary policymaker. Government policymakers also play a role at a broader level, as both regular users and Twitter, Inc. are subject to prevailing state, national, and international laws. Together, these policy markers establish the rules that set the boundaries of users' day-to-day operations.

Rules-In-Use

In examining rules-in-use, the degree of openness and nature of control is a focus of GKC (Frischmann, Madison, and Strandburg 2014). Except in countries where Twitter is banned, it is relatively open to access and participation. IAD identified three levels of rulemaking: operational, policy, and constitutional (Ostrom and Hess 2007). In the context of this study, the operational level is where Twitter users make day-to-day decisions on how they act and interact with each other on the platform. This level is the focus of the current study. Regarding the policy level, it is where people, such as the management of Twitter, Inc., make rules concerning users' day-to-day operations. While Twitter is quite open to public participation, as noted above, control is ultimately vested in Twitter, Inc. Specifically, users' actions on Twitter are subjected to the Twitter User Agreement, including the *Terms of Services*, *Privacy Policy*, and *Twitter Rules and Policies* set forth by Twitter, Inc. (Twitter 2022). Beyond policy, the technology itself can constrain users through system design. For example, tweets were intended to be short, with an original 140 characters and a current 280 characters limit per tweet. Critics have observed that this design demands ideas to be simplified, rendering Twitter ill-suited for nuanced discussions (Ott 2017). The brevity also means that contextual information is often absent, making credibility assessment difficult (Sankaranarayanan et al. 2009; Sin 2016).

The challenges in credibility assessment are not helped by the fact that most US social media companies traditionally have not set strict rules on mis/dis-information (Wardle and Singerman 2021). It was only in 2020 when Twitter began taking more visible actions on mis/dis-information on specific topics, including labeling misleading information related to COVID-19 (Roth and Pickles 2020), the US 2020 election (Gadde and Beykpour 2020), and, more recently, Russia's invasion of Ukraine (Roth 2022).

Legal structures affect knowledge commons (Frischmann, Madison, and Strandburg 2014), which is salient to the current study as both Twitter users and Twitter, Inc., are bounded by relevant legislation. In the US, Section 230 of the federal code has become an issue of contention. Section 230 states:

> "No provider or user of an interactive computer service shall be treated as the publisher or speaker of any information provided by another information content provider" (United States Communications Decency Act 1996) and that they should not be liable for "any action voluntarily taken in good faith to restrict access to or

> availability of material that the provider or user considers to be obscene, lewd, lascivious, filthy, excessively violent, harassing, or otherwise objectionable, whether or not such material is constitutionally protected . . ." (United States Communications Decency Act 1996)

This gives companies such as Twitter, Inc. the power to make their own decisions on what content to keep on their site or to take down. Laws such as Section 230 operate at IAD's constitutional level of rulemaking. Section 230 defines that in the US, it is primarily Twitter, Inc., and not the government, who makes decisions regarding the handling of mis/dis-information on Twitter.

Individual Differences: Demographics and Use Behavior

GKC highlights attention to the attributes of the community and their influences on the *patterns of interaction*. We thus posit that Twitter users' attributes/characteristics may be associated with different behaviors and preferences in the action arena. This aligns with the views and findings from information behavior research, which has often observed various individual and group differences (Case and Given 2016). As a beginning exploration, the current study first examined basic demographic attributes, specifically, age, gender, educational attainment, and frequency of Twitter use.

Although the findings are not unanimous, gender differences are often found in various aspects of social media usage. These include differences in strategies used for evaluating social media information (Kim and Sin 2015), motivations for sharing misinformation (Chen et al. 2015b), and perception of online risk (Lim and Kwon 2010). Furthermore, while men and women share some similarities in how they perceive online trolls, their actions differ. Men were more likely to confront or block trolls, whereas women were more likely to ignore the trolling and discourage confrontation (Fichman and Sanfilippo 2014). A meta-analysis also found that women had greater concerns about social media privacy issues than men. Unlike men, women were more likely to activate privacy settings and less likely to disclose personal information in their profiles (Tifferet 2019). Studies of gender differences in social media have primarily focused on men and women. Less is known about how different groups in the entire gender spectrum compare to each other in their social media use.

Most discussions on age differences in social media focus on the prevalence of use. For example, Twitter is found to be more popular among younger individuals (McClain et al. 2021). Beyond prevalence, some researchers have posited that older users may act differently, such as exercising less self-disclosure (Taddicken 2014) or having different norms regarding emotional expression on social media (Waterloo et al. 2017). The findings have been mixed, however.

Educational differences are less often explored. However, education level was found positively related to the support for social media content reviews (Riedl, Whipple, and Wallace 2021). Microblog users who were college underclassmen reported more difficulties in finding everyday information compared to upperclassmen, master's, and doctoral students (Sin and Kim 2014).

Frequency of use can also have influences on users' attitudes and behaviors. For Twitter, a Pew study found several differences. For example, infrequent users followed fewer accounts than frequent users. They were also more interested in using Twitter to learn about different viewpoints rather than to express their own opinions (Odabaş 2022a). Another study found nonlinear, u-shape patterns between frequency of microblog use and everyday information seeking. Occasional microblog users reported a higher level of difficulties in finding everyday information than non- or frequent users and were less satisfied with the quality of the information found (Sin and Kim 2014). Frequent social media users tended to be less supportive of government regulation of platforms (Riedl, Whipple, and Wallace 2021).

RESEARCH METHODS

Data Collection

Study participants were Twitter users in the US who found misinformation on Twitter problematic. Data were collected using an online questionnaire. Participants answered questions on their demographics, what actions were taken or should be taken to manage problems on Twitter, and what groups should take responsibility. The survey was conducted in Fall 2018. The sample was recruited from Amazon Mechanical Turk (MTurk). MTurk is among the most frequently used channels for recruiting participants online (Aguinis, Villamor, and Ramani 2021), as it is known to facilitate the recruitment of diverse samples (Casler, Bickel, and Hackett 2013).

The current study focused on participants who considered misinformation on Twitter to be problematic. This is operationalized by participants' responses to two questions, "Regarding tweets/retweets posted by Twitter users outside your social network, how problematic is fake news?" and "Regarding tweets/retweets posted by Twitter users outside your social network, how problematic is inaccurate information?" To account for differences in how individuals perceive social boundaries, participants were to decide who was considered "outside [their] social network" instead of following a universal definition set by the researchers. Participants who have an average score of four or above on a five-point scale (indicating "Very problematic" to "Extremely problematic") on the two questions were included in the current analysis. The questionnaire was hosted on Qualtrics. A pilot test was done with ten MTurkers before data collection.

Data Analysis

Descriptive statistics, including means and standard deviations, were used to address RQ1. Robust multiway ANOVAs were used for RQ2. Robust multiway ANOVA is an inferential test for comparing group means. Particularly, it analyzes the relationship between a continuous dependent variable (DV) and multiple categorical independent variables (IVs). Robust multiway ANOVA has the advantage of not requiring the assumptions of regular multiway ANOVA (Field 2009). In this study, the DVs are the top five answers from RQ1a, RQ1b, and RQ1c, respectively. There are four IVs: age, gender, educational attainment, and frequency of Twitter use. Analyses were done using SPSS.

RESULTS

Participant Characteristics

The sample consisted of 400 US participants on MTurk who were registered users of Twitter and found misinformation on Twitter "very" or "extremely problematic." In terms of age, participants in their thirties constituted the majority of the sample (n = 202, 50.5%), followed by those in their forties (n = 80, 20%) and twenties (n = 60, 15%), fifties (n = 40, 10%) and sixty and above (n = 18, 4.5%) (Table 10.1). There were slightly more women (n = 205, 51.3%) than men (n = 193, 48.3%). Two participants (0.5%) selected "Other." Both identified themselves as genderqueer. A plurality of the

TABLE 10.1 *Participant characteristics*

	Count	Percentage (%)
Age group		
20s	60	15.0
30s	202	50.5
40s	80	20.0
50s	40	10.0
60 or above	18	4.5
Gender		
Man	193	48.3
Woman	205	51.3
Other	2	0.5
Educational attainment		
High school	41	10.3
Some college	152	38.0
Bachelor's degree	161	40.3
Master's degree or above	45	11.3
(Missing)	1	0.3
Frequency of use		
Less than monthly	68	17.0
Monthly	74	18.5
Weekly	116	29.0
Daily	116	29.0
Hourly	26	6.5

TABLE 10.2 *Actions taken by participants: frequency of taking actions*

Rank	Actions	M	SD
1	Unfollow the problematic user	3.76	1.19
2	Verify/fact check the information yourself	3.46	1.12
3	Mute the account of the problematic user	3.31	1.39
4	Block the account of the problematic user	3.30	1.35
5	Check the bio and other posts of the problematic user to learn more about them	3.19	1.17
6	Wait and see what happens	2.80	1.13
7	Avoid reading tweets posted by someone outside of your social network	2.74	1.21
8	Spend less time on Twitter	2.48	1.20
9	Report the problematic user to Twitter	2.45	1.35
10	Spend more time on other social media platforms	2.37	1.21

Scales used: 1: Almost never; 2: Occasionally; 3: Sometimes; 4: Frequently; 5: Almost always.

participants held a bachelor's degree (n = 161, 40.3%). Participants with some college education were the second largest group (n = 152, 38.0%), followed by those with a master's or above (n = 45, 11.3%) and those with a high school diploma (n = 41, 10.3%). Most of the participants indicated that they used Twitter daily or weekly.

Actions Taken by Participants (RQ1a)

When encountering problems on Twitter, "Unfollow the problematic user" was the action participants reported taking most frequently (Table 10.2). The second most frequently used action was "Verify/fact check the information yourself." The top five results fall broadly into two categories. One category includes actions that can reduce future encounters with problematic tweets and users (e.g., unfollowing, muting, or blocking). The second group is related to information seeking/verification (e.g., verifying the information and checking the bio and posts of problematic users).

More assertive actions, such as reporting problematic users to Twitter, were used less frequently (ranked ninth out of ten actions). Participants also did not seem too deterred from using Twitter due to problems there. "Spend less time on Twitter" and "Spend more time on other social media platforms" were used only occasionally (ranked eighth and tenth, respectively).

Actions to Be Taken by Others (RQ1b)

Regarding the actions that participants want others to take (Table 10.3), an institutional effort, "Ban problematic users," received the highest average score (M = 4.04, SD = 1.24). This is followed by actions in system design, including

TABLE 10.3 *Actions to be taken by others: extent to which participants want others to take them*

Rank	Actions	M	SD
1	Institutional efforts: Ban problematic users from Twitter	4.04	1.24
2	System design: Provide better functions for users to filter things easily	4.01	0.99
3	System design: Provide better functions for users to report problematic users/tweets easily	3.89	1.16
4	System design: Provide functions for moderators to issue warnings to problematic users	3.81	1.23
5	Institutional efforts: Employ people to monitor and problematic user/tweet	3.69	1.27
6	System design: Provide functions for users to downvote or dislike problematic users/tweets	3.55	1.33
7	System design: Provide functions to automatically and silently block problematic users/tweets without requiring users' inputs	3.44	1.43
8	Institutional efforts: Remind users regularly of Twitter's rules, policies and terms of services	3.18	1.28
9	System design: Provide functions to reward users who have rightfully flagged a problematic user/tweet	3.01	1.42
10	Institutional efforts: Remind users regularly of microblogging etiquettes	2.95	1.32
11	Institutional efforts: Enact and enforce laws to deter and punish problematic behavior on Twitter	2.91	1.47
12	Institutional efforts: Provide training and resources to users on how to use Twitter properly and safely	2.80	1.38
13	Users' efforts: Twitter users to form volunteer groups to monitor and flag problems	2.71	1.24
14	Users' efforts: Individual Twitter user to rebut and deter other users' problematic behaviors	2.69	1.22
15	Institutional efforts: Require all tweets to be posted with real names	2.57	1.57

Scales used: 1: Not at all; 2: To a small extent; 3: To a moderate extent; 4: To a great extent; 5: To a very great extent.

providing better functions for individual users "to filter things easily" (M = 4.01, SD = 0.99) and "to report problematic users/tweets easily" (M = 3.89, SD = 1.16). The fifth highest score went to an institutional effort, "Employ people to monitor and problematic user/tweet" (M = 3.69, SD = 1.27).

Overall, actions related to the system-design category tend to score higher, with average scores ranging from three (*to a moderate extent*) to four (*to a great extent*). In contrast, participants did not prefer actions in the users' efforts category. "Twitter users to form volunteer groups to monitor and flag problems" and "Individual Twitter user to rebut and deter other users' problematic behaviors" scored the second and the third lowest out of fifteen strategies. Actions in the institutional efforts category saw wider variations. For example, while the highest-scoring effort

TABLE 10.4 *Actors to take responsibility: extent to which participants want them to be responsible*

Rank	Actors	M	SD
1	The Twitter company	4.55	0.93
2	Individual Twitter user	3.32	1.27
3	Prominent Twitter users/influencers	3.02	1.40
4	Search engines, social media, and internet companies	2.93	1.39
5	Social media experts, subject specialists, and scholars	2.86	1.35
6	All members of the public	2.73	1.31
7	The media and journalists	2.46	1.27
8	Social activists	2.43	1.31
9	Law enforcement agencies	2.32	1.18
10	Schools, educators, and librarians	2.21	1.30
11	The government, politicians, and elected officials	1.91	2.15
12	Nonprofit organizations	1.62	1.75

Scales used: 1: Not at all; 2: To a small extent; 3: To a moderate extent; 4: To a great extent; 5: To a very great extent.

(i.e., "Ban problematic users") was an institutional effort, the lowest-scoring was also an institutional effort (i.e., "Require all tweets to be posted with real names").

Actors to Take Responsibility (RQ1c)

Regarding the extent to which different groups should take responsibility for managing problems on Twitter, the answer for "The Twitter company" (Twitter, Inc.) stood out ($M = 4.55$, $SD = 0.93$), indicating an average score ranging from *a great extent* to *a very great extent*. Notably, Twitter, Inc. scored considerably higher than any other groups on the list (Table 10.4).

Results on the top four highest-scoring groups show that participants centered the responsibility on companies and users. Beyond Twitter, Inc., mentioned above, "Search engines, social media, and internet companies" ranked fourth. The second- and third-highest groups were related to users: "Individual Twitter user" and "Prominent Twitter users/influencers." Experts ("Social media experts, subject specialists, and scholars") rounded out the fifth. In contrast, governmental groups ("Law enforcement agencies" and "The government, politicians, and elected officials") were seen as having responsibilities to only *a small extent*. Their average scores ranked ninth and eleventh out of twelve groups.

Demographic Differences in Actions Taken by Participants (RQ2a)

"Unfollow the problematic user" was the action most frequently taken by participants (in RQ1a). The robust multiway ANOVA analysis revealed a statistically

	Unfollow the problematic user		Verify/fact check the information yourself		Mute the account of the problematic user		Block the account of the problematic user		Check the bio and other posts of the problematic user to learn more about them	
	B	partial η2	B	partial η2	B	partial η2	B	partial η2	B	partial η2
Intercept	3.99 ***	0.17	3.34 ***	0.14	4.01 ***	0.15	3.95 ***	0.16	3.64 ***	0.17
Age group										
20s	0.08	0.00	-0.06	0.00	-0.16	0.00	-0.44	0.00	-0.28	0.00
30s	0.03	0.00	-0.11	0.00	0.06	0.00	-0.18	0.00	-0.07	0.00
40s	-0.08	0.00	-0.06	0.00	0.06	0.00	-0.40	0.00	-0.32	0.00
50s	0.37	0.00	0.06	0.00	0.18	0.00	0.10	0.00	0.13	0.00
60 or above (reference group)										
Gender										
Man	**-0.32 ***	0.02	0.02	0.00	**-0.34 ***	0.01	**-0.31 ***	0.01	0.02	0.00
Woman (reference group)										
Educational attainment										
High school	0.27	0.00	0.17	0.00	-0.06	0.00	-0.06	0.00	-0.20	0.00
Some college	0.13	0.00	0.05	0.00	-0.04	0.00	0.11	0.00	0.14	0.00
Bachelor's degree	0.15	0.00	0.02	0.00	-0.05	0.00	0.01	0.00	0.15	0.00
Master's degree or above (reference group)										
Frequency of use										
Less than monthly	-0.43	0.01	-0.02	0.00	**-0.93 ***	0.02	-0.56	0.01	**-0.79 ****	0.02
Monthly	-0.38	0.00	0.24	0.00	**-0.93 ***	0.02	**-0.75 ***	0.01	-0.53	0.01
Weekly	-0.30	0.00	0.09	0.00	-0.53	0.01	-0.31	0.00	**-0.62 ***	0.02
Daily	-0.05	0.00	0.26	0.00	-0.15	0.00	0.04	0.00	-0.08	0.00
Hourly (reference group)										

* $p < .05$ ** $p < .01$ *** $p < .001$

FIGURE 10.2 Robust multiway ANOVA results: demographic differences in actions taken by participants.

significant demographic difference (Figure 10.2). Specifically, men tended to use this action less frequently than women. In contrast, the second most used action, "Verify/fact check the information yourself," saw no significant differences.

Gender differences were also found in the third and fourth most frequently taken actions: Men used "Mute the account of the problematic user" and "Block the account of the problematic user" less. Frequency of Twitter use was also statistically significant. Compared to the reference group (hourly users), participants who tweeted infrequently (e.g., less than monthly or monthly) tended to mute or block others less.

For "Check the bio and other posts of the problematic user," the frequency of use was again statistically significant. Infrequent users (e.g., less than monthly) and weekly users applied the above action less than the reference group.

Demographic Differences in Actions to Be Taken by Others (RQ2b)

In contrast to the above results, where gender and frequency of use differences were found, no such difference was found regarding the top actions participants wanted others to take.

Demographic Differences in Actors Considered Responsible (RQ2c)

"The Twitter company," the highest-ranked group that participants consider most responsible, saw a statistically significant difference in frequency of Twitter use

	The Twitter company		Individual Twitter user		Prominent Twitter users/influencers		Search engines, social media, and Internet companies		Social media experts, subject specialists, and scholars	
	B	partial η2	B	partial η2	B	partial η2	B	partial η2	B	partial η2
Intercept	4.18 ***	0.26	4.30 ***	0.23	4.02 ***	0.14	3.94 ***	0.14	3.77 ***	0.15
Age group										
20s	0.00	0.00	-0.55	0.01	-0.13	0.00	-0.59	0.01	-0.13	0.00
30s	0.07	0.00	-0.36	0.01	-0.21	0.00	-0.49	0.01	-0.20	0.00
40s	0.14	0.00	-0.44	0.01	-0.62	0.01	**-0.68 ***	0.01	-0.51	0.01
50s	0.29	0.00	-0.44	0.01	-0.31	0.00	-0.34	0.00	-0.30	0.00
60 or above (reference group)										
Gender										
Man	-0.12	0.00	-0.04	0.00	**-0.33 ***	0.01	**-0.36 ***	0.02	**-0.29 ***	0.01
Woman (reference group)										
Educational attainment										
High school	0.14	0.00	0.15	0.00	0.01	0.00	-0.35	0.00	-0.41	0.00
Some college	0.20	0.00	-0.11	0.00	-0.15	0.00	-0.13	0.00	-0.27	0.00
Bachelor's degree	-0.13	0.00	0.08	0.00	-0.04	0.00	-0.05	0.00	-0.34	0.01
Master's degree or above (reference group)										
Frequency of use										
Less than monthly	**0.46 ***	0.01	**-0.61 ***	0.01	-0.65	0.01	-0.15	0.00	-0.07	0.00
Monthly	0.25	0.00	**-0.75 ** **	0.02	**-0.75 ***	0.01	-0.56	0.01	-0.55	0.01
Weekly	0.26	0.00	**-0.58 ***	0.01	-0.44	0.01	-0.30	0.00	-0.35	0.00
Daily	0.32	0.01	**-0.52 ***	0.01	-0.37	0.00	-0.03	0.00	-0.09	0.00
Hourly (reference group)										

* $p < .05$ ** $p < .01$ *** $p < .001$

FIGURE 10.3 Robust multiway ANOVA results: demographic differences in actors to take responsibility.

(Figure 10.3). That is, infrequent users (e.g., less than monthly) tended to rate the company as having higher responsibilities than did the reference group (hourly users). Frequency of use was again significant for the second-highest response – "Individual Twitter user." This time, however, infrequent users rated "Individual Twitter user" lower than the reference group.

"Prominent Twitter users/influencers" also saw a significant difference in frequency of Twitter use. Similar to the "Individual Twitter user," the infrequent users (e.g., monthly) rated "Prominent Twitter users/influencers" as having less responsibility than did the reference group. In addition, gender differences were found. Women tended to rate "Prominent Twitter users/influencers" as having higher responsibilities than men did.

For "Search engines, social media, and internet companies," a similar gender difference was observed: Women rated them higher than men did. Furthermore, age differences emerged. Compared to younger participants, particularly those in their forties, participants who were sixty or above (the reference group) saw these companies as having higher responsibilities.

For the fifth-highest group, "Social media experts, subject specialists, and scholars," a gender difference was found. Women rated the responsibilities of this group higher than men did.

To summarize, women, more than men, tended to view the following three groups as having higher responsibilities in handling Twitter problems: "Prominent Twitter users/influencers"; "Search engines, social media, and internet companies"; "Social media experts, subject specialists, and scholars." Infrequent Twitter users rated "The Twitter company" highly while rating "Individual Twitter user" and "Prominent Twitter users/influencers" lower than did the reference group (hourly users). Age difference was found only in "Search engines, social media, and Internet companies."

DISCUSSION

Action Arena

Participants' top actions taken when encountering problems on Twitter (RQ1a) can be categorized in two groups: information filtering (including unfollowing, muting, and blocking) and information verification (including fact-checking and checking the bio and posts of problematic users). From the perspective of information literacy education, it is promising that fact-checking, one of the highly recommended strategies, is among the top actions. Interestingly, while all participants considered fake news and inaccurate information on Twitter "very problematic" or "extremely problematic," they did not frequently take actions. Even the top action – unfollowing – averaged only 3.76, and fact-checking averaged 3.46, which translates to between *sometimes* and *frequently* used. A possible reason for this may be that the information filtering strategies used by participants may have helped reduce the chance of encountering problematic users and tweets. Twitter's algorithms (Newberry and Sehl 2022) may also have reinforced the users' filtering actions and reduced unpreferred/unwanted encounters. Even for earlier information systems, "the need for quality filtering of information" was identified as an essential need (Faibisoff and Ely 1976, 6). Quality filtering is no less vital today. For social media platforms such as Twitter, how to ensure necessary filtering does not exacerbate current problems of "echo chamber" and polarization (Cinelli et al. 2020) would require the attention of policymakers and platform designers.

Banning problematic users emerged as the top action that participants wanted others to take. Whether social media companies should ban (or be allowed to ban) problematic users has become increasingly politicized. For example, Republicans and Democrats differed significantly on whether it was right or wrong for some social media companies to ban Donald Trump from their platforms. Among Republicans and Republican leaners, 78 percent said the ban was wrong. In contrast, 89 percent of Democrats and Democrat leaners said the ban was right. Republicans were especially more likely than Democrats to believe that social media platforms intentionally censor political viewpoints (McClain and Anderson 2021). Two states, Texas and Florida, passed laws barring large social media companies from banning people based on their viewpoints. Currently, these laws are on hold as they are being contested in courts, but the debate is expected to continue (Robertson 2022). Unlike the ongoing legal contests, participants' views on the topic have been quite similar. They notably favored such bans *to a large extent* (mean score of 4.04 out of 5). Future studies may explore if this preference holds and if respondents' political affiliation plays a role.

This result on which participants wanted Twitter, Inc. to ban problematic users can be viewed in tandem with the findings from RQ1a about participants' own actions. Participants often unfollowed, muted, or blocked problematic users, but they did not frequently report problematic users to Twitter. Reporting to Twitter ranked only ninth, with a mean score between *occasionally* and *sometimes*. The findings suggest that participants did not take the power of banning users lightly. They appear to be using the function judiciously while recognizing the need to do something to prevent problematic users from disrupting the community further.

Participants saw Twitter, Inc. as most responsible for handling Twitter issues. The mean score (4.55, *to a great extent*) of their level of responsibility was considerably higher than those for other actors such as individual Twitter users or influencers. This finding confirms the calls from the public, scholars, and policymakers for social media companies to take more responsibility (Cusumano, Gawer, and Yoffie 2021; Suzor et al. 2019; Wakabayashi 2019). On the other end of the spectrum, participants did not emphasize the government's role. "The government, politicians, and elected officials" ranked eleventh out of twelve actors to take responsibility. Law enforcement ranked ninth. The option of enacting and enforcing laws also drew little interest (eleventh out of fifteen actions to be taken by others). More studies can be conducted to investigate the reasons behind these preferences. Areas to explore are whether users' preferences are affected by their understanding of internet laws, users' familiarity with the roles different actors play in social media governance, and what users think different actors can do.

The US public has complicated views on government regulation of social media. While the view that the government should restrict false information online has gained more support recently (from 39 percent in 2018 to 48 percent in 2021), a larger share of the public considered that tech companies should take care of the issue (56 percent in 2018 and 59 percent in 2021) (Mitchel and Walker 2021). A more recent Pew study shows a steep decline in the preference for government regulation of tech companies, however. The share of support dropped from 56 percent in 2021 to 44 percent in 2022. This drop was observed not only among Republicans but also among Democrats (Vogels 2022). The issue is likely made more complex by what different levels of governments would regulate. Now, government regulation may not be limited to restricting false or harmful materials. It could also include government regulation that restricts a company's power to handle false information (e.g., the Texas and Florida state laws mentioned above). The *constitutional level of rules making*, which defines "who must, may, or must not participate in making collective choices" (Ostrom and Hess 2007, 50), would become the space to watch. If the center of action indeed shifts to the fights between governments and big tech companies in the legal arena – and away from the day-to-day operations and relationships between a company and its users – whether users may feel sidelined and subsequently disengaged could become a matter of concern. With GKC's attention to the character of control and the "nestness" (Frischmann, Madison, and Strandburg 2014, 32) of knowledge commons in broader sociocultural and institutional settings, future studies may apply GKC to further probe whether exogenous institutions such as the government may gain more control of platform governance over internal governing mechanisms within the Twitter community.

Participants' preference for content moderation is also worth noting. Twitter uses commercial content moderation, including algorithmic moderation. Such moderations are often found to be opaque (Gorwa, Binns, and Katzenbach 2020; Roberts 2018), especially when compared to content moderation done by user volunteers (such as on Reddit and Twitch) (Cook, Patel, and Wohn 2021). In the current study, participants indeed prefer human moderation (fifth) over algorithmic moderation (seventh out of fifteen actions to be taken by others). However, they did not like involving volunteers to monitor or rebuke problems (ranked thirteenth and fourteenth). Instead, they prefer Twitter to employ people to monitor (ranked fifth) and issue warnings to problematic users (ranked fourth). This preference for the social media company (instead of users) to do more on content moderation is also observed in Cook and her colleagues' cross-platform study about content moderation for dealing with toxicity on social media (Cook, Patel, and Wohn 2021).

Unlike their study concluding that users seem to "want to be taken care of when it comes to content moderation as opposed to engaging themselves" (Cook, Patel, and Wohn 2021, 1), participants of the current study seem to want others to provide more ways to engage. The top actions they wanted others to take involved system functions that give them the power to filter things (ranked second), to report users or tweets (third), and to downvote them (sixth out of fifteen actions to be taken by others). It appears that participants do not want to burden users with the *responsibility* of wrangling Twitter problems, as noted above. Instead, they want system features and tools at their disposal, which would allow them to participate in tackling problematic users and tweets whenever they *choose* to.

Differences between previous and current study findings could be due to the differences in study focus. Cook and her colleagues' study, for example, was on social media toxicity, and their data covered multiple platforms, not just Twitter. Another possibility is due to potential differences in motivation levels. The current study focused on participants who considered dis/misinformation on Twitter very or extremely problematic. The perceived seriousness of the problem may have contributed to a higher interest in having the tools available to combat problems themselves. While any user engagement features could potentially be targeted for manipulation by malicious actors (Lee, Tamilarasan, and Caverlee 2021), companies should appreciate users' desire for more agency in their day-to-day interactions on social media.

Demographic Differences

Among the four user attributes, frequency of Twitter use and gender yielded more significant differences. Frequent users tended to take actions more frequently and think various actors should take responsibility to a larger extent than infrequent users. The exception is related to Twitter, Inc. Compared to infrequent Twitter users, frequent users rated the company's responsibility lower (while rating individual users' and influencers' responsibilities higher). A possible reason may be that frequent users are more familiar with the dynamics of the platform. They may be more cognizant of the impacts of individual users and influencers on compounding (or lessening) social media problems, such as rumor propagation (Mikhaeil and El Mougy 2020).

The study found that women tended to take actions more frequently and rated several actors with higher responsibilities. One contributing factor may be that women face more gender-based harassment than men on social media (Simons 2015; Winkelman et al. 2015); they have been the targets of disinformation and character attack (Bradshaw and Henle 2021). This may have led women to take more frequent actions to filter out problematic users and believe more strongly that various actors should take more active measures. It is also observed that women of color and other minorities often face situations where they have to provide unpaid labor to actively call out misogyny and racism on social media and end up facing harassment and threats as a result (Nakamura 2015). Along this line, more in-depth research may be needed to understand the lived experience and action arenas of user groups who have been found more targeted on social media, such as racial/ethnic minorities, LGBTQIA+, and people with disabilities (Silva et al. 2016).

Limitations and Further Studies

The study participants were a self-selected sample. Studies suggested that US MTurkers tend to be younger, have a higher female-to-male ratio, are more educated, and have lower median household income than the overall US population (Aguinis, Villamor, and Ramani 2021; Difallah, Filatova, and Ipeirotis 2018). They have also been found to be more ethnically diverse than other US internet panels (Smith et al. 2016). Overall, the sample may not represent all US adult Twitter users who considered mis/dis-information problematic. For instance, a limitation of this study sample is that there are too few gender-nonconforming participants. Further studies on different samples will shed light on whether they share similar patterns of behavior and preferences. As an initial step, this study focused on only a few user attributes. Future research may include more user attributes such as race/ethnicity, political ideology, and problem-solving styles. Studies may also expand the analysis to include other parts of GKC (e.g., patterns of interaction and evaluation criteria). A more diverse range of methods can also be employed. More inductive and qualitative methods such as interviews can be used to reveal the reasons behind participants' actions and preferences and to investigate how the action arena in turn influences users' thoughts and experiences. Social media data may also be crawled and collected for triangulation. As the socioecological, political, and legal milieus may evolve, longitudinal research would be beneficial to capture changes in the underlying factors and action arena over time. Regarding theoretical development, this study suggests that using GKC to study the Twitter community is promising. More research about this platform with GKC would contribute to (1) understanding the challenges of such research, and (2) identifying areas of development for using GKC to study Twitter and other social media as knowledge commons.

CONCLUSION

The current study is an initial exploration of the actions and preferences of Twitter users who found mis/dis-information on Twitter problematic. The findings have implications for Twitter platform governance, system design, and everyday management of mis/dis-information on social media. For instance, the study found that participants took more information filtering and fact-checking steps than assertive actions such as reporting problematic users or tweets. Participants wanted problematic users/contents to be banned, as well as having Twitter, Inc. employ people to moderate problematic users/contents. They also sought improvements in system design to provide users with better tools to engage, such as easier reporting and upvoting/downvoting capabilities. Participants considered Twitter, Inc. most responsible for managing problems on Twitter. There was less appetite for actions by governments, however. Some demographic differences were also found in the study. While age showed few differences, frequency of Twitter use and gender differences were more pronounced.

By collecting data from Twitter users who are concerned about mis/dis-information, this study sheds light on the actions they take, and their views on how to and who should take responsibility to confront these issues. Social media policymakers, system designers, researchers, and Twitter users may find insights from the study results. As challenges such as mis/dis-information continue to rage, and the governance of social media appears more and more politicized, it is of paramount importance to understand the everyday decision-making of diverse users and amplify their voices, so as to inform social media policies and system designs to take into consideration users' well-being and agency.

REFERENCES

Acar, Adam, and Yuya Muraki. 2011. "Twitter for Crisis Communication: Lessons Learned from Japan's Tsunami Disaster." *International Journal of Web Based Communities* 7 (3): 392–402.

Aguinis, Herman, Isabel Villamor, and Ravi S. Ramani. 2021. "MTurk Research: Review and Recommendations." *Journal of Management* 47 (4): 823–837.

Alothali, Eiman, Nazar Zaki, Elfadil A Mohamed, and Hany Alashwal. 2018. "Detecting Social Bots on Twitter: A Literature Review." In *Proceedings of the 2018 International Conference on Innovations in Information Technology (IIT)*, 175–180. Piscataway, NJ: IEEE.

Auxier, Brooke. 2020. "64% of Americans Say Social Media Have a Mostly Negative Effect on the Way Things Are Going in the US Today." Pew Research Center. www.pewresearch.org/fact-tank/2020/10/15/64-of-americans-say-social-media-have-a-mostly-negative-effect-on-the-way-things-are-going-in-the-u-s-today/.

Auxier, Brooke, and Monica Anderson. 2021. "Social Media Use in 2021." Pew Research Center. www.pewresearch.org/internet/2021/04/07/social-media-use-in-2021/.

Baccarella, Christian V., Timm F. Wagner, Jan H. Kietzmann, and Ian P. McCarthy. 2018. "Social Media? It's Serious! Understanding the Dark Side of Social Media." *European Management Journal* 36 (4): 431–438.

Beskow, David M., and Kathleen M. Carley. 2020. "Characterization and Comparison of Russian and Chinese Disinformation Campaigns." In *Disinformation, Misinformation, and Fake News in Social Media: Emerging Research Challenges and Opportunities*, ed. Kai Shu, Suhang Wang, Dongwon Lee, and Huan Liu, 63–81. Cham: Springer.

Bovet, Alexandre, and Hernán A. Makse. 2019. "Influence of Fake News in Twitter during the 2016 US Presidential Election." *Nature Communications* 10 (7): 1–14.

Bradshaw, Samantha, and Amélie Henle. 2021. "The Gender Dimensions of Foreign Influence Operations." *International Journal of Communication* 15: 4598–4618.

Breu, Anthony C. 2019. "Why Is a Cow? Curiosity, Tweetorials, and the Return to Why." *New England Journal of Medicine* 381 (12): 1097–1098.

Bruns, Axel. 2007. "Produsage: Towards a Broader Framework for User-Led Content Creation." In *Proceedings of the 6th ACM SIGCHI Conference on Creativity and Cognition*, 99–106. New York: ACM.

Case, Donald Owen, and Lisa M. Given. 2016. *Looking for Information: A Survey of Research on Information Seeking, Needs, and Behavior*. 4th ed. Bingley: Emerald.

Casler, Krista, Lydia Bickel, and Elizabeth Hackett. 2013. "Separate but Equal? A Comparison of Participants and Data Gathered Via Amazon's MTurk, Social Media, and Face-to-Face Behavioral Testing." *Computers in Human Behavior* 29 (6): 2156–2160.

Chaffey, Dave. 2022. "Global Social Media Statistics Research Summary 2022." Smart Insights. www.smartinsights.com/social-media-marketing/social-media-strategy/new-global-social-media-research/.

Chamberlain, P. R. 2010. "Twitter as a Vector for Disinformation." *Journal of Information Warfare* 9 (1): 11–17.

Chandler, Daniel, and Rod Munday. 2016a. "Microblogging." In *A Dictionary of Social Media*. Oxford: Oxford University Press.

2016b. "Social Media." In *A Dictionary of Social Media*. Oxford: Oxford University Press.

Chen, Xinran, Sei-Ching Joanna Sin, Yin-Leng Theng, and Chei Sian Lee. 2015a. "Why Do Social Media Users Share Misinformation?" In *Proceedings of the 15th ACM/IEEE-CS Joint Conference on Digital Libraries*, 111–114. New York: ACM.

2015b. "Why Students Share Misinformation on Social Media: Motivation, Gender, and Study-Level Differences." *Journal of Academic Librarianship* 41 (5): 583–592.

Cinelli, Matteo, Gianmarco De Francisci Morales, Alessandro Galeazzi, Walter Quattrociocchi, and Michele Starnini. 2020. "Echo Chambers on Social Media: A Comparative Analysis." *arXiv preprint*. https://doi.org/10.48550/arXiv.2004.09603.

Cook, Christine L., Aashka Patel, and Donghee Yvette Wohn. 2021. "Commercial Versus Volunteer: Comparing User Perceptions of Toxicity and Transparency in Content Moderation across Social Media Platforms." *Frontiers in Human Dynamics* 3: 1–8.

Cusumano, Michael, Annabelle Gawer, and David Yoffie. 2021. "Social Media Companies Should Self-Regulate. Now." *Harvard Business Review*. https://hbr.org/2021/01/social-media-companies-should-self-regulate-now.

Difallah, Djellel, Elena Filatova, and Panos Ipeirotis. 2018. "Demographics and Dynamics of Mechanical Turk Workers." In *Proceedings of the Eleventh ACM International Conference on Web Search and Data Mining*, 135–143. New York: ACM.

Faibisoff, Sylvia G., and Donald P. Ely. 1976. "Information and Information Needs." *Information Reports and Bibliographies* 5: 2–16.

Fichman, Pnina, and Madelyn Rose Sanfilippo. 2014. "The Bad Boys and Girls of Cyberspace: How Gender and Context Impact Perception of and Reaction to Trolling." *Social Science Computer Review* 33 (2): 163–180.

Field, Andy P. 2009. *Discovering Statistics Using SPSS*. 3rd ed. Los Angeles: SAGE.

Frischmann, Brett M, Michael J. Madison, and Katherine Jo Strandburg. 2014. "Governing Knowledge Commons." In *Governing Knowledge Commons*, ed. Brett M. Frischmann, Michael J. Madison, and Katherine Jo Strandburg, 1–43. Oxford: Oxford University Press.

Gadde, Vijaya, and Kayvon Beykpour. 2020. "Additional Steps We're Taking Ahead of the 2020 US Election." Twitter. https://blog.twitter.com/en_us/topics/company/2020/2020-election-changes.

Gorwa, Robert, Reuben Binns, and Christian Katzenbach. 2020. "Algorithmic Content Moderation: Technical and Political Challenges in the Automation of Platform Governance." *Big Data and Society* 7 (1): 1–15.

Hess, Charlotte, and Ostrom Elinor. 2007. "Introduction: An Overview of the Knowledge Commons." In *Understanding Knowledge as a Commons: From Theory to Practice*, 3–26. Cambridge, MA: MIT.

Hindman, Matthew, and Vlad Barash. 2018. "Disinformation, and Influence Campaigns on Twitter." *Knight Foundation: George Washington University*. www.issuelab.org/resources/33163/33163.pdf.

Internet Live Stats. 2022. "Twitter Usage Statistics." www.internetlivestats.com/twitter-statistics/.

Kim, Kyung-Sun, and Sei-Ching Joanna Sin. 2011. "Selecting Quality Sources: Bridging the Gap between the Perception and Use of Information Sources." *Journal of Information Science* 37 (2): 178–188.

2015. "Use of Social Media in Different Contexts of Information Seeking: Effects of Sex and Problem-Solving Style." *Information Research* (online) 20 (1). http://informationr.net/ir/20-1/isic2/isic24.html.

2016. "Use and Evaluation of Information from Social Media in the Academic Context: Analysis of Gap between Students and Librarians." *Journal of Academic Librarianship* 42 (1): 74–82.

Kim, Kyung-Sun, Sei-Ching Joanna Sin, and EunYoung Yoo-Lee. 2014. "Undergraduates' Use of Social Media as Information Sources." *College and Research Libraries* 75 (4): 442–457.

2021. "Use and Evaluation of Information from Social Media: A Longitudinal Cohort Study." *Library and Information Science Research* 43 (3): 1–11.

Lee, Kyumin, Prithivi Tamilarasan, and James Caverlee. 2021. "Crowdturfers, Campaigns, and Social Media: Tracking and Revealing Crowdsourced Manipulation of Social Media." In *Proceedings of the International AAAI Conference on Web and Social Media*, 331–340. Palo Alto: AAAI.

Leeder, Chris. 2019. "How College Students Evaluate and Share 'Fake News' Stories." *Library and Information Science Research* 41 (3): 1–11.

Lim, Sook, and Nahyun Kwon. 2010. "Gender Differences in Information Behavior Concerning Wikipedia, an Unorthodox Information Source?" *Library and Information Science Research* 32 (3): 212–220.

Linvill, Darren L., and Patrick L. Warren. 2020. "Troll Factories: Manufacturing Specialized Disinformation on Twitter." *Political Communication* 37 (4): 447–467.

Lokot, Tetyana, and Nicholas Diakopoulos. 2016. "News Bots." *Digital Journalism* 4 (6): 682–699.

MacAllister, Julia M. 2016. "The Doxing Dilemma: Seeking a Remedy for the Malicious Publication of Personal Information." *Fordham Law Review* 85 (5): 2451–2483.

McClain, Colleen, and Monica Anderson. 2021. "Republicans, Democrats at Odds over Social Media Companies Banning Trump." Pew Research Center. https://pewrsr.ch/3pnZ9d7.

McClain, Colleen, Regina Widjaya, Gonzalo Rivero, and Aaron Smith. 2021. "The Behaviors and Attitudes of U.S. Adults on Twitter." Pew Research Center. www.pewresearch.org/internet/2021/11/15/the-behaviors-and-attitudes-of-u-s-adults-on-twitter/.

Madison, Michael J., Brett M. Frischmann, and Katherine J. Strandburg. 2010. "Constructing Commons in the Cultural Environment." *Cornell Law Review* 95 (4): 657–710.

Mikhaeil, Merna, and Amr El Mougy. 2020. "The Effects of Personality Traits on the Lifetime of a Rumor." In *Highlights in Practical Applications of Agents, Multi-Agent Systems, and Trust-worthiness. The PAAMS Collection*, 108–117. Cham: Springer.

Mitchel, Amy, and Mason Walker. 2021. "More Americans Now Say Government Should Take Steps to Restrict False Information Online Than in 2018." https://pewrsr.ch/37TynlF.

Mohammadi, Ehsan, Mike Thelwall, Mary Kwasny, and Kristi L. Holmes. 2018. "Academic Information on Twitter: A User Survey." *PLoS ONE* 13 (5): e0197265. https://doi.org/10.1371/journal.pone.0197265.

MSNBC. 2022. "Social Media Is Rewiring Our Brains and Changing Our Brain Chemistry." www.msnbc.com/american-voices/watch/social-media-is-rewiring-our-brains-and-changing-our-brain-chemistry-148741189504.

Murthy, Dhiraj. 2011. "Twitter: Microphone for the Masses?" *Media, Culture and Society* 33 (5): 779–789.

Nakamura, Lisa. 2015. "The Unwanted Labour of Social Media: Women of Colour Call out Culture as Venture Community Management." *New Formations* 86: 106–112. www.muse.jhu.edu/article/604492.

Newberry, Christina, and Katie Sehl. 2022. "How the Twitter Algorithm Works in 2022 and How to Make It Work for You." Hootsuite. https://blog.hootsuite.com/twitter-algorithm/.

Odabaş, Meltem. 2022a. "5 Facts About Twitter 'Lurkers'." Pew Research Center. https://pewrsr.ch/3u5R54f.

2022b. "10 Facts About Americans and Twitter." Pew Research Center. https://pewrsr.ch/3vM1gga.

Osborne, Miles, and Mark Dredze. 2014. "Facebook, Twitter and Google Plus for Breaking News: Is There a Winner?" In *Proceedings of the International AAAI Conference on Web and Social Media*, 611–614. Palo Alto: AAAI.

Ostrom, Elinor, and Charlotte Hess. 2007. "A Framework for Analyzing the Knowledge Commons." In *Understanding Knowledge as a Commons: From Theory to Practice*, ed. Charlotte Hess and Elinor Ostrom, 41–81. Cambridge, MA: MIT.

Ott, Brian L. 2017. "The Age of Twitter: Donald J. Trump and the Politics of Debasement." *Critical Studies in Media Communication* 34 (1): 59–68.

Picard, Robert G. 2015. "The Humanisation of Media? Social Media and the Reformation of Communication." *Communication Research and Practice* 1 (1): 32–41.

Riedl, Martin J., Kelsey N. Whipple, and Ryan Wallace. 2021. "Antecedents of Support for Social Media Content Moderation and Platform Regulation: The Role of Presumed Effects on Self and Others." *Information, Communication and Society* 25 (11): 1632–1649.

Roberts, Sarah T. 2018. "Digital Detritus: 'Error' and the Logic of Opacity in Social Media Content Moderation." *First Monday* (online) 23 (3). https://journals.uic.edu/ojs/index.php/fm/article/view/8283.

Robertson, Adi. 2022. "Supreme Court Blocks Texas Social Media Moderation Ban." *The Verge*. www.theverge.com/2022/5/31/23124281/supreme-court-texas-hb20-social-media-ban-response.

Roth, Yoel. 2022. "Introducing Our Crisis Misinformation Policy." Twitter. https://blog.twitter.com/en_us/topics/company/2022/introducing-our-crisis-misinformation-policy.

Roth, Yoel, and Nick Pickles. 2020. "Updating Our Approach to Misleading Information." Twitter. https://blog.twitter.com/en_us/topics/product/2020/updating-our-approach-to-misleading-information.

Sandoval-Almazan, Rodrigo, and J. Ramon Gil-Garcia. 2014. "Towards Cyberactivism 2.0? Understanding the Use of Social Media and Other Information Technologies for Political Activism and Social Movements." *Government Information Quarterly* 31 (3): 365–378.

Sankaranarayanan, Jagan, Hanan Samet, Benjamin E. Teitler, Michael D. Lieberman, and Jon Sperling. 2009. "Twitterstand: News in Tweets." In *Proceedings of the 17th ACM SIGSPATIAL International Conference on Advances in Geographic Information Systems*, 42–51. New York: ACM.

Shao, Chengcheng, Giovanni Luca Ciampaglia, Onur Varol, Kai-Cheng Yang, Alessandro Flammini, and Filippo Menczer. 2018. "The Spread of Low-Credibility Content by Social Bots." *Nature Communications* 9: 4787. https://doi.org/10.1038/s41467-018-06930-7.

Shin, Jieun, Lian Jian, Kevin Driscoll, and François Bar. 2018. "The Diffusion of Misinformation on Social Media: Temporal Pattern, Message, and Source." *Computers in Human Behavior* 83: 278–287.

Silva, Leandro, Mainack Mondal, Denzil Correa, Fabrício Benevenuto, and Ingmar Weber. 2016. "Analyzing the Targets of Hate in Online Social Media." In *Proceedings of the 10th International Conference on Web and Social Media, ICWSM 2016*, 687–690. Palo Alto: AAAI.

Simons, Rachel Noelle. 2015. "Addressing Gender-Based Harassment in Social Media: A Call to Action." *iConference 2015 Proceedings*. http://hdl.handle.net/2142/73743.

Sin, Sei-Ching Joanna. 2016. "Social Media and Problematic Everyday Life Information-Seeking Outcomes: Differences across Use Frequency, Gender, and Problem-Solving Styles." *Journal of the Association for Information Science and Technology* 67 (8): 1793–1807.

Sin, Sei-Ching Joanna, and Kyung-Sun Kim. 2014. "Impacts of Social Media Usage on the Outcomes of Students' Everyday Life Information Seeking." In *Proceedings of the American Society for Information Science and Technology*, 1–4. Silver Spring, MD: Association for Information Science and Technology.

Smith, Scott M., Catherine A. Roster, Linda L. Golden, and Gerald S. Albaum. 2016. "A Multi-Group Analysis of Online Survey Respondent Data Quality: Comparing a Regular USA Consumer Panel to MTurk Samples." *Journal of Business Research* 69 (8): 3139–3148.

Statista. 2022. "Number of Twitter Users Worldwide from 2019 to 2024." www.statista.com/statistics/303681/twitter-users-worldwide/.

Suzor, Nicolas, Molly Dragiewicz, Bridget Harris, Rosalie Gillett, Jean Burgess, and Tess Van Geelen. 2019. "Human Rights by Design: The Responsibilities of Social Media Platforms to Address Gender-Based Violence Online." *Policy and Internet* 11 (1): 84–103.

Taddicken, Monika. 2014. "The 'Privacy Paradox' in the Social Web: The Impact of Privacy Concerns, Individual Characteristics, and the Perceived Social Relevance on Different Forms of Self-Disclosure." *Journal of Computer-Mediated Communication* 19 (2): 248–273.

Tifferet, Sigal. 2019. "Gender Differences in Privacy Tendencies on Social Network Sites: A Meta-Analysis." *Computers in Human Behavior* 93: 1–12.

Twitter. 2022. "Terms of Service." https://twitter.com/en/tos.

United States Communications Decency Act. 1996. "U.S. Code 47 (1996), § 230." https://uscode.house.gov/view.xhtml?req = (title:47%20section:230% 20edition:prelim).

Vogels, Emily A. 2022. "Support for More Regulation of Tech Companies Has Declined in U.S., Especially among Republicans." Pew Research Center. https://pewrsr.ch/3l8NYEs.

Vosoughi, Soroush, Deb Roy, and Sinan Aral. 2018. "The Spread of True and False News Online." *Science* 359 (6380): 1146–1151.

Wakabayashi, Daisuke. 2019. "Legal Shield for Websites Rattles under Onslaught of Hate Speech." *The New York Times*. www.nytimes.com/2019/08/06/technology/section-230-hate-speech.html.

Walker, Mason, and Katerina Eva Matsa. 2021. "News Consumption across Social Media in 2021." Pew Research Center. www.pewresearch.org/journalism/2021/09/20/news-consumption-across-social-media-in-2021/.

Wardle, Claire, and Eric Singerman. 2021. "Too Little, Too Late: Social Media Companies' Failure to Tackle Vaccine Misinformation Poses a Real Threat." *British Medical Journal* 372 26: 1–3.

Waterloo, Sophie F., Susanne E. Baumgartner, Jochen Peter, and Patti M. Valkenburg. 2017. "Norms of Online Expressions of Emotion: Comparing Facebook, Twitter, Instagram, and WhatsApp." *New Media and Society* 20 (5): 1813–1831.

Wathen, C. Nadine, and Jacquelyn Burkell. 2002. "Believe It or Not: Factors Influencing Credibility on the Web." *Journal of the American Society for Information Science and Technology* 53 (2): 134–144.

Whittaker, Elizabeth, and Robin M. Kowalski. 2015. "Cyberbullying Via Social Media." *Journal of School Violence* 14 (1): 11–29.

Winkelman, Sloane Burke, Jody Oomen-Early, Ashley D. Walker, Lawrence Chu, and Alice Yick-Flanagan. 2015. "Exploring Cyber Harassment among Women Who Use Social Media." *Universal Journal of Public Health* 3 (5): 194–201.

11

Accepting and Expecting Deception

Community Governance of False, Fabricated, Omitted, and Out of Context Claims on Instagram

Madelyn Rose Sanfilippo and Smita Katherine Nair

INTRODUCTION

Mis- and mal-information pervades our everyday lives, ranging from dangerous to harmless, malicious to well intentioned. Intentional deception serves as one form of misinformation that is wide reaching, including examples that are both harmful or socially positive or acceptable. As one such example of acceptable everyday deception, when people ask how others are doing, they expect responses on some variation of "fine" or "great." This is not only innocuous but is widely socially accepted, much akin to putting one's "best face forward" for job interviews or on resumes, dating apps, or first dates. Misrepresentation, in these instances, may be misleading or false but are normatively expected as standard practice. People are startled when someone doesn't respond with superficial pleasantries or positive misrepresentations. In contrast, there are many other instances in which deception is met with outrage, such as outright lies about heritage á la Rachel Dolezal, past behavior á la Hershel Walker, or professional innovation á la Elizabeth Holmes, for personal gain. In the middle is collective incredulity, as with fake product reviews or the recent phenomenon of in-person job postings labeled as remote to generate more clicks. Thus, we ask: What aspects of deception produce different perceptions? How can community governance of misinformation adjust based on social perceptions and norms?

Deception, as the various behaviors that obscure the truth, mislead, or promote falsehoods (Whaley 1982), is a multifaceted concept in the social sciences (Zhou 2005), reflecting psychological (Hyman 1989), sociological (Meltzer 2003), communicative (Buller, Daly, and Wiemann 1994), and legal (Klass 2011) dimensions. So,

Madelyn Rose Sanfilippo is Assistant Professor, School of Information Sciences, University of Illinois at Urbana-Champaign, with Affiliations in the Informatics Institute and the Center for Global Studies; Affiliate Scholar, The Vincent and Elinor Ostrom Workshop in Political Theory and Policy Analysis, Indiana University, Bloomington; PhD., Indiana University, Bloomington; MIS, Indiana University, Bloomington; BS, University of Wisconsin-Madison.
Smita Katherine Nair is an Undergraduate Student, University of Illinois at Urbana-Champaign.

FIGURE 11.1 Visual themes from community governance of false, fabricated, omitted, and out of context claims on Instagram.

too, is deception multifaceted in experience. We recognize that the concept of "deception" has different implications for different scholarly and everyday communities; we draw on an interdisciplinary perspective in this chapter to understand the ways in which everyday engagement with deception as a form of misinformation does and does not map onto academic conceptualization in specific contexts.

BACKGROUND

Social Media and Deception

Social media platforms are widely trusted and integral information environments in everyday life (Rubin, Burkell, and Quan 2010), though they often lack meaningful institutionalization as associated with more traditional communication channels (Obar and Wildman 2015) or information resources (e.g., Mansoor 2021). Because social platforms both possess a degree of decentralization (Blanke and Pybus 2020) and facilitate immediate conversation (Couldry 2015), they lend themselves easily to everyday knowledge commons for collection and dissemination of digital resources (Sanfilippo and Strandburg 2021a, 2021b). As communities emerge around popular topics, members generate, use, and share information co-created within these commons. Within these sociotechnical systems, communities and individual users produce and curate valuable knowledge resources, yet they lack enforcement structures to enforce norms or rules, and, as such, typical information behavior in these environments involves generating and passively consuming information, as well as regulating mis- and disinformation on an individual scale. Community members hold a personal stake in identifying acts of deception perpetuated by misinformed or mal-intentioned actors to maintain the integrity of knowledge generation.

As knowledge commons grow, certain contributors and content creators build informal authority as "influencers," entrusted by their communities to share products and tips that will accurately serve informational needs. These actors can be professionals within their fields or amateur content creators; as community members relate to influencers, situated strategically and visibly within networks, trust in them grows and their behaviors and preferences are normalized in the eyes of the community. Because engagement drives influencer profit, influencers are incentivized not to violate this trust. However, influencers do not always abide by social and contextual norms; often they develop strategies to achieve their own aims, contrary to social norms (Wellman, Stoldt, Tully, and Ekdale 2020). Among the most prevalent norm violations influencers commit are acts of deception; yet not all deception is viewed as equivalent and where there is a lack of consensus, there are opportunities to deceive without repercussion or else to salvage trust in the long term. It is important to develop better understanding of types of deception, social perceptions of these acts and associated information, and community responses to deception as misinformation.

Past scholarship of deception highlights multiple social and psychological dimensions of misrepresentation (e.g., Hyman 1989; Meltzer 2003; Whaley 1982; Zhou 2005), emphasizing lies of omission and rational means of obfuscation (Brunton and Nissenbaum 2015), as well as motivations for fabrication or falsification (e.g., Kumar and Shah 2018). Yet, most of this research does not explore either governance of these behaviors or the characteristics of the information itself produced by deceptive behaviors. In synthesizing this research, along with literature on misinformation, information quality, and social perceptions of information, including expectations regarding contextual integrity, we propose a typology of four distinct misinformation types associated with deception: omitted information, false information, fabricated information, and information out of context. We will explore everyday misinformation, assessing the validity of these suggested constructs throughout knowledge commons associated with food, dating, and retail on Instagram, so associated for their clear subcontexts and inherent nature as co-producing community and information. The next section provides conceptual background, applying the knowledge commons frame to understand social media and associated communities on social platforms, such as Instagram.

Conceptualizing Social Media as a Knowledge Commons

Social media manifests as a polycentric arrangement of everyday knowledge commons (Sanfilippo and Strandburg 2021b), wherein individuals self-arrange based on networked and algorithmically recommended points of connection to communicate, generate, and exchange content, as well as to collaboratively produce norms of interactions concurrent with the emergence of the community built upon their interactions (Sanfilippo, Frischmann, and Strandburg 2021). This is a key attribute of knowledge commons, wherein information generation, sharing, and use are co-created with the community (Frischmann, Madison, and Strandburg 2014), illustrating how meaningful the framing of everyday information environments as knowledge commons is in analysis.

A *knowledge commons* refers to shared knowledge resources, wherein sharing is institutionalized via knowledge commons governance (Frischmann, Madison, and Strandburg 2014) as the socially collaborative production of strategies, norms, and rules that specify how particular information and information types can be shared, among whom, for what purposes, and under what conditions (Sanfilippo, Frischmann, and Strandburg 2021). Note that strategies, norms, and rules represent a hierarchy, drawing on the institutional grammar (Sanfilippo, Frischmann, and Strandburg 2021) in which strategies formalize functions to achieve specific aims, norms reflect consensus about the best strategies for the conditions, and rules enforce norms (Shvartzshnaider, Sanfilippo, and Apthorpe 2022). This perspective reflects the Governing Knowledge Commons (GKC) Framework, which has been employed to analyze action arenas around a variety of communities and information types (e.g., Frischmann, Madison, and Strandburg 2014; Sanfilippo, Frischmann, and Strandburg 2021).

Institutionalization of knowledge production, sharing, and use is evident throughout online communities, both those built as sites of knowledge sharing, such as Wikipedia (Morell 2014) or Galaxy Zoo (Madison 2014), and those built for digital mediation of communication and other social interactions, such as Facebook or Twitter (e.g., Sanfilippo and Strandburg 2021a, 2021b). It is not necessary for the platform or infrastructure itself to reflect commons principles to support knowledge commons. Instead, commons arrangements amongst participants build within exogenously governed sociotechnical systems; intersections between platform and community governance are often complex, but communities and individuals often innovate to arrange information flows, resources, and participation in ways that best reflect their values and objectives (Sanfilippo, Frischmann, and Strandburg 2021). Further, the knowledge resources produced span many types of information; commons arrangements often govern personal information (Sanfilippo, Frischmann, and Strandburg 2021), in addition to more generalizable scientific data (e.g., Geary et al 2019; Strandburg, Frischmann, and Madison 2017), historical information (e.g., Murray 2014), instructions (e.g., Meyer 2012; Shvartzshnaider, Sanfilippo, and Apthorpe 2022), or other impersonal information. It is also notable that knowledge commons may build resources that center around everything from facts to advice, opinions, and beliefs, making many of these resources subjective and context specific (Sanfilippo, Frichmann, and Strandburg 2018). Knowledge commons arrangements often address issues of information quality and participation or exclusion as key parameters, making this framework useful to structure analysis of misinformation, given the relevance of information quality to the concept, and social interactions around misinformation.

METHODS

This chapter empirically focuses on deception as a type of misinformation. To understand how online communities respond to this form of everyday misinformation, we employed qualitative and computational content analysis of public social media posts to assess how groups come to consensus about different examples of deception on Instagram. We focus on group dynamics and processes to form consensus, or deliberate by examining their language and interactions. We aim to determine what types of deception are socially acceptable and unacceptable, anticipated or expected and unanticipated, looking at three contexts: posts about food, dating, and shopping. While these three cases do not reflect the totality of everyday interactions on Instagram, they span cultures and geography and are emblematic of popular experiences with misinformation. It is of particular interest to understand how groups collectively determine what type of deception a particular post is. We also provide understanding regarding how Instagram subcommunities respond to unacceptable forms of deception. This research is significant in helping online communities develop strategies to cope with everyday misinformation, based on past success.

Criteria for inclusion are based primarily on public posts and comments on Instagram that pertain to the keywords deception, fraud, and misinformation as they intersect with the thirty most common shopping, dating, and food keywords as ranked by Best-Hashtags in 2022 (see Table 11.1). We use these keywords to identify relevant posts, and associated comments, as both hashtags and via full-text search supported by natural-language processing (NLP) via the application programming interface (API). We collected posts by a variety of user account types, from individuals with small followings to famous influences and celebrities, as well as corporate accounts, excluding only those posts made by users under the age of eighteen. These details were screened prior to collection. To appropriately anonymize the posts considered, we did not collect nor consider any user profile details during this study, including profile pictures or usernames.

Using a plugin to the Instagram Graph API that exclusively captured text without media, we downloaded Instagram posts, and subsequent associated comments, based on the keyword sampling design at the intersection of deception, fraud, and misinformation and shopping, dating, and food keywords in Table 11.1. We saved only the text of individual posts and comments, which we numbered in chronological order. We did not include any pseudonyms or links between comments or posts by the same user. Via the API, it is not possible to download posts without including usernames and timestamps. Thus, these details and all associated metadata were deleted immediately upon download. Upon download, posts were numbered to maintain the sequence of communication, and to assess consensus development over time. No indirect identifiers indicated whether any posts were written by the same user.

A significant concern in this research is to ensure that users cannot be identified and will not be adversely impacted in this work; we minimized data retention and dissemination, beyond aggregate qualitative analysis and communication of common words people use to describe the issues we are investigating. We do not retain or analyze usernames, profile details, photos, or metadata. We also recognize that direct quotes of public social media posts in publications could identify users and as such will not include anything more than examples of words used to describe misinformation (e.g., "deceptive" or "fraud") which cannot be used without context to reidentify users. In addition to data minimization to exclude direct identifiers, care will be taken to avoid the use of any direct quotations in publication or presentation, which may identify users due to the public nature of the social media posts considered. As such, we used Python to strip out personally identifiable fields and converted to plain text upon download, prior to retention. A summary of the sample is provided in Table 11.2.

TABLE 11.1 *Contextual hashtags for sample selection*

Food	Dating	Shopping
#food #foodporn #foodie	#dating #love #relationships	#shopping #fashion #style
#instafood #foodphotography	#onlinedating	#onlineshopping #shop
#foodstagram #yummy	#relationshipgoals	#love #shoppingonline
#foodblogger #foodlover	#datingadvice #single	#instagood #outfit #moda
#instagood #love #delicious	#relationship #marriage	#instafashion #ootd
#follow #like #healthyfood	#romance #datingapp	#fashionblogger #dress
#homemade #dinner	#datingtips #tinder	#fashionista #sale #like
#foodgasm #tasty	#relationshipadvice	#shoes #instagram #beauty
#photooftheday #foodies	#couples #couplegoals	#fashionstyle #follow #online
#restaurant #cooking #lunch	#datinglife #datingcoach	#beautiful #design
#picoftheday #bhfyp	#singles #datingapps	#onlineshop #shoponline
#foodpics #instagram	#instagood #couple #follow	#clothes #model #stylish
#healthy #chef	#datingmemes #date	
	#lovequotes #bhfyp	
	#singlelife	
	#relationshipquotes #life	

TABLE 11.2 *Threads, posts collected and coded by context*

	Threads	Posts collected	Posts coded
Food	35	3,214	2,906
Dating	26	2,056	1,980
Shopping	17	1,824	1,721

The text was analyzed via a mixed-method content analysis approach. Qualitative analysis was structured by the Governing Knowledge Commons Framework (presented in the Introduction) and based on misinformation literature; content analysis was structured over multiple phases. First, both investigators co-coded a subsample, discussing what concepts from the literature applied and what other common themes emerge, in a grounded manner, to finalize the codebook (Table 11.3) and begin to gauge validity via inter-rater reliability. We also used this process to identify duplicate posts, which were not coded, nor were any posts not in English. Second, both investigators independently coded a second subsample and inter-rater reliability was again assessed, iterating until adequate agreement was established. Krippendorf's alpha measures of reliability ranged from 0.81 to 0.97. Third, the remaining content was divided in two and each investigator coded independently. Themes and patterns were assessed based on this manual, qualitative coding. We also employed a computational linguistic approach to count terms in the sample that apply to specific codes of interest, in order to ascertain how users describe acts of

TABLE 11.3 *Codebook for content analysis*

Concept	Definition	Reference
Falsification	Deliberate manipulation or misrepresentation of information	Kumar and Shah (2018); Simmons and Lee (2020); Whaley (1982)
Fabrication	Deliberate invention or synthetic creation of information	Kumar and Shah (2018)
Miscontext	Implying, insinuating, or framing something as true because it is true in another context, though it is either definitively false or uncertain in this context	Shivangi, Bregler, and Nießer (2021); Strathern, Crick, and Marcus (1987)
Omission	Exclusion of relevant data/ information	Brunton and Nissenbaum (2015)
Influencer	A social media content creator and/ or brand promoter with a following typically >100k	Jin, Muqaddam, and Ryu (2019)
Cultural heritage	Communal customs, expressions, and values, passed down from generation to generation	Brown (2009)
Personal attributes	Individual qualities or features, Typically innate	Latour (2011)
Fictitious entry	Deliberately incorrect information, typically associated with humor, consumer deception, and/or honeypots	Banerjee and Chua (2017)
False advertising or claims	Deliberate or negligent use of dubious, untrue, or misleading information to encourage beliefs, uses, or sales	Mangus (2008)
Misrepresentation	Intentionally or negligently false depiction of innovation, service, good, or procedure	Boutron and Ravaud (2018)
Trade secret	Information that is commercially valuable, known to a limited group of people, and actively protected as a secret	Graves (2007)
Skepticism	Doubting or questioning the veracity of information, based on observations or facts	Williams (2017)
Uncertainty	Unknown or undecided about information or judgment, often based on incomplete information	FeldmanHall and Shenhav (2019)
Trust	Confidence in truthfulness or accuracy	Hirschmann (1970); Sanfilippo, Frischmann, and Strandburg (2021)

deception and what types of deception are perceived more positively or negatively, as measures of sentiment. All protocols were approved by the UIUC IRB.

COMPARING GOVERNANCE OF EVERYDAY DECEPTION IN THREE INSTAGRAM CASES

Contextual Background

Three broad subcontexts within Instagram pervasively provide examples of deception and community responses to this form of misinformation: food, dating, and shopping. In this section, we provide contextual background for each of these three cases, delineating challenges regarding the status of deception as misinformation in each, as well as examples.

Do You Mind Sharing the Recipe?

Dinner parties, potlucks, and local fairs – among other social interactions – often offer opportunities to share homemade food, from experiments without recipes to trusted, and in some cases closely guarded, family recipes. It is considered perfectly normal and appropriate to ask the cook or baker for the recipe, though some will engage a right of refusal in response. More common than flat refusal are examples in which people acquiesce to share, yet don't share key ingredients or procedural details for a variety of reasons. For example, some steps might be hard to explain or else someone may assume tacit knowledge or generalities are understood, based on cultural backgrounds. Many share lists of ingredients without steps, a practice exacerbated by apps and platforms that gatekeep directions. Further, ingredient lists may include general references to complex mixtures, such as "Cajun spices" or garam masala. Beyond assumptions that someone else will understand, people often try to obfuscate family secrets, leaving out a key spice or method, to protect their heritage or their status as a better cook. We might describe all of these as practices of omission.

Alternatively, others may share complete information in such a way as to manage boundaries via contextually specific terminology. "Sure you can have the recipe!" Yet, they provide it in illegible photocopies of century-old handwritten Polish or Russian recipe cards, or with direction that simply say "assemble as per [name of the dish]" (e.g., a favorite family recipe handed down through my family instructs the baker to "assemble as per butterhorns"), which presumes that one can only make the recipe if they can read it, or translate, or know enough about the food to imagine the order and process of assembly. Many of these examples illustrate boundary setting, intentionally or unintentionally including and excluding based on whether others have enough domain or cultural knowledge to understand the references or make appropriate inferences from incomplete information.

Yet, foodie communities on Instagram and other social platforms, which encompass large populations with diverse interests (e.g., Calefato, La Fortuna, and Scelzi 2016), experience other issues of misinformation beyond omission. In cases outside heritage recipes or traditions, issues of experimentation or substitution may contribute to deceptive instructions in intentional or unintentional ways. For example, people may genuinely not remember the order of steps taken, duration of specific steps, or ingredient details, particularly if their food philosophy is more artistic than scientific (MacGovern 2021). More problematic are instances in which posts take credit for recipes or foods that are not their own; subsequent falsification and fraudulence in communication of the recipe are to be expected, when they don't know how it was prepared or understand the nuance of culturally relevant ingredients or processes. These cases gain the most visibility around celebrity chefs and prominent influencers, rather than individuals who try to pass off takeout food as homemade and represent the more significant misinformation problem.

Food and human relationships to food reflect a complex context in which identity, culture, and social structures are strongly interrelated (Almerico 2014). Religion, ethnicity, and geography significantly influence our understanding of, experiences with, and relationships to food, in addition to the ways they are prepared. Food reflects a facet of everyday life that is both complex and all too seldom analyzed in a serious, scholarly manner (e.g., Aspray, Royer, and Ocepek 2013), yet it tells us what people prioritize, their traditions, their socioeconomic status, and much more about them. This is a context in which to explore everyday misinformation that importantly highlights the tensions between expertise and experience, importance of belongingness in communities and their associated knowledge commons, and distinctions between tacit and recorded knowledge. Further, it emphasizes the importance of variability and tradition in influencing information quality throughout everyday knowledge commons.

Looking for Love

Dating and courtship have historically been built on optimizing one's assets and representing one's "best self," if not a "better you." It is commonplace for people to round up their height, or lie about how tall they are, or to enhance their physical features or descriptive profile to gain attention. The use of photoshop and filters is ubiquitous on social media beyond dating profiles, but especially so in this space, in ways that build on in-real-life (IRL) enhancements, such as cosmetics, haircoloring, or surgical procedures. There are real incentives to mimic advertising strategies and produce clickbait, rather than a high-fidelity and honest profile, whether one aims for a hookup or a long-term relationship.

Some of the tactics to achieve these aims are transparently manipulative, such as posts that serve as "thirst traps," integrating misrepresentation and sexualization on social media (Davis 2018). They are institutionalized via clear aims and conditions, as well as normative consensus around practice in the context of dating. Popular culture and relevant communities understand their purpose and can identify them, yet they are embraced as a strategy to meet romantic and sexual needs, as well as attract attention more generally (Davis 2018). Provocative and enhanced photos, depicting the most attractive and/or positive features are not viewed negatively or sanctioned in the context of dating. This aligns with the ways in which public relations and the concept of "spin" have influenced dating norms as commercial actors increasingly play a role in how people meet and interact (Di Domenico, Sit, Ishizaka, and Nunan 2021; Ewen 1996); modern dating profiles mimic retail descriptions for products, more so than the historic ways in which communities and matchmakers have depicted eligible singles across cultures. While there has always been a tendency toward representation of the individual as favorably as possible, truth is treated much more malleably in modern circumstances.

Technology plays major roles in these trends, around the world. Not only is the use of dating apps extremely common (Hobbs, Owen, and Gerber 2017; Wu and Trottier 2022), but social media documents and brokers many relationships as they take their course, with communities emerging around specific relationships and the context more broadly to process interactions and experiences (e.g., Hobbs, Owen, and Gerber 2017). Dating and romantic partnerships represent a significant sub community on Instagram, with influencers posting considerable details and many accounts dedicated to commentary on dating apps, culture, first dates observed, and humor. This also builds on online culture built around older platforms, such as Tumblr, which were so much a part of dating experiences that various countries that did not yet have dedicated dating apps in their countries used Tumblr as dating profiles (Byron, Robards, Hanckel, Vivienne, and Churchill 2019; Hart 2015).

A key issue that intersects with deception in the context of dating discussion subcommunities on Instagram is that of catfishing, which engages outright lies via a false personal profile on social media to engage in fraud or deceive other users (Simmons and Lee 2020). This is distinct from phishing, in that the aim is not personal information or financial gain, typically, but rather social and psychological harm. Research on this issue shows a high degree of overlap in impact with misinformation in the context of dating, but motivations are identified to be quite distinct (e.g., Chen 2016; Hancock, Toma, and Ellison 2007; Simmons and Lee 2020).

Where Can I Get One?

As digital communities form around popular themes – over any number of hobbies, such as makeup or sneaker collecting – so do "influencers" gain popularity as opinion leaders. Communities trust influencers to share tips and recommendations as normal, relatable people who seemingly have more in common with community members than brands or celebrities; emotional bond-building behaviors (directly interacting with their audience; referring to their fanbase as a "family") drive influencer popularity and engagement. Full-time influencers earn their income through their content creation. Brand partnerships, sponsorships, and promotional deals are all typical among influencers and acceptable to community members with proper disclosure and transparency (Wellman, Stoldt, Tully, and Ekdale 2020).

However, community members can be sensitive to the possibility that profit – financial or social – would lead an influencer to misrepresent the truth to their audience. For instance, an influencer might endorse a product for payment, but neglect to disclose this transaction so that their endorsement ("I'm obsessed with this serum, it saved my skin!") appears more credible or encourage fans to shop for the product through a specific URL without revealing that it's an affiliate link.

More common are examples where deception is arguably present, but not clearly defined. Influencers may be transparent about their intent to profit yet misdirect consumers by sharing false or incomplete information to represent their brand more favorably (or, to make another influencer's brand appear worse), "#sponsored," yet lying about the product having officially recognized certifications. Or they may create a post to demonstrate the effects of a product while filtering or editing the image to make the product seem more effective. These examples could be described as fabrication and falsification respectively.

The bond between influencer and community can be such that, in cases unlike the first examples where deception is clearly present, community members who enjoy an influencer's content may be more likely to contest claims of deception on their behalf. Fans may also acknowledge the deception yet try to minimize the severity ("we all make mistakes!") or excuse the deception on external circumstances, for instance claiming that an influencer's age or cultural background should exempt them from consequences. The relationship fans develop to an influencer can produce "an increased level of trust ... akin to that felt for one's personal friends" (Wellman, Stoldt, Tully, and Ekdale 2020). For this reason, fans may trust an influencer's opinions over those of someone with relevant credentials or expertise, even when the two are incompatible.

Influencers can leverage this trust to manipulate narratives around deception. They may try to miscontextualize the consequences they face because of deception (attributing consequences to "cancel culture," for instance, or claiming that they're being targeted as a social minority) or question the trust of fans who believe the influencer would intentionally deceive their audience, in some cases engaging in gaslighting behaviors. Heterogeneity of personal experiences community-wide with opinion leadership, opinion seeking, and parasocial relationships fractures subcommunities and prevents consensus from forming with respect to public opinions on influencer deception or how to govern these behaviors and information.

Key Attributes

Instagram subcommunities, spanning topics from food to dating to consumer products (often fashion and cosmetic retail), have few barriers to participation. While an Instagram and/or Facebook account is required to post, content is public facing with opportunities for the average user to comment on content from celebrities, influencers, or other users, as well as to engage multiple audiences. A major issue that arises in this social environment is that of power differentials and the differences in ways that nonpowerful, nonfamous, or infamous users are governed or expected to behave, in comparison to brands or more powerful individuals.

Yet how do subcommunities arise or persist amid these everyday contexts? The answer centers on shared interests and collective negotiation of values. Not only does the topic matter, but so also do priorities. Aspiration, as a motivation, centers focus in retail communities on consumption of bags or shoes, skincare routines, or other consumer products endorsed by celebrities and influencers. It produces very different interactions than, say, those with a shared interest in socially curated and visible shopping, whose primary motivation is brand loyalty. Similarly, differences in the motivation to participate in discussion of or consumption of dating content produce very different communities; while some care primarily about advice, others care about commiseration or humor.

Intersecting with cognitive, emotional, and value-oriented motivations, are the attributes of the actors themselves. Status and reputation play a significant role in the relative visibility and influence of individuals within these communities, reflecting related research on social capital, reputation, and Instagram (e.g., Zulli 2018). Negotiation of social position is a complex assemblage of strategies and norms regarding representation of self (e.g., Bullingham and Vasconcelos 2013) and interaction, thereby incrementally shaping how the community arranges into groups, as well as information content and personal status. In this sense, boundary management reflects informal strategies and norms, as well as being shaped via externalities associated with cultural capital and hierarchies. These behavioral and incentive attributes also contribute to the quality of information resources within the community, reflecting gossip, hearsay, and marketing that stray from any semblance of objective truth, yet are still socially valuable and often trusted.

Overall, a broad array of misinformation challenges these interactions and are ubiquitous, with communities grappling with distinctions between information quality and "messaging," as well as social versus informational harms. Importantly, these communities, regardless of topic and motivation, perceive lies to be a form of manipulation distinct from marketing, and their discussions suggest that pervasive approaches to deal with misinformation via automated detection or user flagging are unlikely to succeed in everyday contexts.

Governing Everyday Misinformation Action Arenas

Exploring deception as misinformation in food, dating, and retail subcommunities on Instagram, we find common action arenas around different types of deception, yet some divergence in community governance in response. Norms are contextually specific yet evaluated attributes and general perceptions appear to be shared.

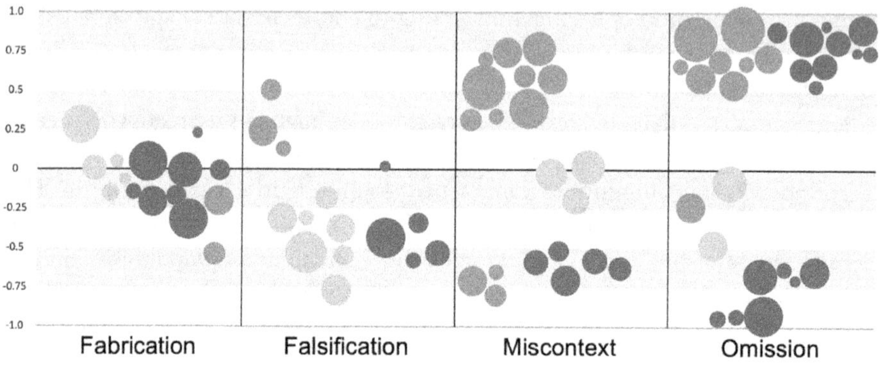

FIGURE 11.2 Sentiment averages by thread, weighted by posts.

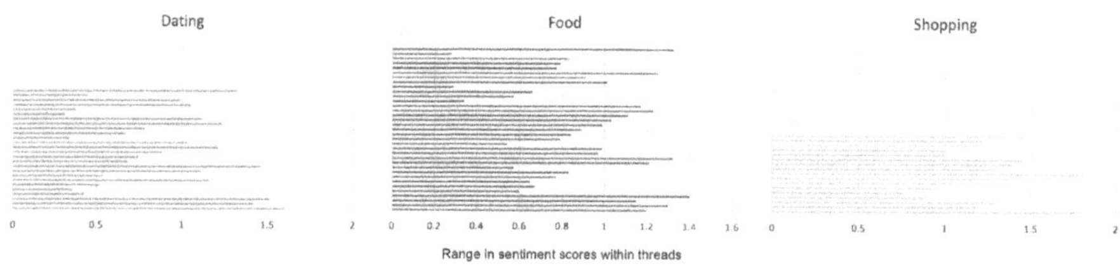

FIGURE 11.3 Consensus within and between threads.

We discuss action arenas around falsification, fabrication, miscontext, and omission in subsequent subsections, drawing on qualitative coding and computational assessment of sentiment within given threads (Figure 11.2), to understand disagreement and consensus formation 8 (Figure 11.3) within action arenas.

Falsification

Falsification, which we define as instances of deliberate manipulation or misrepresentation of information in communication, presents nuanced governance challenges. First, it is necessary to determine whether exogenous, legal governance parameters were violated. For example, issues of false claims may merit intervention at state or federal levels in the United States, based on consumer protection, or in some cases, where contention over ownership and innovation come into play, intellectual property rights. The latter is evident amongst influencers who seek to market their own supposed innovations, when they are actually knockoffs of others' products, processes, or devices.

Second, it is necessary for communities to determine whether the false information led to social or psychological harms to the community, as in cases of cultural appropriation and product releases or recipes on the part of major brands, celebrities, or influencers. In instances where lasting harms were identified, discussions focused on breaches of "trust" and growing "skepticism" and scrutiny about apologies and past behavior. This form of misinformation seems to pose the greatest challenge to the communities, with significant discussion about how to proceed and how actors behave or post in good faith. The average sentiment expressed across threads addressing issues of falsification was −0.51, bifurcating on issues of psychological harm, to −0.17 (no lasting harm) and −0.78 (harms).

In contrast, a form of falsification that was viewed neutrally and not perceived to be a problem, on average, were threads centered on humor driving false information in dating profiles, such as implausible career/location combinations or achievements relative to age or eloquence. Here, posters or users who falsified their information were mocked and discredited, but there was no meaningful harm to the community. This contrasts with instances of financial harms or outright catfishing.

Fabrication

Fabrication refers to the deliberate invention or creation of information as described and depicted in our typology (see Table 11.3). Examples of fabrication would include instances where brands promote information about a product that is not only untrue but entirely invented (as opposed to manipulated in cases of falsification). For example, there are many cases of brands claiming cruelty-free or other lifestyle-conscious credentials for products that have not only never received such certification but demonstrably do not meet their criteria; claiming Leaping Bunny certification yet still retailing in countries that would require their products be tested on animals, for instance, or using nonvegan ingredients. We see numerous instances of brand or influencing accounts attempting to increase the perception of information credibility within their target community in such a way that highlights a lack of respect for community values. There are many examples of brands and influencers capitalizing on the collaborative nature of digital information environments, for instance such as by posing as community members to leave fake reviews for products or otherwise endorsing a brand under explicitly false premises.

Often in threads discussing acts of deception in sales, we see discussion of community retribution as well that may indicate greater faith in communal policing than exogenous administration; for instance, commenters may question the lack of involvement from the Federal Trade Commission (FTC) or confer with others in the thread over whether an FTC-enforced penalty is appropriate for the severity of the deception. Many responses indicate that individuals have reported these issues to the Better Business Bureau (BBB) or encourage others to do so. Predictably, commenters are less likely to endorse a penalty if they find the offending account's actions acceptable, but acceptability on an individual scale isn't always dependent on whether deception is believed to be present. For instance, a commenter may perceive but disregard deception if they plan to continue purchasing from the offending brand.

Threads for exchanging opinions on deception indicate that commenters may perceive deceptive marketing of credible information as a distinct (and not necessarily equally unforgivable) circumstance separate from deceptive information. Commenters may be more inclined to accept the former if they perceive the behaviors separately, with arguments that deceptive marketing behavior is "common practice" within the digital sales industry and unrelated to the quality of the product. This is supported by an average sentiment measure of 0.35 across four threads that illustrate this pattern is generally viewed near neutrally across individuals. More specifically, given the range of sentiment, as distributed throughout these four threads (see Figures 11.2 and 11.3), consensus is reached around neutrality for three of four, through the course of the thread; while the latter is more mixed in attitudes, perceptions lack significant amplitude, illustrating minor disagreement.

Omission

Another key form of deceptive misinformation is omission, which we define as exclusion of relevant data or information. Omission is diverse across all three contexts on Instagram, ranging from the omission of sensitive characteristics in volatile or especially public discussions, to protect privacy and promote safety, as well as the omission of key facts or information pertinent to socially relevant knowledge, for purposes of personal gain.

In many examples, when individuals leave out sensitive characteristics, this is perceived to be perfectly rational and advisable behavior. We see social consensus around the appropriateness of leaving out race, gender, sexual orientation, or other characteristics that may denote an individual as vulnerable or stigmatized in certain situations. Such examples are most prominent in threads discussing dating and other romantic norms. Keywords associated with safety, reasonableness, rationality, and justification describe these perceptions, for which the average sentiment measurements of these posts is 0.92. There is notably also a high degree of certainty expressed in these posts, ranging from 0.71 to 0.94. Similar behaviors and favorable interpretation of the omissions are also evident in threads stemming from food posts in which cooks leave out details that could be relevant to the cultural authenticity of shared recipes or food images, but reflect contentious geopolitics, such as not sharing details relative to ancestry in Palestine or Israel in posts related to Mediterranean and Middle Eastern cuisine. Sentiment in responding posts measured 0.87, on average.

These examples are sharp contrasts with the few threads sampled that demonstrated omission to obscure nonbelongingness into a given community or cultural identity; in those instances, it was perceived to be deception via implicit communication as opposed to omission for the purpose of safety. For example, Caucasian American or European influencers implying BIPOC ancestry via implicit communication, without specifying their race or ethnicity, was viewed negatively; responses to these posts measured −0.69. Few responses gave the original poster the benefit of the doubt in these cases, with the only exceptions tying into social capital dynamics, wherein powerful and esteemed individuals might be forgiven.

Another recurring issue around omission pertains to examples in which key details or facts have been left out of processes. For example, there are many recipes shared on Instagram without complete instructions or with relative ambiguity regarding culturally specific ingredients such as garam masala, Chinese 5 spice, or a culturally nonspecific phrasing of season to taste. This is generally not viewed negatively, except for those users who post as detractors and display their xenophobia or particular biases or prejudices against relevant communities. This represents a problem for public opinion that is not inherently about the misinformation problem. In these instances of discrimination, the average sentiment expressed in comments responding to these posts are the most extremely negative included in this study, measuring −.97. Here, the intersection of misinformation and racial discrimination presents an action arena that associated communities take more seriously than any other instances of misinformation examined in this research. Dialogue emerges around how to appropriately address these issues, removing misinformation, while not allowing the problematic poster to simply move on unscathed. The social consensus that we should not simply let this go contrasts with other examples such as controversies around product endorsement that miss information associated with celebrities, as in the Pharrell case, where fans allow the celebrity to move on without consequence.

Out of Context

Deception via information out of context is evidenced by implying, insinuating, or framing something as true because it is true in another context, though it is either definitively false or uncertain in this context, as described and depicted in our typology (see Table 11.3).

One distinct set of examples in which deception via miscontextual information is evident includes those instances when public figures, publications, or other influential accounts communicate culturally, ethnically, or religiously relevant information that is true for those within the community but does not reflect the heritage or community membership of the poster. For example, there are various instances in which the lack of nuance or respect shown for recipes or food associated with religious holidays highlights the nonbelongingness of the poster. We see examples in which ethnically ambiguous influencers or celebrity chefs share what are supposedly family recipes associated with cultural heritage or religious feasts, yet are appropriated from cultural traditions or communities to which they do not belong. Examples range from framing Juneteenth or Kwanzaa celebrations and family traditions by white influencers or publicly Christian chefs claiming family Passover traditions. These examples are distinctly viewed as inappropriate forms of deception, by and large, with an average sentiment of responding posts at −0.63, which indicates the conversation following the initial post is negative in tone.

Similarly, we see various threads discussing the appropriateness or inappropriateness of individuals who create profiles for themselves on culturally specific dating websites or apps, wherein they represent themselves accurately but for their membership in the relevant community group. Sometimes these practices are perceived to be acceptable, such as when individuals have not misrepresented themselves per se but are found to be out of context based on algorithmic inferences about missing attributes, which may be justified as with gender identity or sexual orientation. These threads measure average sentiment of 0.54, indicating an overall slightly positive tone. There are times these practices are perceived to be unacceptable, as evidenced by an average sentiment measure of −0.71, such as when individuals intentionally and knowingly impersonate other cultural identities, as with individuals creating J date profiles when they are not, in fact, Jewish.

We also see forms of deception through information out of context with respect to dating profiles and behaviors that are viewed neutrally to positively, with threads scoring from 0.35 to 0.76 in average sentiment, wherein posters comment on sharing accomplishments, education, or employment details in ways that are deceptive based on context, such as individuals labeling themselves as alumni of prestigious universities or programs, though they did not graduate or perhaps only attended a semester-long program, rather than a full degree program. Here, many posters argue these practices are "par for the course" or to be "expected" in the modern dating culture.

Patterns/Outcomes

A spectrum of acceptability, from very acceptable to very unacceptable emerges from these action arenas, overall, as depicted in Figure 11.4.

We see deception via miscontext and omission as viewed benignly, along with most instances of fabrication. The exceptions here are a few comparatively rare instances in which harms are more lasting, which aligns with the overwhelmingly negative responses to cases of falsification. While exogenous governance can be useful to deal with instances in which consumer deception takes place, communities must also reach consensus to deal with the local implications within their communities, more so for the regulation of their community boundaries and future participation norms than for information quality. In this sense, while examples of medical and political misinformation spread within existing social networks, the examples of everyday misinformation studied herein are contextually dependent on social capital, norms, and values. Institutionalization regarding appropriate, or at the very least acceptable, manipulation of content really needs to address power dynamics, platform visibility, and enforcement mechanisms, more so than content standards themselves.

Another key pattern is that people are willing to accept deception if they agree with it or if they see individual benefits to overlooking deception, as with various

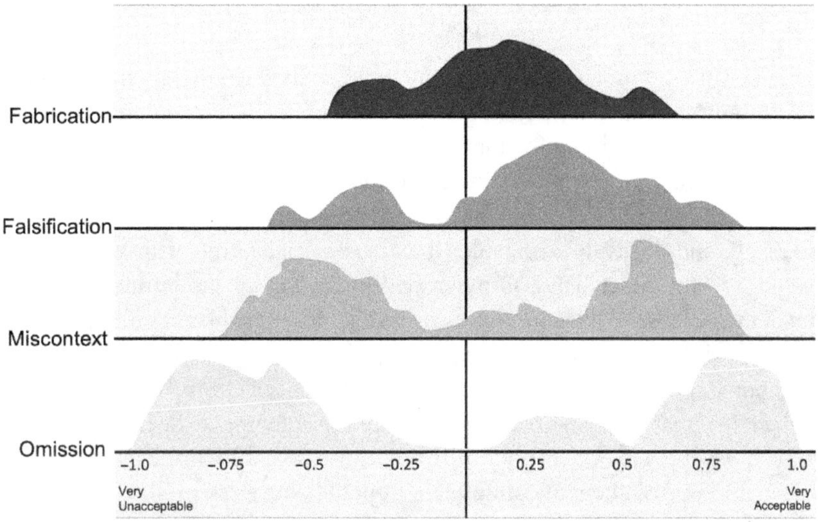

FIGURE 11.4 Sentiment distributions by action arena, weighted by number of posts.

celebrities they prefer not to "cancel." Fans might either deny wrongdoing on an influencer's behalf, or attribute wrongdoing to cultural differences or other factors that could excuse deception. For example: Fans of influencers claim an influencer's cultural background excuses their deception, frame an influencer's purposeful deception is a mistake or learning opportunity, or negate influencers' deceptions by pointing out that critics have likely also been dishonest at some point themselves. Under an influencer's apology post, comments are largely forgiving, echoing sentiment that making mistakes is part of human nature and encouraging influencer to learn and grow. However, people appear to love people more than they love brands. Responses to brand deception are on average much less forgiving; at best, commenters may remark on the triviality of the deception.

That commenters who defend influencers are much more favorable and fervent reflects the sort of cult of personality that emerges around the personas portrayed. Often, the behaviors of famous people are excused even though others find the behavior unacceptable for the average person. Commenters also mention trust at a far greater rate when commenting on influencer deception versus brand deception, regardless of in a positive or negative context ("I still trust them" versus "they betrayed our trust"), indicates higher emotional stakes in influencer deception versus brand or celebrity.

Deception as misinformation experienced in everyday knowledge commons is most problematic in terms of issues of trust, fragmenting communities or damaging relationships between influencers and followers, rather than the information itself posing the harm. In this sense, the most serious objections expressed are about social and psychological harms, rather than information quality.

CONCLUSIONS

Despite negative connotations of deception as associated with manipulation, deception with respect to misinformation generates diverse social perceptions. On the one hand, various forms of deception that are associated with issues of identity and values of safety or privacy – including issues of gender identity, sexuality, religious or philosophical beliefs, and ethnicity or cultural identity – are perceived to be normatively and logically justifiable. This is especially true with deception via omission. On the other hand, forms of deception that aim to enrich the original poster at the expense of the community or that demonstrate bias or cultural illiteracy are largely viewed negatively. Social acceptance along the lines of normative justifications and social harms help to explain this spectrum of perceptions, as described in greater detail in Section "Patterns/Outcomes." Social expectations align with acceptance of deception that minimizes social harms, in that we expect people not to make others uncomfortable in polite conversation, thus deception to frames information or communication in a normative format is not only desirable, but also anticipated as the default for interaction.

A key takeaway, thus, is that misinformation in everyday knowledge commons is as much a quality of valued knowledge resources to these communities as privacy is to contact information or location data as knowledge resources. The series of governance challenges, or action arenas, that arise around everyday misinformation reflect intersections with other social challenges. This certainly raises the question: Is misinformation an externality of social issues, such as mistrust in science or political contention, in other contexts, as well?

We also identify specific implications of this work for community and platform governance regarding misinformation as a dimension of information quality in everyday knowledge commons. First, it is important that mechanisms by which community members can exercise voice and consensus can build around evaluation of knowledge quality are essential to the construction of healthy, sustainable everyday knowledge commons (e.g., Hirschmann 1970; Sanfilippo, Frischmann, and Strandburg 2021). Second, it is important to recognize the nuance of social dimensions in evaluating deception as a form of misinformation. It is not possible to address many of these instances of misinformation via automated or other primarily technological means. Instead, it is critical that the approach to deal with these forms of misinformation be sociotechnical in nature and oriented at the level of communities. Infrastructure that communities can use to appropriately address issues of misinformation they deem to be unacceptable is a much more worthwhile investment at the platform level. Third, both prior recommendations also illustrate the intersection of platform level design and community level actions; we must recognize that there is no panacea to address misinformation and communities must have the flexibility to address problematic misinformation in context.

Recognizing the parallels between this key takeaway and Ostromian institutional analysis, there are meaningful implications from this work for the GKC framework. It is absolutely necessary to recognize that attributes of information, including metadata, context, and perception, are equally as important to governance of misinformation in everyday knowledge commons as the category of information itself. A more granular analysis of these attributes can help us to understand why patterns diverge around deception, among other types of misinformation, as well as how to compare contextual action arenas around similar information types.

REFERENCES

Allcott, Hunt, Matthew Gentzkow, and Chuan Yu. 2019. "Trends in the Diffusion of Misinformation on Social Media." *Research and Politics* 6 (2): 2053168019848554.

Almerico, Gina M. 2014. "Food and Identity: Food Studies, Cultural, and Personal Identity." *Journal of International Business and Cultural Studies* 8 (1): 1–8.

Aspray, William, George Royer, and Melissa G. Ocepek. 2013. *Food in the Internet Age*. New York: Springer.

Banerjee, Snehasish, and Alton Y. K. Chua. 2017. "Theorizing the Textual Differences between Authentic and Fictitious Reviews: Validation across Positive, Negative and Moderate Polarities." *Internet Research*. https://doi.org/10.1108/IntR-11-2015-0309.

Banerjee, Shubhomoy, and Abhijit Ghosh. 2022. "Till Death Do Us Part – Customer Commitment after Negative Publicity: The Role of Relational Variables and Cognitive Dissonance." *Asia-Pacific Journal of Business Administration* 15 (5): 649–672.

Blanke, Tobias, and Jennifer Pybus. 2020. "The Material Conditions of Platforms: Monopolization through Decentralization." *Social Media + Society* 6 (4): 2056305120971632.

Bode, Leticia, and Emily K. Vraga. 2015. "In Related News, that Was Wrong: The Correction of Misinformation through Related Stories Functionality in Social Media." *Journal of Communication* 65 (4): 619–638.

Bond Jr, Charles F., and Bella M. DePaulo. 2006. "Accuracy of Deception Judgments." *Personality and Social Psychology Review* 10 (3): 214–234.

Boutron, Isabelle, and Philippe Ravaud. 2018. "Misrepresentation and Distortion of Research in Biomedical Literature." *Proceedings of the National Academy of Sciences* 115 (11): 2613–2619.

Brown, Michael F. 2009. *Who Owns Native Culture?* Cambridge, MA: Harvard University Press.

Brunton, Finn, and Helen Nissenbaum. 2015. *Obfuscation: A User's Guide for Privacy and Protest*. Cambridge, MA: MIT.

Buller, David B., and Judee K. Burgoon. 1996. "Interpersonal Deception Theory." *Communication Theory* 6 (3): 203–242.

Buller, David B., Judee K. Burgoon, J. A. Daly, and J. M. Wiemann. 1994. "Deception: Strategic and Nonstrategic Communication." *Strategic Interpersonal Communication*: 191–223.

Bullingham, Liam, and Ana C. Vasconcelos. 2013. "'The Presentation of Self in the Online World': Goffman and the Study of Online Identities." *Journal of Information Science* 39 (1): 101–112.

Byron, Paul, Brady J. Robards, Benjamin Hanckel, Son Vivienne, and Brendan Churchill. 2019. "'Hey, I'm Having these Experiences': Tumblr Use and Young People's Queer (Dis) Connections." *International Journal of Communication*: 2239–2259.

Calefato, Patrizia, Loredana La Fortuna, and Raffaella Scelzi. 2016, "Food-ography: Food and New Media." *Semiotica* 211: 371–388.

Chen, Xinran. 2016. "The Influences of Personality and Motivation on the Sharing of Misinformation on Social Media." *IConference 2016 Proceedings*.

Chen, Xinran, Sei-Ching Joanna Sin, Yin-Leng Theng, and Chei Sian Lee. 2015. "Why Do Social Media Users Share Misinformation?" In *Proceedings of the 15th ACM/IEEE–CS Joint Conference on Digital Libraries*, 111–114.

Cinotto, Simone. 2013. *The Italian American Table: Food, Family, and Community in New York City*. University of Illinois Press.

Cooper, Harris. 2003. "Cues to Deception." *Psychological Bulletin* 129 (1): 74.

Couldry, Nick. 2015. "Illusions of Immediacy: Rediscovering Hall's Early Work on Media." *Media, Culture and Society* 37 (4): 637–644.

Cullen, Rowena. 2009. "Culture, Identity and Information Privacy in the Age of Digital Government." *Online Information Review*.

Cundy, Linda. 2021. "Guess Who's Coming to Dinner: Culture, Community, Identity . . . and Food." In *Attachment, Relationships and Food*. New York: Routledge. 150–171.

D'Sylva, Andrea, and Brenda L. Beagan. 2011. "'Food is Culture, but It's Also Power': The Role of Food in Ethnic and Gender Identity Construction among Goan Canadian Women." *Journal of Gender Studies* 20 (3): 279–289.

Davis, Stefanie E. 2018. "Objectification, Sexualization, and Misrepresentation: Social Media and the College Experience." *Social Media + Society* 4 (3): 2056305118786727.

DePaulo, Bella M., James J. Lindsay, and Daniel B. Wright et al. 2004. "Say It to My Face: Examining the Effects of Socially Encountered Misinformation." *Legal and Criminological Psychology* 9 (2): 215–227.

Di Domenico, Giandomenico, Jason Sit, Alessio Ishizaka, and Daniel Nunan. 2021. "Fake News, Social Media and Marketing: A Systematic Review." *Journal of Business Research* 124: 329–341.

Ewen, Stuart. 1996. *PR!: A Social History of Spin*. New York: Basic Books.

FeldmanHall, Oriel, and Amitai Shenhav. 2019. "Resolving Uncertainty in a Social World." *Nature Human Behaviour* 3 (5): 426–435.

Frischmann, Brett M., Michael J. Madison, and Katherine Jo Strandburg, eds. 2014. *Governing Knowledge Commons*. Oxford University Press.

Graves, Charles Tait. 2007. "Trade Secrets as Property: Theory and Consequences." *Journal of Intellectual Property* 15 (1): 39–89.

Guerra, Gerardo A., Daniel John Zizzo, William H. Dutton, and Malcolm Peltu. 2003. "Economics of Trust in the Information Economy: Issues of Identity, Privacy and Security." OECD Information Security and Privacy Working Paper No. JT00142557.

Hancock, Jeffrey T. 2007. "Digital Deception." *Oxford Handbook of Internet Psychology* 61 (5): 289–301.

Hancock, Jeffrey T., Catalina Toma, and Nicole Ellison. 2007. "The Truth about Lying in Online Dating Profiles." In *Proceedings of the SIGCHI Conference on Human Factors in ComputingSsystems*, 449–452.

Hart, Matt. 2015. "Youth Intimacy on Tumblr: A Pilot Study." *Young* 23 (3): 193–208.

Hirschman, Albert O. 1970. *Exit, Voice, and Loyalty: Responses to Decline in Firms, Organizations, and States*. Vol. 25. Cambridge, MA: Harvard University Press.

Hobbs, Mitchell, Stephen Owen, and Livia Gerber. 2017. "Liquid Love? Dating Apps, Sex, Relationships and the Digital Transformation of Intimacy." *Journal of Sociology* 53 (2): 271–284.

Hyman, Ray. 1989. "The Psychology of Deception." *Annual Review of Psychology* 40 (1): 133–154.

Gabbert, Fiona, Amina Memon, Kevin Allan, and Daniel B. Wright. 2004. "Say It to My Face: Examining the Effects of Socially Encountered Misinformation." *Legal and Criminological Psychology* 9 (2): 215–227.

Geary, Janis, Trish Reay, and Tania Bubela. 2019. "The Impact of Heterogeneity in a Global Knowledge Commons: Implications for Governance of the DNA Barcode Commons." *International Journal of the Commons* 13 (2): 909–930.

Gneezy, Uri. 2005. "Deception: The Role of Consequences." *American Economic Review* 95 (1): 384–394.

Jacobs, Janet Liebman. 1996. "Women, Ritual, and Secrecy: The Creation of Crypto-Jewish Culture." *Journal for the Scientific Study of Religion* 35 (2): 97–108.

Jin, S. Venus, Aziz Muqaddam, and Ehri Ryu. 2019. "Instafamous and Social Media Influencer Marketing." *Marketing Intelligence and Planning* 37 (5): 567–579.

Klass, Gregory. 2011. "Meaning, Purpose, and Cause in the Law of Deception." *Georgetown Law Journal* 100: 449.

Kumar, Srijan, and Neil Shah. 2018. "False Information on Web and Social Media: A Survey." *arXiv preprint arXiv*:1804.08559.

Latour, Bruno. 2011. "Network Theory | Networks, Societies, Spheres: Reflections of an Actor–Network Theorist." *International Journal of Communication* 5: 15.

MacGovern, Kyle. 2021. "Food Narrative: Storytelling, Memories, Art Expression, and Recipes–A Community Engagement Project." Expressive Therapies Capstone Theses. 486. https://digitalcommons.lesley.edu/expressive_theses/486.

Madison, Michael J. 2014. "Commons at the Intersection of Peer Production, Citizen Science, and Big Data: Galaxy Zoo." In *Governing Knowledge Commons*, ed. Brett M. Frischmann, Michael J. Madison, and Katherine J. Strandburg, 209–215.

Magnus, P. D. 2008. "Early Response to False Claims in Wikipedia." *First Monday* 13 (9). https://doi.org/10.5210/fm.v13i9.2115.

Mansoor, Mahnaz. 2021. "Citizens' Trust in Government as a Function of Good Governance and Government Agency's Provision of Quality Information on Social Media during COVID-19." *Government Information Quarterly* 38 (4): 101597.

Marwick, Alice E. 2013. "Online Identity." *A Companion to New Media Dynamics*: 355–364. www.wiley.com/en-us/A+Companion+to+New+Media+Dynamics-p-978.1444332247.

Meltzer, Bernard M. 2003. "Lying: Deception in Human Affairs." *International Journal of Sociology and Social Policy* 23 (6–7): 61–79. https://doi.org/10.1108/01443330310790598.

Meyer, Peter B. 2012. "Open Technology and the Early Airplane Industry." In *Economic History Association 2012 Annual Meeting*, Vancouver, Canada.

Morell, Mayo Fuster. 2014. "Governance of Online Creation Communities for the Building of Digital Commons: Viewed through the Framework of Institutional Analysis and Development." In *Governing Knowledge Commons*, ed. Brett M. Frischmann, Michael J. Madison, and Katherine J. Strandburg, 281–312, Oxford University Press.

Murray, Laura J. 2014. "Exchange Practices among Nineteenth-Century US Newspaper Editors: Cooperation in Competition." In *Governing Knowledge Commons*, ed. Brett M. Frischmann, Michael J. Madison, and Katherine J. Strandburg, 365–390, Oxford University Press.

Narayan, Uma. 2018. "Eating Cultures: Incorporation, Identity and Indian Food." In *Food and Culture*, 67–87. New York: Routledge.

Nelson, Michelle R., and Jiwoo Park. 2015. "Publicity as Covert Marketing? The Role of Persuasion Knowledge and Ethical Perceptions on Beliefs and Credibility in a Video News Release Story." *Journal of Business Ethics* 130 (2): 327–341.

Nissenbaum, Helen. 2009. "Privacy in Context." In *Privacy in Context*. Redwood City, CA: Stanford University Press.

Nguyen, Lilly Uyen, Patrick Keilty, Colin Doty, and Leah Lievrouw. 2014. "In Information's Shadow: Considering Failure as Noise, Misinformation, Error and Breakdown." *iSchools. IConference 2014 Proceedings*. 1238–1239. https://core.ac.uk/download/pdf/19961166.pdf.

Obar, Jonathan A., and Steven S. Wildman. 2015. "Social Media Definition and the Governance Challenge: An Introduction to the Special Issue." *Telecommunications Policy* 39 (9): 745–750.

Papacharissi, Zizi, ed. 2010. A *Networked Self: Identity, Community, and Culture on Social Network Sites*. New York: Routledge.

Patterson, Patrick Hyder. 2003. "Truth Half Told: Finding the Perfect Pitch for Advertising and Marketing in Socialist Yugoslavia, 1950–1991." *Enterprise and Society* 4 (2): 179–225.

Rubin, Victoria L., Jacquelyn Burkell, and Anabel Quan-Haase. 2010. "Everyday Serendipity as Described in Social Media." *Proceedings of the American Society for Information Science and Technology* 47 (1): 1–2.

Sanfilippo, Madelyn, Brett Frischmann, and Katherine Strandburg. 2018. "Privacy as Commons: Case Evaluation through the Governing Knowledge Commons Framework." *Journal of Information Policy* 8 (1): 116–166.

Sanfilippo, Madelyn Rose, Brett M. Frischmann, and Katherine J. Strandburg, eds. 2021. *Governing Privacy in Knowledge Commons*. Cambridge, UK: Cambridge University Press.

Sanfilippo, Madelyn Rose, and Katherine J. Strandburg. 2021a. "Privacy Governing Knowledge in Public Facebook Groups for Political Activism." *Information, Communication and Society* 24 (7): 960–977.

2021b. "Participatory Privacy: How Privacy Governs Community Boundaries and Inclusion in Online Social Movements." *Journal of Social Media in Society* 10 (2): 102–135.

Shivangi, Anja, Chris Bregler, and Matthias Nießner. 2021. "Cosmos: Catching Out-of-Context Misinformation with Self-Supervised Learning." *arXiv preprint arXiv:2101.06278*.

Short, Frances. 2006. *Kitchen Secrets: The Meaning of Cooking in Everyday Life*. Oxford: Berg.

Shvartzshnaider, Yan, Madelyn Rose Sanfilippo, and Noah Apthorpe. 2022. "GKC-CI: A Unifying Framework for Contextual Norms and Information Governance." *Journal of the Association for Information Science and Technology* 73 (9): 1297–1313.

Simmons, Mariah, and Joon Suk Lee. 2020. "Catfishing: A Look into Online Dating and Impersonation." In *International Conference on Human–Computer Interaction*, 349–358. Cham: Springer.

Srinivas, Tulasi. 2006. "'As Mother Made It': The Cosmopolitan Indian Family, 'Authentic' Food and the Construction of Cultural Utopia." *International Journal of Sociology of the Family* 32 (2): 191–221.

Strandburg, Katherine J., Brett M. Frischmann, and Michael J. Madison, eds. 2017. *Governing Medical Knowledge Commons*. Cambridge, UK: Cambridge University Press.

Strathern, Marilyn, M. R. Crick, and George E. Marcus et al. 1987. "Out of Context: The Persuasive Fictions of Anthropology [and Comments and Reply]." *Current Anthropology* 28 (3): 251–281.

Susser, Daniel, Beate Roessler, and Helen Nissenbaum. 2019. "Technology, Autonomy, and Manipulation." *Internet Policy Review* 8 (2): 1–22.

Wang, Hong, and Xin-An Lu. 2007. "Cyberdating: Misinformation and (Dis) trust in Online Interaction." *Informing Science* 10 (online). https://inform.nu/Articles/Vol10/ISJv10p001-015Wang66.pdf.

Wellman, Mariah L., Ryan Stoldt, Melissa Tully, and Brian Ekdale. 2020. "Ethics of Authenticity: Social Media Influencers and the Production of Sponsred Content." *Journal of Media Ethics* 35 (2): 68–82.

Whaley, Barton. 1982. "Toward a General Theory of Deception." *Journal of Strategic Studies* 5 (1): 178–192.

Williams, Michael. 2017. "Skepticism." In *The Blackwell Guide to Epistemology*, ed. John Greco, and Ernest Sosa, 33–69.

Wu, Liang, Fred Morstatter, Kathleen M. Carley, and Huan Liu. 2019. "Misinformation in Social Media: Definition, Manipulation, and Detection." *ACM SIGKDD Explorations Newsletter* 21 (2): 80–90.

Wu, Shangwei, and Daniel Trottier. 2022. "Dating Apps: A Literature Review." *Annals of the International Communication Association* 46 (2): 91–115.

Zhou, Lina. 2005. "An Empirical Investigation of Deception Behavior in Instant Messaging." *IEEE Transactions on Professional Communication* 48 (2): 147–160.

Zulli, Diana. 2018. "Capitalizing on the Look: Insights into the Glance, Attention Economy, and Instagram." *Critical Studies in Media Communication* 35 (2): 137–150.

12

Co-constructing Misinformation and Community

Some Conclusions

Madelyn Rose Sanfilippo and Melissa G. Ocepek

Misinformation is no more nor less than a category of information that diverges in some way from objectivity or truth; like its veracious counterparts, it is produced, saved, processed, and shared among individuals and groups of people, for better or for worse. The predominant assumption, however, is that misinformation is always bad or dangerous, disregarding little white lies and satire. As such, the broader sphere of misinformation research tends to focus on automatic detection or moderation of misinformation, the negative impacts of such content on political dynamics and polarization, and social and psychological effects that impact susceptibility to misinformation or conspiracy.

In addition to all these social and psychological factors producing misinformation and increasing our susceptibility, there are a variety of structural and institutional facets that support and constrain our engagement with misinformation, as well as our ability to effectively manage, mitigate, and remediate associated harms. We have developed a variety of social norms over time that encourage overly positive or optimistic communication of information regardless of whether it aligns with truth, as we saw in Chapter 10; for example, concerning our expectations of positivity and superficiality in greeting others and asking how they are. Further, we have strong social norms across many countries with respect to face saving, as grounded in Hofstede, facets of cultural identity (Hofstede 2011), as well as Goffman's frame analysis (Goffman 1974). We expect people to put their best face forward in a variety of circumstances from dating interviewing to school applications, and it scarcely matters whether they are honest or truthful.

Beyond the social norms, we have a legal and regulatory system in the United States that privileges freedom of speech, regardless of information quality. While we have many institutional restrictions on hate speech or threats, with varying degrees of efficacy in their impact and enforcement, we do very little to restrict speech with respect to information quality in other contexts, or with respect to restrictions outside of government. Given that the modern-day public sphere exists as much, if not more in online environments which are owned, managed, and governed by private entities, rather than representative or public officials, there is much greater discretion in permitting low quality information to flourish, disseminate, and impact society. It is costly, and in fact, cost prohibitive, to manage your impact and the content produced by users via moderation to the degree necessary to fully eliminate every embellishment, falsehood, or misinformation.

Yet this does not mean that we should give up. Rather, it means that we need to differentiate between those instances in which this information is truly harmful, and those instances in which misinformation occurs. Aligning rules with social norms can have this affect. For this to be effective, it is critical that researchers and policymakers effectively gauge social perceptions in different contexts, and with respect to different populations. Toward this aim, we find that the Governing Knowledge Commons (GKC) framework serves to understand how different concerns and policy objectives align with different groups of people in order to institutionalize strategies as norms. Building on the GKC research tradition, which has historically focused on aspects of common resources such as intellectual property or biomedical research contexts, personal information as a resource, smart cities, and markets (e.g., Frischmann et al. 2014; Sanfilippo et al. 2021; Strandburg et al. 2017), this collection of case studies moves beyond that to specifically inquire about aspects of information and data quality. This approach helps us to understand how the GKC framework can more practically support research and governance arrangements, as well as the ways in which it ought to be extended and modified to better address aspects of misinformation as a social product call produced with community.

One of the key aspects that needs to be addressed in amendments to the GKC framework is that of information quality. Historically there has been an assumption that knowledge products produced by communities were inherently valuable. This series of case studies regarding misinformation highlights the ways in which many community knowledge products have a fine edge between valuable and costly, normatively acceptable or appropriate, and normatively frowned upon. It is imperative that we tease apart these tensions and identify the contextually specific line between good and bad, useful and harmful, in order to fully understand the nature of the resource and develop appropriate governance arrangements around it.

COMPLEXITY OF MISINFORMATION IN EVERYDAY LIFE

Consider a child caught in a lie. Whatever the falsehood or inconsistency, whether innocuous or serious, regardless of motivation, the child has only a few options when confronted in their lie. On the one hand, the child may own up to their lie, embarrassed or abashed. On the other hand, the child might double down, digging in their heels, or even upping the ante. Well, of course, there are middle-of-the-road options wherein they acknowledge exaggeration or walk back parts of the lie, and we can imagine many children who don't want to give in, cede ground, or lose face. It is a pattern that happens every day, in every culture, in every context.

While we often recognize these lies for what they are, as in the dog ate my homework scenarios, there are certainly instances in which we take some untruth at face value, building future decisions and social constructions of reality upon poor quality information. More importantly, in many instances a child's peers will buy in to the lie, basing their own understanding or expectations upon falsehoods. Little by little, every day, normative assumptions about honesty allow, and even facilitate, tiny bits of misinformation and common nonsense to enmesh themselves in our perceived reality.

Further, it is our perceptions of reality, rather than some ideal neutral or objective reality, that often matter the most to our behaviors, expectations, and interactions. Every time we misremember an experience or detail, it is the interactions between our fallible memories and socially, emotionally influenced perceptions that fill in the blanks of uncertainties. Just like the children, we may or may not acknowledge uncertainty or inaccuracy in our memories. Many adults also double down on untruths, perhaps based on the little bits of truth they are certain of within the assemblage of ideas or memories, perhaps based on their perceived expertise in the domain, perhaps based on any number of contributing factors. For example, worldviews and prior knowledge color human perceptions and information behaviors, shaping our understanding of the world around us in ways to make us more or less susceptible to misinformation that aligns with our preconceptions and epistemology. In this sense, just as beauty lies in the eye of the beholder, so too does truth.

Information asymmetries also often align with social structures, hierarchies, and the status quo, thereby allowing ample opportunity for manipulation of information or framing to bend representation and reflections of reality to social dynamics in subjective and favorable ways. We are more likely to accept misinformation when it aligns with our interests, and to reject or suspect it when it is misaligned or infringes on our status, safety, or interests. It takes a more critical eye to recognize misinformation or falsehood when to do so counters one's interests.

Ubiquitous misinformation in our everyday lives poses numerous challenges, to be sure, but more than that, the spectrum of information quality throughout our information environments requires nuance in how we address it. While misinformation is often created and propagated without necessarily an intent to deceive, or with malicious motivations, it nonetheless makes us all more susceptible in cases where that is the aim via desensitization, inescapability, and an intersection between information quality problems and information overload. However, it doesn't make us powerless to address these challenges. We need to learn from the innocuous to appropriately identify when it is noxious or controversial, and how we can best handle those instances when we encounter them.

Through the case studies in this volume, we recognize the complexity of misinformation and its origins. Not all, or even in a majority of misinformation that we encounter is malicious, but those few malicious instances can have disproportionate impact. As such, the emergent patterns regarding norms, expectations, and behaviors associated with misinformation engagement in the everyday world offer insight into what separates acceptable and unacceptable misinformation, and how we might address those problematic aspects and instances.

EMERGING PATTERNS ACROSS CASES

Narrative Frames and Rhetoric

In considering the patterns of discussion surrounding informational framing and rhetoric with respect to misinformation, which has emerged in our multiple case studies and has been built on in conceptual chapters, a few key elements arise that highlight both the unique aspects of misinformation as a resource and the inherent challenges in governing these resources.

Partial Truth

First among them, *the most convincing pieces of misinformation couple truth with untruth*. Regardless of intention, deception is especially impactful via this structure. From the principles of storytelling explored in Chapters 1 and 2, to Instagram influencers in Chapter 10, this was a key pattern providing evidence both for how people influence dialogue via misinformation and why people are susceptible. If we recognize that most information is analogous to an onion with many layers, how do we peel apart the layers of truth from those layers that are problematic? How do we ensure that interpretation of information is correct or contextual? In this sense, problems of governing misinformation are not merely about enforcement of standards of quality. At a much more complex level, the real challenge is in how to differentially enforce standards or norms when some inaccuracies are tolerable and some are not, or when untruths are tightly coupled to truths.

Chapters 2 and 3 provide qualitative and conceptual understanding as to why this is particularly effective, as well as to the duality of this practice. For example, Chapter 3 contained an analysis of why misremembering who won the Oscar for best picture in a particular year might layer fact with honest mistake, although only causing problems as that misinformation gets passed on, is certainly without serious consequences. In other instances, it's not a matter of a simple mistake but rather opinion that layers with fact; Chapter 6 illustrates how Google's Local Guides and crowdsourcing of business and service reviews via digital platforms embed "hot takes" in what otherwise may be factual recounting of individuals' experiences in particular places. The compelling nature of truth and conspiracy coupled together is illustrated in Chapter 9, via analysis of QAnon. Taken as a whole, we see a spectrum from the innocuous to quite serious misinformation that leverages this strategy.

How might we deal with this practice? Wikipedia is one place we might wish to look for lessons, as their content moderation strategies reflecting knowledge commons arrangements do not throw out entire articles over bits of unsupported opinions or inaccuracies, but rather iteratively approach some consensus over truth.

Audience and Receptiveness

Second, *audiences are especially receptive to messaging*, whether with valuable information or with misinformation, *when it is delivered by or framed with respect to trusted sources or individuals.* Whether those sources are public figures, as in Chapter 10, or family or friends, as in Chapters 2 and 7 though 9, or sources of authority, as in Chapter 4, there is considerable evidence that framing with respect to legitimate, credible, trusted, and recognizable entities leads people to share and believe misinformation.

Chapters 6 and 10 offered how platform design features indicating contributions, even when not directly known, to Local Guides or influence on Twitter respectively via metrics and engagement, serve as signposts for other users and imply credibility as a source. Other cases, as with Chapters 4 and 11 illustrate how triangulation, in the form of signposting in course syllabi or to official policies on social media can be used to communicate implicit authority and reinforce false statements with policies, which are not necessarily aligned with that misinformation. In this sense, institutionalization is compelling to audiences, even when they do not understand or attempt to interpret the details. This aligns with the discussion of epistemology, as presented relative to the complexity of everyday misinformation; signifiers make us more receptive to false content.

Storytelling

Third, *technology eases storytelling and knowledge construction patterns*, providing social and networked mechanisms to communicate in ways that appear polished, compelling, and are more far-reaching than traditional analog writing or oral storytelling traditions. In lowering barriers to content production, reducing distance between those who share content and those who consume it, and providing feedback mechanisms, technologies make many forms of storytelling engaging without requiring us to be authentic (or BeReal). In fact, it facilitates more social construction of misinformation while providing the appearance of authenticity by supporting multiple emotional layers and enhancing contextual narratives in ways that might enhance or downplay the messiness of reality, depending on the objectives of the storyteller, to provide salience.

Every chapter in this collection highlighted the intersection between technology and storytelling, with Chapter 2 and those cases that engaged with social media providing some of the clearest illustrations. However, Chapter 5 illustrated this in one of the most surprising ways; password misinformation and misperceptions regarding security practices are not a problem born of attitudes or a lack of knowledge, but rather an often technologically mediated system of delivering best practices or rules to users that leads to nonexperts absorbing the wrong lessons.

Socialization in Context

As quality knowledge and misinformation are coproduced with community, socialization in those communities is key to discerning fact from fiction, as well as understanding the line of what is and what is not appropriate. Certainly, humans are fallible, making genuine mistakes, minor errors, or miscommunications. However, without meaningful socialization in a community, we are much more likely to be susceptible to deception, as observed in many cases.

Belongingness

Whoever belongs and perceives their belongingness, not merely those who participate in a community, is the key to outcomes and governance patterns in many cases.

At a deeper level than active or passive participation or secondary impacts in interacting with active participants in other relationships, as will be explored in relationship to issues of community boundaries in the next subsection, issues of acculturation and belongingness are also tightly coupled to evaluation of appropriateness of misinformation or governance interventions regarding this, as with Chapters 6, 7, 8, and 11. Subjective judgments about intentions to deceive versus novice misunderstandings are related to perceptions of belongingness in online communities, as well as educational contexts – as in Chapter 4 – and interactions between experts and novices around security, as in Chapter 5. Understanding of contextual norms about truthfulness, or perceived understanding, also appears associated with strength of enforcement and the types of consequences violators of expectations might face.

Power and Incentives

Power and the incentives to produce and share and engage with content within many of the cases explored tells us a lot about the relative impact of influencers on spreading misinformation, the pressures to accept misinformation in order to belong, and inequities with respect to enforcement of governance.

Many of the cases examined instances in which misinformation sharing and production actually empowered the powerless. From marginalized voices in Brazil and India – Chapters 7 and 8, respectively – to US undergraduates simply looking to fit in to computer science programs, people often reinforce misinformation when they get past community boundaries so as to reinforce their own belongingness. Power was also tightly coupled with influence on norm formation processes, setting community standards via influence and esteem within communities, such as with reinforcement logic on Facebook QAnon groups, in Chapter 8, and the power of Instagram influencers to move on from violations of community expectations, as in Chapter 11. Perceptions about the appropriateness of misinformation were tightly coupled in many chapters to power, as power was also an incentive to produce misinformation by producing more engagement via sensational or manipulative content.

Echoing Other Applications of the GKC Framework

The final two patterns importantly echo key aspects of other collections of knowledge commons case studies, including case studies on governance privacy (Sanfilippo et al. 2021) and smart cities (Frischmann et al. 2023). Issues regarding infrastructure and the margins of participation, as associated with boundaries or gatekeeping or outsiders, were prevalent and impactful on outcomes in many of the cases in this volume.

Infrastructure

As with other knowledge resources, *we are dependent on infrastructure at material, social, knowledge, and technological levels in order to produce, share, and engage with misinformation*. Infrastructure is often easy to overlook, but its importance to the lifecycle of knowledge resources, including misinformation, is especially evident when examining the coproduction of ephemeral communities on social media and prominent content of the moment. Across many of the cases in this collection, the layers of information and support for that were important in shaping the unique concerns in each case, including contextual norms about what was and was not appropriate in terms of sharing misinformation, as with Chapter 11.

Platforms documentation, including moderation policies, were clearly identified as providing support for the proliferation of misinformation, as opposed to mechanisms to appropriately govern or moderate if and when it gets out of hand, in many other cases, such as Chapters 7, 9, and 10. This highlights where platforms fall short in supporting the needs of their communities, in contrast with the example provided in Chapter 8 regarding WhatsApp introducing limits on forwarding content to reduce the reach of misinformation.

Boundaries, Gatekeeping, and Impacted Outsiders

Participation dynamics also contributed to the production, reach, and impact of misinformation throughout every case study. As we considered *who are the relevant communities* as researchers, again and again we were reminded that it is necessary to look more broadly. What does it really mean to participate? Who is eligible? If someone didn't participate in misinformation production, but was exposed or passed that along to others, what is their responsibility?

Various issues regarding boundaries, gatekeeping, and impacted outsiders arose in each case study, echoing a key pattern evident in many other applications of the GKC framework. Those with power over knowledge – and misinformation – production and governance are not necessarily the only people impacted. While at the surface, the impact on lurkers in online communities examined might be the group of people most directly impacted by misinformation, however, the impact on family members, as in Chapter 9's discussion of QAnon and Chapter 2's scenarios regarding family lore highlight the everydayness of these issues and the broad reach in terms of secondary influence.

Further, gatekeeping is often supported by misinformation, as in Chapter 4. Exclusionary practices and proliferation of misinformation as a weapon to maintain control and dominance in an academic discipline or field, as in Chapter 5, depend on bending the truth, layering fact with negative and fearful sentiment, and framing.

BROAD LESSONS

Patterns from this collection of case studies offer several broad lessons, from reminders of unremarkable but often overlooked axioms to more nuanced and actionable insights.

At one end of that spectrum: *Sometimes people simply get things wrong*. This can happen by accident, perhaps compounding other factors. It may not reflect malice, even when it is undoubtedly a reflection of bias. For example, in 2023, the ESPN credited Vladimir Guerrero Jr. as the second Cuban-born winner of the home run derby, after Yeonis Cespedes, despite the fact that he's the son of Dominican MLB hall of famer Vladimir Guerrero (not coincidentally also a winner of the Home Run Derby) and was born in Montreal, making him Canadian-Dominican. Yet his name led to an assumption by the network research team and their claim went unchecked prior to inclusion in an internationally broadcast chyron. This incident both illustrates a simple truth – mistakes happen – and an uglier reality – complex social factors contribute to those mistakes in undeniable ways that cannot go unremarked upon. As such, a broad lesson is that we need to be able to correct mistakes, and those who produced them are responsible for that.

However, *responsibility to correct erroneous content is not necessarily a responsibility to correct misinterpretation*. Beyond the cases published in this volume, there are many examples of satire that are misinterpreted as sincere news or communication, from politicians who participated in Stephen Colbert's Better Know a District (2005–2014) parody interview series to those who quote The Onion as if it were journalism. The comedians behind that are not responsible for correcting misconceptions and in fact the humor elicited by a dividing line between who is and is not in on the joke is key to the success of these examples. In this way, parody as a light-hearted and certainly appropriate example of misinformation illustrates the analogy to political trolling, which despite its extremely negative connotations to the general public given recent election cycles and issues of election interference, is pervasive among innocuous social media interactions between users and elected officials, such as rickrolling or pseudo-sincere engagement to draw attention to policy positions that the public feels are absurd; for example, asking former Vice President Mike Pence for reproductive health advice (e.g., Connor 2020). Both embody the truthiness issues that continue to grow.

The *malleability of truth* is another key take away from this collection of case studies. While we have long known about the social construction of reality in the social sciences, broader perceptions of quantitative indicators, whether measurements, or apparently "neutral" statistics as objective, have long delineated certain types of information as unassailable truths. Yet mounting distrust of experts, authorities, and professionals has eroded the trust and perceived legitimacy of many of these facts. Is the earth flat or round? This question was answered correctly at least as early as the sixth century BCE by Pythagoras (Arif et al. 2019), with increasing acceptance over time, yet, recent years have seen the number of flat earthers grow again, perhaps foremost among them professional basketball player Kyrie Irving (Paolillo 2018). Politicization and polarization over these issues, coupled with socio-economic dynamics, and technologies that enhance and exacerbate these processes explain the unfortunate trend.

However, *eradication of misinformation is not only impossible, but also undesirable*. Beyond satire, there are instances as in Chapter 11 where misinformation is socially normative. We need to reduce the spread and impact and availability of harmful and antinormative misinformation, but not all of it. Many communities are good at identifying what is right or wrong for them. Contextual norms should stipulate enforcement and prioritization. What communities need are meaningful tools to address and to reduce pervasiveness and spread as with both WhatsApp cases in Chapters 7 and 8, wherein simple changes regarding limits on forwarding had significant impacts. Beyond these simple design changes that can introduce useful friction to reduce the spread of misinformation, the space where we most clearly require additional support is with respect to manipulation, as it reflects a more challenging problem for communities to address without support or meaningful enforcement mechanisms, whether technical or legal.

As we think about the increasing ubiquity and nonplussing nature of misinformation in everyday life, we might consider the lyrics from Michael Jackson's (1982) song "Billie Jean" "and the lies become the truth." Not only does this provide an excellent example of how socially constructed truthiness may be perceived as reality, it also reflects the inherent polarization around truth and lie, fact and fiction. Michael Jackson, as the singer and songwriter in question, is the embodiment of social disagreement regarding truth, with some revering the art and reviling the man, while others embrace or reject both. If there is a high degree in of uncertainty around him, and the reality of his life. what do we really know? What do we believe? What do we want to believe? How do we know it? What do we trust? Yet despite all these questions, we feel strongly about our position.

Finally, as a pair of academic parents editing this book who both try to varying degrees of success to pass along a love of reading to our children, we saw many these insights echoed in children's books moralizing and satirizing values around truth and honesty. For example, Ian Falconer, author of *Olivia* (an anthropomorphic pig with a colorful personality) wrote:

MOTHER: *"Well, Olivia, what have you learned by eavesdropping?"*
OLIVIA: *"Partial truths and misinformation – "*
MOTHER: *"And how did that make you feel?*
OLIVIA: *"Insecure and suspicious – "*

(Falconer 2017)

Olivia speaks for us all in this moment. There are certainly costs to misinformation and, as we saw in the study of QAnon on Facebook in Chapter 9, conspiracy. It is important to confront these issues in our everyday lives and for families to have engagement with reality, not merely digitally mediated constructions of it.

AMENDING THE GKC FRAMEWORK

Many cases illustrated the coproduction of information resources and community, exemplifying the typical institutional structure of the knowledge commons. In this sense, the GKC framework is useful for exploring institutionalization processes to manage misinformation resources, as well as to mitigate harms from communities' misinformation. Yet we also gain significant insights into the ways in which the GKC framework might be expanded in more nuanced ways to explore the quality of information resources, and the nuanced processes that allow for distinct governance patterns of similar misinformation differentiated upon intent and perception by the community.

The patterns that emerged from these misinformation cases suggest the following concepts are key to knowledge commons processes:

1. *Infrastructure*: Shared infrastructure has long been explored with respect to commons, both in relationship to physical resources (McGinnis 1999; Ostrom 1990) and knowledge commons (Frischmann et al. 2014), yet control of the infrastructure has not directly been explored. Rather, endogeneity has been assumed via a question regarding shared infrastructure. As such, we suggest it would be fruitful to ask: *Is exogenous infrastructure required to engage with resources? Who controls infrastructure?*
2. *Incentives for knowledge production:* These cases indicated that there are often financial and power incentives to create misinformation or sensationalize relevant information to capitalize on modern recommendation systems and surveillance capitalism, as when we benefit from advertising dynamics, there is power in a click or share, or other forms of engagement. As such, it is important to interrogate: *What incentives exist to produce knowledge? To participate in the commons?*
3. *Contextual complexity*: Analysis of misinformation in the everyday importantly illustrated how complex aspects of context unintentionally influence governance arrangements and knowledge production dynamics. Intersections between the everyday and politics and speech protections, for example, complicate efforts to enforce governance and appropriately address misinformation. However, the implications suggest applications beyond this context: *What parallel action arenas might inadvertently or unintentionally impact governance?*
4. *Implicit institutional dynamics*: Past cases have explored informal norms and their formation as implicit institutional forces on commons arrangements, yet did not directly examine other implicit issues of political influence or power. We suggest asking: *What implicit power dynamics influence the commons?*
5. *Information quality*: Falsehood or deception are key attributes of misinformation, both indicating information quality issues for information behavior in the everyday and for information professionals. Based on these issues, we offer the following questions: *Who evaluates dimensions of resource quality? Are there criteria or standards to delineate important dimensions of quality?*

As with other collections of case studies, the questions raised based on these concepts were appended to illuminate future cases, as indicated in Table 12.1 via italicization.

TABLE 12.1 *An appended GKC framework (emerging questions based on everyday misinformation cases are indicated via* italicization)

Knowledge commons framework and representative research questions	
Background environment	
	What is the background context (legal, cultural, etc.) of this particular commons?
	What normative values are relevant for this community?
	What is the "default" status of the resources involved in the commons (patented, copyrighted, open, or other)?
	How does this community fit into a larger context? What relevant domains overlap in this context?
Attributes	
Resources	What resources are pooled and how are they created or obtained?
	What are the characteristics of the resources? Are they rival or nonrival, tangible or intangible?
	Is there shared infrastructure? *Is exogenous infrastructure required to engage with resources? Who controls infrastructure?*
	What is personal information relative to resources in this action arena?
	What technologies and skills are needed to create, obtain, maintain, and use the resources?
	What are considered to be appropriate resource flows? How is appropriateness of resource use structured or protected?
	Who evaluates dimensions of resource quality? Are there criteria or standards to delineate important dimensions of quality?
Community members	Who are the community members and what are their roles, including with respect to resource creation or use, and decision making?
	Are community members also information subjects?
	What are the degree and nature of openness with respect to each type of community member and the general public?
	What noncommunity members are impacted?
Goals and objectives	What are the goals and objectives of the commons and its members, including obstacles or dilemmas to be overcome?
	Who determines goals and objectives?
	What values are reflected in goals and objectives?
	What are the history and narrative of the commons?
	What is the value of knowledge production in this context?
	What incentives exist to produce knowledge? To participate in the commons?
Governance	
Context	What are the relevant action arenas and how do they relate to the goals and objective of the commons and the relationships among various types of participants and with the general public?
	Are action arenas perceived to be legitimate?
	What parallel action arenas might inadvertently or unintentionally impact governance?
Institutions	What legal structures (e.g., intellectual property, subsidies, contract, licensing, tax, antitrust) apply?
	What other external institutional constraints are imposed? What government, agency, organization, or platform established those institutions and how?

	How is institutional compliance evaluated?
	What are the governance mechanisms (e.g., membership rules, resource contribution or extraction standards and requirements, conflict resolution mechanisms, sanctions for rule violation)?
	What are the institutions and technological infrastructures that structure and govern decision making?
	What informal norms govern the commons? *What implicit power dynamics influence the commons?*
	What institutions are perceived to be legitimate? Illegitimate? How are institutional illegitimacies addressed?
Actors	What actors and communities: are members of the commons, participants in the commons, users of the commons, and/or subjects of the commons?
	Who are the decision makers and how are they selected? Are decision makers perceived to be legitimate? Do decision makers have an active stake in the commons?
	How do nonmembers interact with the commons? What institutions govern those interactions?
	Are there impacted groups that have no say in governance? If so, which groups?
Patterns and outcomes	
	What benefits are delivered to members and to others (e.g., innovations and creative output, production, sharing, and dissemination to a broader audience, and social interactions that emerge from the commons)?
	What costs and risks are associated with the commons, including any negative externalities?
	Are outcomes perceived to be legitimate by members? By decision makers? By impacted outsiders?
	Do governance patterns regarding participation provide exit and/or voice mechanisms for participants and/or community members?
	Which rules-in-use are associated with exit-shaped, voice-shaped, or imposed governance? Are there governance patterns that indicate the relative impact of each within the commons overall?

FINAL REFLECTIONS

As the opportunity to make final revisions came in summer 2023, it seemed imperative to reflect on how what we'd learned from these cases applied to broader conversations about truth (and Truth™), misinformation, and politics as wave after wave of indictments associated with a former US president rolled in. What about misinformation in everyday life and the innocuous and contentious ways in which people govern that might help us to make sense of the surreality we find ourselves in? Issues of trust around channels of communication and specific messages across cases and throughout history point to why this scenario is so challenging; long-standing digital literacy principles to look toward experts and authority for quality information are undermined when the legitimacy of the role is in question, and when those in power have the motivation and behavioral patterns of deception. This is not merely a problem of media or political communication, but a much broader one.

Perhaps the answer is in something as relatable as a childhood game of telephone? The average person does not find it difficult to see how the message becomes distorted as one giggling child whispers to another. It also maps intuitively onto longstanding information theory, supporting Shannon and Weaver's mathematical theory of communication, regarding noise as it impacts communication channels and information quality (Shannon 1948; Weaver 1953). Each time the message is forwarded, opportunities arise for the message to become distorted, whether intentionally or unintentionally. Beyond that there is also incentive to begin with a silly or attention-grabbing or sensational message to keep and hold engagement. If this happens on the scale of the everyday, of course it also happens in all contexts regardless of impact or seriousness. If we look at how this game maps onto patterns of sharing on social media, in an era of filter bubbles, we see how algorithms may reduce noise, both in the form of detraction from and diversity of content; this too poses a problem, as we are increasingly exposed to echo chambers that are susceptible to misinformation via partial truth or rhetoric that seems analogous to legitimate content, from the perspective of the recommendation system.

With or without modern technology, there is also a long history demonstrating human motivation to ignore ugly facts or hypothesize unfounded alternative explanations for reality, as with numerous conspiracies about the Kennedy family or holocaust deniers. These conspiracies are sometimes emotional reactions or efforts to reconcile facts with alternative epistemologies, while others are born of malice or bias or bigotry. Sometimes they reflect concerted efforts to control the historical record, as with friends of presidents writing their official biographies to proclaim their significance or greatness (Aldrich 1992), when the reality was that they ranged from unremarkable to terrible. They offer a contrast to instances of error in modern history, such as with a headline of "Dewey defeats Truman."

And thus, the takeaway from these examples and comparisons is that while the fact pattern might change, as we see in the many scandals and challenges around Trump – or Theranos or FTX, as in the Introduction – providing an absurd combination of issues, none of the issues are new. For example, Octavian leveraged misinformation to gain support and undermine Mark Anthony in the Roman Empire between 42 and 23 BC (Fraser 2020); he certainly was not the first to do so.

These examples and the cases presented in this book illustrate that while the same behaviors occur across time and contexts, reflecting universal behavioral patterns about information quality, engagement, and flow, norms about appropriateness vary widely. When we compare half-truths in dating profiles to half-truths in the personal narratives of politicians, it's a comparison between something that is par for the course and "lies and the lying liars who tell them" (Franken 2004). Ironically, NY Representative George Santos is the embodiment of this dichotomy in a single person, illustrating the lack of concern about patterns of untruth when it did not affect democracy, as opposed to widespread bipartisan concern when it did (Guest and Wild 2023).

We collectively produce (mis)information, community, and governance of that (mis)information and community in the everyday. They are necessarily linked and sociotechnically constructed in the modern information environment. Just as misinformation is quotidian, so too are the knowledge commons arrangements with the potential to appropriately govern misinformation.

REFERENCES

Aldrich, Elizabeth Kaspar. 1992. "Necessary Saints: Some Reflections on the Varieties of American Hagiography." *Yale Journal of Criticism* 5 (3): 1.

Arif, Faiz, Abdul Aziz Ab Rahman, Khairul Nizam Abdul Maulud, and Amir Husni Kamaludin. 2019. "Debunking Flat Earth: From Geomatics Perspective." In *2019 6th International Conference on Space Science and Communication (IconSpace)*, 150–153. IEEE.

Colbert, Stephen. 2005–2014. "Better Know a District." The Colbert Report. Comedy Central.

Conner, Berkley D. 2020. "Menstrual Trolls: The Collective Rhetoric of Periods for Pence." *The Palgrave Handbook of Critical Menstruation Studies*: 885–899.

Falconer, Ian. 2017. *Olivia the Spy*. New York: Atheneum/Caitlyn Dlouhy.

Franken, Al. 2004. *Lies – and the Lying Liars Who Tell Them: A Fair and Balanced Look at the Right*. New York: Penguin.

Fraser, Matthew. 2020. *In Truth: A History of Lies from Ancient Rome to Modern America*. Lanham, MD: Rowman & Littlefield.

Frischmann, Brett M., Michael J. Madison, and Madelyn Rose Sanfilippo, eds. 2023. *Governing Smart Cities as Knowledge Commons*. Cambridge, UK: Cambridge University Press.

Frischmann, Brett M., Michael J. Madison, and Katherine Jo Sanfilippo, eds. 2014. *Governing Knowledge Commons*. Oxford, UK: Oxford University Press.

Goffman, Erving. 1974. *Frame Analysis: An Essay on the Organization of Experience*. Cambridge, MA: Harvard University Press.

Guest, Michael, and Susan Wild. 2023. "Statement of the Chairman and Ranking Member of the Committee on Ethics Regarding Representative George Santos." Committee on Ethics, US House of Representatives, March 2.

Hofstede, Geert. 2011. "Dimensionalizing Cultures: The Hofstede Model in Context." *Online Readings in Psychology and Culture* 2 (1): 8.

Jackson, Michael. 1982. "Billie Jean." Epic Records.

McGinnis, Michael Dean. 1999. *Polycentricity and Local Public Economies: Readings from the Workshop in Political Theory and Policy Analysis*. Ann Arbor, MI: University of Michigan Press.

Ostrom, Elinor. 1990. *Governing the Commons: The Evolution of Institutions for Collective Action*. Cambridge, MA: Cambridge University Press.

Paolillo, John C. 2018. "The Flat Earth Phenomenon on YouTube." *First Monday*. https://doi.org/10.5210/fm.v23i12.8251.

Sanfilippo, Madelyn Rose, Brett M. Frischmann, and Katherine J. Strandburg, eds. 2021. *Governing Privacy in Knowledge Commons*. Cambridge, UK: Cambridge University Press.

Shannon, Claude Elwood. 1948. "A Mathematical Theory of Communication." *Bell System Technical Journal* 27 (3): 379–423.

Strandburg, Katherine J., Brett M. Frischmann, and Michael J. Madison, eds. 2017. *Governing Medical Knowledge Commons*. Cambridge, UK: Cambridge University Press.

Weaver, Warren. 1953. "Recent Contributions to the Mathematical Theory of Communication." *ETC: A Review of General Semantics* 10 (4), Special Issue on Information Theory (Summer): 261–281.